ATMOSPHERIC POLLUTION

Atmospheric Pollution: History, Science, and Regulation provides a comprehensive introduction to the history and science of major air pollution issues. It begins with an introduction to the basic atmospheric chemistry and the history of discovery of chemicals in the atmosphere, then moves on to a discussion of the evolution of the earth's atmosphere and the structure and composition of the present-day atmosphere. Subsequently, a comprehensive and accessible discussion of the five major atmospheric pollution topics – urban outdoor air pollution, indoor air pollution, acid deposition, stratospheric ozone reduction, and global climate change – is provided. Each chapter discusses the history and science behind these problems, their consequences, and the effort made through government intervention and regulation to mitigate them. The book contains numerous student examples and problems, more than 200 color illustrations, and is international in scope.

Atmospheric Pollution: History, Science, and Regulation forms an ideal introductory textbook on atmospheric pollution for undergraduate and graduate students taking courses in atmospheric chemistry and physics, meteorology, environmental science, earth science, civil and environmental engineering, chemistry, environmental law and politics, and city planning and regulation. It also forms a valuable reference text for researchers and an introduction to the subject for general audiences.

Mark Z. Jacobson is an Associate Professor of Civil and Environmental Engineering at Stanford University. He has published over 40 peer-reviewed papers and another textbook, *Fundamentals of Atmospheric Modeling* (1998, Cambridge University Press), that has been rated by students in the top 5% of textbooks in the School of Engineering at Stanford for nine consecutive quarters. Professor Jacobson is a recipient of the National Science Foundation Career Award, the Powell Foundation Award, a Frederick Terman Fellowship, and a NASA New Investigator Award. In addition to these awards and scholarships, in 1985, 1986, and 1987, he received an NCAA–ITCA scholar–athlete of the year award at Stanford University.

To Yvonne and William

ATMOSPHERIC POLLUTION

HISTORY, SCIENCE, AND REGULATION

Mark Z. Jacobson

Stanford University

CAMBRIDGE
UNIVERSITY PRESS

PUBLISHED BY THE PRESS SYNDICATE OF THE UNIVERSITY OF CAMBRIDGE
The Pitt Building, Trumpington Street, Cambridge, United Kingdom

CAMBRIDGE UNIVERSITY PRESS
The Edinburgh Building, Cambridge CB2 2RU, UK
40 West 20th Street, New York, NY 10011-4211, USA
477 Williamstown Road, Port Melbourne, VIC 3207, Australia
Ruiz de Alarcón 13, 28014 Madrid, Spain
Dock House, The Waterfront, Cape Town 8001, South Africa

http://www.cambridge.org

First published 2002

Printed in the United Kingdom at the University Press, Cambridge

Typeface Times Roman 10/12 pt.   *System* QuarkXPress® [TB]

A catalog record for this book is available from the British Library.

Library of Congress Cataloging in Publication Data
Jacobson, Mark Z. (Mark Zachary)
      Atmospheric pollution : history, science, and regulation / Mark Z. Jacobson.
        p.   cm.
      Includes bibliographical references and index.
      ISBN 0-521-81171-6 – ISBN 0-521-01044-6 (pb.)
      1. Air – Pollution. 2. Atmospheric chemistry. 3. Air – Pollution – Law and legislation.
   I. Title.
   TD883 .J37 2002
   363.739'2 – dc21   2001037645

ISBN 0 521 81171 6 hardback
ISBN 0 521 01044 6 paperback

CONTENTS

Preface   *page* ix
Acknowledgments   xi

1. **BASICS AND HISTORY OF DISCOVERY OF
    ATMOSPHERIC CHEMICALS**   1

    1.1.   *Basic Definitions*   2
    1.2.   *History of Discovery of Elements and Compounds of
             Atmospheric Importance*   4
    1.3.   *Chemical Structure and Reactivity*   21
    1.4.   *Chemical Reactions and Photoprocesses*   24
    1.5.   *Lifetimes of Chemicals*   26
    1.6.   *Summary*   26
    1.7.   *Problems*   26

2. **THE SUN, THE EARTH, AND THE EVOLUTION OF
    THE EARTH'S ATMOSPHERE**   29

    2.1.   *The Sun and Its Origin*   30
    2.2.   *Spectra of the Radiation of the Sun and the Earth*   33
    2.3.   *Primordial Evolution of the Earth and Its Atmosphere*   36
    2.4.   *Summary*   47
    2.5.   *Problems*   48

3. **STRUCTURE AND COMPOSITION OF THE PRESENT-DAY
    ATMOSPHERE**   49

    3.1.   *Air Pressure and Density Structure*   50
    3.2.   *Processes Affecting Temperature*   52
    3.3.   *Temperature Structure of the Atmosphere*   54

3.4.   Equation of State   58
3.5.   Composition of the Present-Day Atmosphere   62
3.6.   Characteristics of Selected Gases and
       Aerosol Particle Components   63
3.7.   Summary   79
3.8.   Problems   79

4.   URBAN AIR POLLUTION   81

4.1.   History and Early Regulation of Urban Air Pollution   82
4.2.   Chemistry of the Background Troposphere   93
4.3.   Chemistry of Photochemical Smog   99
4.4.   Pollutant Removal   111
4.5.   Summary   111
4.6.   Problems   112

5.   AEROSOL PARTICLES IN SMOG AND
     THE GLOBAL ENVIRONMENT   115

5.1.   Size Distributions   116
5.2.   Sources and Compositions of New Particles   118
5.3.   Processes Affecting Particle Size   128
5.4.   Summary of the Composition of Aerosol Particles   138
5.5.   Aerosol Particle Morphology and Shape   139
5.6.   Health Effects of Aerosol Particles   140
5.7.   Summary   142
5.8.   Problems   142

6.   EFFECTS OF METEOROLOGY ON AIR POLLUTION   145

6.1.   Forces   146
6.2.   Winds   147
6.3.   Global Circulation of the Atmosphere   150
6.4.   Semipermanent Pressure Systems   154
6.5.   Thermal Pressure Systems   155
6.6.   Effects of Large-Scale Pressure Systems on Air Pollution   156
6.7.   Effects of Local Meteorology on Air Pollution   168
6.8.   Summary   175
6.9.   Problems   176

7.   EFFECTS OF POLLUTION ON VISIBILITY,
     ULTRAVIOLET RADIATION, AND ATMOSPHERIC OPTICS   179

7.1.   Processes Affecting Solar Radiation in the Atmosphere   180
7.2.   Visibility   197
7.3.   Colors in the Atmosphere   202
7.4.   Summary   205
7.5.   Problems   206
7.6.   Project   207

8. INTERNATIONAL REGULATION OF URBAN SMOG SINCE THE 1940s
209

8.1. Regulation in the United States   210
8.2. Pollution Trends and Regulations Outside the United States   225
8.3. Summary   238
8.4. Problems   239

9. INDOOR AIR POLLUTION
241

9.1. Pollutants in Indoor Air and Their Sources   242
9.2. Sick Building Syndrome   251
9.3. Regulation of Indoor Air Pollution   251
9.4. Summary   252
9.5. Problems   252

10. ACID DEPOSITION
253

10.1. Historical Aspects of Acid Deposition   254
10.2. Causes of Acidity   257
10.3. Sulfuric Acid Deposition   260
10.4. Nitric Acid Deposition   263
10.5. Effects of Acid Deposition   263
10.6. Natural and Artificial Neutralization of Lakes and Soils   266
10.7. Recent Regulatory Control of Acid Deposition   270
10.8. Summary   271
10.9. Problems   272

11. GLOBAL STRATOSPHERIC OZONE REDUCTION
273

11.1. Structure of the Present-Day Ozone Layer   274
11.2. Relationship between the Ozone Layer and UV Radiation   277
11.3. Chemistry of the Natural Ozone Layer   278
11.4. Recent Changes to the Ozone Layer   283
11.5. Effects of Chlorine on Global Ozone Reduction   286
11.6. Effects of Bromine on Global Ozone Reduction   293
11.7. Regeneration Rates of Stratospheric Ozone   294
11.8. Antarctic Ozone Depletion   295
11.9. Effects of Enhanced UV-B Radiation on Life and Ecosystems   301
11.10. Regulation of CFCs   303
11.11. Summary   306
11.12. Problems   307

12. THE GREENHOUSE EFFECT AND GLOBAL WARMING
309

12.1. The Temperature on the Earth in the Absence of a Greenhouse Effect   310
12.2. The Greenhouse Effect and Global Warming   316
12.3. Recent and Historical Temperature Trends   323

12.4. Feedbacks and Other Factors That May Affect
       Global Temperatures   337
12.5. Possible Consequences of Global Warming   342
12.6. Regulatory Control of Global Warming   345
12.7. Summary   349
12.8. Problems   350
12.9. Essay Questions   351

Appendix: Conversions and Constants   353
References   355
Photograph Sources   371
Index   377

PREFACE

Natural air pollution problems on the Earth are as old as the Earth itself. Volcanoes, fumaroles, natural fires, and desert dust have all contributed to natural air pollution. Humans first emitted air pollutants when they burned wood and cleared land (increasing windblown dust). More recently, the burning of coal; chemicals; oil, gasoline, kerosene, diesel, jet and alcohol fuels; natural gas; and waste and the release of chemicals have contributed to several major air pollution problems on a range of spatial scales. These problems include outdoor urban smog, indoor air pollution, acid deposition, Antarctic ozone depletion, global ozone reduction, and global warming.

Urban smog is characterized by the outdoor buildup of gases and particles emitted from vehicles, smokestacks, and other human sources, or formed chemically in the air from emitted precursors. Smog affects human and animal health, structures, and vegetation. Urban smog occurs over scales of tens to hundreds of kilometers.

Indoor air pollution results from the emission of pollutant gases and particles in enclosed buildings and the transport of pollutants from outdoors into buildings. Indoor air pollutants cause a variety of human health effects. Indoor air pollution occurs over scales of meters to tens of meters.

Acid deposition occurs when sulfuric acid, nitric acid, or hydrochloric acid in the air deposits to the ground as a gas or dissolved in rainwater, fogwater, or particles. Acids harm soils, lakes, forests, and structures. In high concentrations, they can harm humans. Acid deposition occurs over scales of meters to thousands of kilometers.

Antarctic ozone depletion and **global ozone reduction** are caused, to a large extent, by human-produced chlorine and bromine compounds that are emitted into the air and break down only after they have traveled to the upper atmosphere. Ozone reduction increases the intensity of ultraviolet (UV) radiation from the sun reaching the ground. Intense UV radiation destroys microorganisms on the surface of the Earth and causes skin cancer in humans and animals. Antarctic ozone depletion occurs over a region the size of North America. Global ozone reduction occurs globally.

Global warming is the increase in global temperatures, rainfall patterns, and sea level due to human emission of carbon dioxide, methane, nitrous oxide, other gases, and particulate black carbon. Global warming is a global problem with regional impact.

Air is not owned privately; instead, it is common property (accessible to all individuals). As a result, air has historically been polluted without limit. This is the classic **tragedy of the commons**. The only known mechanism of limiting air pollution, aside from volunteerism, is government intervention. Intervention can take the form of setting up economic markets for the rights to emit pollution, limiting emissions from specific sources, requiring certain emission control technologies, or setting limits on pollutant concentrations and allowing the use of any emission reduction method to meet those limits.

Because government action usually requires consensus that a problem exists, the problem is severe enough to warrant action, and action taken will not have its own set of adverse consequences (usually economic), national governments did not act aggressively to control global air pollution problems until the 1970s and 1980s. For the most part, action was not taken earlier because lawmakers were not always convinced of the severity of air pollution problems. Even when problems were recognized, action was often delayed because industries used their political strength to oppose government intervention. Even today, government intervention is opposed by many industries and politicians out of often-misplaced concern that intervention will cause adverse economic consequences. In many developing countries, intervention is sometimes opposed because of the concern that developed countries are trying to inhibit economic expansion of the less-developed countries. In other cases, pollution is not regulated strictly due to the perceived cost of emission-control technologies and enforcement.

Despite the opposition to government intervention and although work still needs to be done, government intervention has proved effective in mitigating several of the major air pollution problems facing humanity. The problems mitigated but not eliminated include urban air pollution (in some countries), acid deposition (in some countries), and stratospheric ozone reduction. The problem of global climate change has not been controlled to date, and only recently has it been addressed on a global scale.

The purpose of this book is to discuss the history and science of major air pollution problems, the consequences of these problems, and efforts to control these problems through government intervention. Such a study involves the synthesis of chemistry, meteorology, radiative processes, particle processes, cloud physics, soil sciences, microbiology, epidemiology, economics, and law. The field of air pollution is a true interdisciplinary field.

This book is directed at students in the environmental, Earth, and atmospheric sciences. It was designed to be detailed enough to be used as a reference text as well. Chemical symbols and chemical equations are used, but all chemistry required is introduced in Chapter 1 – no previous knowledge of chemistry is needed. The text also describes a handful of physical laws; however, no calculus, geometry, or high math is needed.

ACKNOWLEDGMENTS

I thank several colleagues who reviewed different sections of this text. In particular, I am indebted to: (in alphabetic order) Joe Cassmassi, Frank Freedman, Ann Fridlind, Lynn Hildemann, Jinyou Liang, Cristina Lozej Archer, Gerard Ketefian, Nesrin Osalp, Ana Sandoval, Roberto San Jose, Alfred Spormann, Amy Stuart, and Azadeh Tabazadeh, who all provided comments, suggestions, or corrections relating to the text. I also thank Jill Nomura, William Jacobson, and Yvonne Jacobson for helping with graphics and editing of this text. Finally, I would like to thank the anonymous reviewers and students who used drafts of the text in an air pollution course and provided suggestions and corrections.

BASICS AND HISTORY OF DISCOVERY OF ATMOSPHERIC CHEMICALS

The study of air pollution begins with the study of chemicals that make up the air. These chemicals include molecules in the gas, liquid, or solid phases. Because the air contains so many different types of molecules, it is helpful to become familiar with important ones through the history of their discovery. Such a history also gives insight into characteristics of atmospheric chemicals and an understanding of how much our knowledge of air pollution today relies on the scientific achievements of alchemists, chemists, natural scientists, and physicists of the past. This chapter starts with some basic chemistry definitions, then proceeds to examine historical discoveries of chemicals of atmospheric importance. Finally, types of chemical reactions that occur in the atmosphere are identified, and chemical lifetimes are defined.

1.1. BASIC DEFINITIONS

Air is a mixture of gases and particles, both of which are made of atoms. In this section, atoms, elements, molecules, compounds, gases, and particles are defined.

1.1.1. Atoms, Elements, Molecules, and Compounds

In 1913, **Niels Bohr** (1885–1962), a Danish physicist, proposed that an **atom** consists of one or more negatively charged electrons in discrete circular orbits around a positively charged nucleus. Each **electron** carries a charge of -1 and a tiny mass.* The **nucleus** consists of 1–92 protons and 0–146 neutrons. **Protons** have a net charge of $+1$ and a mass 1,836 times that of an electron. **Neutrons** have zero net charge and a mass 1,839 times that of an electron. For the net charge of an atom to be zero, the number of electrons must equal the number of protons. Positively charged atoms have fewer electrons than protons. Negatively charged atoms have more electrons than protons. Positively or negatively charged atoms are called **ions**.

The average mass of protons plus neutrons in a nucleus is called the **atomic mass**. Electrons are not included in the atomic mass calculation because the summed mass of electrons in an atom is small in comparison with the summed masses of protons and electrons. The number of protons in an atomic nucleus is called the **atomic number**.

An **element** is a single atom or a substance composed of several atoms, each with the same atomic number (the same number of protons in its nucleus). Whereas all atoms of an element have a fixed number of protons, not all atoms of the element have the same number of neutrons. Atoms of an element with the same number of protons but a different number of neutrons are **isotopes** of the element. Isotopes of an element have different atomic masses but similar chemical characteristics.

The **periodic table of the elements**, developed in 1869 by Russian chemist **Dmitri Mendeleev** (1834–1907), lists elements in order of increasing atomic number. Table 1.1 identifies the first ten elements of the periodic table and some of their characteristics. The atomic mass of an element in the periodic table is the sum, over all isotopes of the element, of the percentage occurrence in nature of the isotope multiplied by the atomic mass of the isotope.

---

*Mass is an absolute property of a material. Mass, multiplied by gravity, equals weight, which is a force. Because gravity varies with location and altitude, weight is a relative property of a material. A person who is nearly "weightless" in space, where gravity is small, has the same mass, whether in space or on the surface of the Earth.

Table 1.1. Characteristics of the First Ten Elements in the Periodic Table

Element	Symbol	Number of Protons (Atomic Number)	Number of Neutrons in Main Isotope	Atomic mass (g mol^{-1})	Number of Electrons
Hydrogen	H	1	0	1.00794	1
Helium	He	2	2	4.00206	2
Lithium	Li	3	4	6.941	3
Beryllium	Be	4	5	9.01218	4
Boron	B	5	6	10.811	5
Carbon	C	6	6	12.011	6
Nitrogen	N	7	7	14.0067	7
Oxygen	O	8	8	15.9994	8
Fluorine	F	9	10	18.9984	9
Neon	Ne	10	10	20.1797	10

The simplest element in the periodic table is **hydrogen** (H), which contains one proton, no neutrons, and one electron. Hydrogen occurs in three natural isotopic forms. The most common (one proton and one electron) is that shown in Fig. 1.1. The other two are **deuterium**, which contains one proton, one neutron, and one electron, and **tritium**, which contains one proton, two neutrons, and one electron. **Helium** (He), also shown in Fig. 1.1, is the second simplest element and contains two protons, two neutrons, and two electrons.

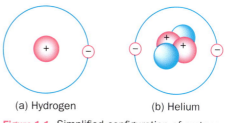

(a) Hydrogen   (b) Helium

Figure 1.1. Simplified configuration of protons, neutrons, and electrons in (a) a hydrogen atom and (b) a helium atom.

When one atom bonds to another atom of either the same or different atomic number, it forms a molecule. A **molecule** is a group of atoms of like or different elements held together by chemical forces. When a molecule consists of different elements, it is a compound. A **compound** is a substance consisting of atoms of two or more elements in definite proportions that cannot be separated by physical means.

1.1.2. Gases and Particles

Gases are distinguished from particles in two ways. First, a **gas** consists of individual atoms or molecules that are separated, whereas a **particle** consists of aggregates of atoms or molecules bonded together. Thus, a particle is larger than a single gas atom or molecule. Second, whereas particles contain liquids or solids, gases are in their own phase state. Particles may be further segregated into aerosol particles and hydrometeor particles.

An **aerosol** is an ensemble of solid, liquid, or mixed-phase particles suspended in air. An **aerosol particle** is a single liquid, solid, or mixed-phase particle among an ensemble of suspended particles. The term *aerosol* was coined by British physicochemist **Frederick George Donnan** (1870–1956) near the end of World War I (Green and Lane, 1969).

A **hydrometeor** is an ensemble of liquid, solid, or mixed-phase water particles suspended in or falling through the air. A **hydrometeor particle** is a single such particle. Examples of hydrometeor particles are cloud drops, ice crystals, raindrops, snowflakes, and hailstones. The main difference between an aerosol particle and a hydrometeor particle is that the latter contains much more water than the former.

Liquids in aerosol and hydrometeor particles may be pure or may consist of a solution. A solution is a homogeneous mixture of substances that can be separated into individual components on a change of state (e.g., freezing). A solution consists of a solvent, such as water, and one or more solutes dissolved in the solvent. Solids may be mixed throughout a solution, but are not part of the solution. In this text, pure water and solutes dissolved in water are denoted with "(aq)" for aqueous (dissolved in water). Gases are denoted with "(g)," and solids are denoted with "(s)."

Gases and aerosol particles may be emitted into the air naturally or anthropogenically or formed chemically in the air. Anthropogenic emissions are human-produced emissions, such as from fossil-fuel combustion or industrial burning. Hydrometeor particles generally form from physical processes in the air. Air pollution occurs when gases or aerosol particles, emitted anthropogenically, build up in concentration sufficiently high to cause direct or indirect damage to humans, plants, animals, other life forms, ecosystems, structures, or works of art.

1.2. HISTORY OF DISCOVERY OF ELEMENTS AND COMPOUNDS OF ATMOSPHERIC IMPORTANCE

In this section, the history of discovery of elements and compounds of atmospheric importance is discussed. Reactive elements that make up most gases are hydrogen (H), carbon (C), nitrogen (N), oxygen (O), fluorine (F), sulfur (S), chlorine (Cl), and bromine (Br). Unreactive elements in the air include helium (He), argon (Ar), krypton (Kr), neon (Ne), and xenon (Xe). Two radioactive elements of importance are polonium (Po) and radon (Rn). Aerosol particles contain the elements present in gases and possibly sodium (Na), magnesium (Mg), aluminum (Al), silicon (Si), potassium (K), calcium (Ca), iron (Fe), lead (Pb), or phosphorus (P). Tables 1.2 and 1.3 summarize the dates of discovery of elements and compounds, respectively, of atmospheric importance.

1.2.1. Solids and Liquids, Ancient World–1690

In this subsection, solids and liquids discovered from ancient times through the seventeenth century are discussed.

1.2.1.1. Iron

The first elements in the periodic table to be identified were the metals gold (Au), silver (Ag), mercury (Hg), copper (Cu), iron (Fe), tin (Sn), and lead (Pb). Many cultures, including the Egyptians and the Chaldeans, were aware of these metals. Of note were the Chaldeans (612–539 B.C.), who connected them with planets, identifying gold as the sun, silver as the moon, mercury as Mercury, copper as Venus, iron as Mars, tin as Jupiter, and lead as Saturn. Of these six metals, iron and lead are the most important in aerosol particles today. Iron (*ferrum* in Latin; *iarn* in Scandinavian) is a dense metal element that is the primary component of the Earth's core and the fourth most abundant element in the Earth's crust. It is emitted into the air in soil–dust particles. It is also the particulate element emitted in the greatest abundance from industrial sources today.

1.2.1.2. Lead

Lead (*plumbum* in Latin) is a dense bluish-white metal element. Lead was referred to in the Books of *Job* and *Numbers* as "biblicalx." The Roman Pliny the Elder (23–79 A.D.)

Table 1.2. Dates of Discovery of Elements of Atmospheric Importance

Element	Origin of Name or Previous Name	Year Discovered	Discoverer
Iron (Fe)	Named after *Iarn*	B.C.	?
Lead (Pb)	Previously *biblicalx, plumbum nigrum*	B.C.	?
Carbon (C)	Named from *carbo*, "charcoal"	B.C.	?
Sulfur (S)	Named from *sulvere, sulphurium*; previously *brimstone*	B.C.	?
Phosphorus (P)	Means "light bearer"	1669	Brand (Sweden)
Hydrogen (H)	Means "water producer"	<1520, 1766	Paracelsus (Switzerland), Cavendish (England)
Fluorine (F)	Named from *fluere*, "flow" or "flux"	1771	Scheele (Sweden)
Nitrogen (N)	Means "nitre maker"	1772	Rutherford (England)
Oxygen (O)	Means "acid maker"	1774, 1772–5	Priestley (England), Scheele (Sweden)
Chlorine (Cl)	Means "green gas"	1774	Scheele (Sweden)
Sodium (Na)	Named from *soda*	1807	Davy (England)
Potassium (K)	Named from *potash*	1807	Davy (England)
Calcium (Ca)	Named from *calx*	1808	Davy (England)
Silicon (Si)	Named from *silex*, "flint"	1823	Berzelius (Sweden)
Bromine (Br)	Means *stench*	1826	Balard (France)
Aluminum (Al)	Found in *alum*	1827	Wöhler (Germany)
Magnesium (Mg)	Named after the city of Magnesia	1830	Bussy (France)
Helium (He)	Named from *Helios*, Greek sun god	1868	Janssen (France), Lockyer (England)
Argon (Ar)	Named from *argos*, "lazy"	1894	Rayleigh (England), Ramsay (Scotland)
Krypton (Kr)	Named from *kryptos*, "concealed"	1898	Ramsey, Travers (Scotland)
Neon (Ne)	Named from *neos*, "new"	1898	Ramsey, Travers (Scotland)
Xenon (Xe)	Named from *xenos*, "guest"	1898	Ramsey, Travers (Scotland)
Polonium (Po)	Named after the country of Poland	1898	Curie, Curie (France)
Radon (Rn)	Originally named *radium emanation*	1900	Dorn (Germany)

called it *plumbum nigrum*. The English word "plumber" describes a person who installs or fixes lead pipes. Beginning in the 1920s, lead was emitted in gasoline. Due to its serious health effects, most countries have since banned leaded gasoline. Lead is also still emitted worldwide during certain industrial processes.

1.2.1.3. Sulfur

Elemental **sulfur** (*sulvere* in Sanskrit; *sulphurium* in Latin) is a nonmetallic, pale yellow, crystalline mineral found in volcanic and hot spring deposits, sedimentary beds, and salt domes. Sulfur was known by ancient Egyptian alchemists (Brown, 1913). It was also mentioned by the Greek Dioscorides and by Pliny the Elder in the first century A.D. The word **brimstone** (or "burn-stone," referring to its combustibility) is an Old English word for sulfur. In the Book of *Genesis*, "brimstone and fire" were said to have rained down on the cities of Sodom and Gomorrah, destroying them. If this event occurred, it may have been due to a volcanic eruption in which various forms of sulfur emanated. Sulfur in the air is primarily in the form of sulfur dioxide gas [$SO_2(g)$] and aqueous sulfuric acid [$H_2SO_4(aq)$].

1.2.1.4. Carbon

Elemental **carbon** (*carbo* in Latin, meaning "charcoal") was well known in the Ancient World, although it is unlikely that alchemists at the time knew that diamonds,

Table 1.3. Dates of Discovery of Compounds of Atmospheric Importance

Molecule	Chemical Formula	Mineral Name	Former Name, Alternate Name, or Meaning	Year Discovered	Discoverer
Calcium carbonate	$CaCO_3(s)$	Calcite, aragonite	Calcspar	B.C.	?
Sodium chloride	$NaCl(s)$	Halite	Common salt	B.C.	?
Potassium nitrate	$KNO_3(s)$	Nitre	Saltpeter, nitrum	B.C.	?
Sulfurous acid	$H_2SO_3(aq)$		Oil of sulfur, acidum volatile	B.C.	?
Sodium carbonate	$Na_2CO_3(s)$	Natrite	Nitrum, nator, nitron, natrum, soda ash, washing soda, salt-cake, calcined soda	B.C.	?
Calcium sulfate dihydrate	$CaSO_4 \cdot 2H_2O(s)$	Gypsum	"Plaster"	315 B.C.	Theophrastus (Greece)
Sulfuric acid	$H_2SO_4(aq)$		Oil of vitriol, acidum fixum, vitriolic acid, spirit of alum, spirit of vitriol	<1264	de Beauvais (France)
Ammonium chloride	$NH_4Cl(s)$	Sal ammoniac		<1400	Geber or later author
Molecular hydrogen	$H_2(g)$		Inflammable air	<1520, 1766	Paracelsus (Switzerland), Cavendish (England)
Nitric acid	$HNO_3(aq)$		Spirit of nitre	1585	Libavius (Germany)
Hydrochloric acid	$HCl(aq)$		Spirit of salt	<1640	Sala (Germany)
Carbon dioxide	$CO_2(g)$		Gas silvestre, fixed air	<1648, 1756	Van Helmont (Belgium), Black (Scotland)
Ammonia	$NH_3(g)$		Gas pingue, alkaline acid air	<1648, 1756	Van Helmont (Belgium), Black (Scotland)
Ammonium nitrate	$NH_4NO_3(s)$	Nitrammite	Nitrum flammans, ammonia–nitre, ammoniak–saltpeter	1648	Glauber (Germany)
Sodium sulfate	$Na_2SO_4(s)$	Thenardite	Sal mirabile, Glauber's salt	1648	Glauber (Germany)
Amonium sulfate	$(NH_4)_2SO_4(s)$	Mascagnite	Secret sal ammoniac	<1648	Glauber (Germany)
Potassium sulphate	$K_2SO_4(s)$	Arcanite	Sal polychrestum glaseri, Arcanum duplicatum	1663	Glaser (France)
Calcium nitrate	$Ca(NO_3)_2 \cdot 4H_2O(s)$	Nitrocalcite	Baldwin's phosphorus	<1669	Baldwin (Germany)
Magnesium sulfate	$MgSO_4 \cdot 7H_2O(s)$	Epsomite	Epsom salt	1695	Grew (England)
Magnesium carbonate	$MgCO_3(s)$	Magnesite	Magnesia alba	c. 1695	?
Nitrogen dioxide	$NO_2(g)$		Nitrous gas, red nitrous vapor	<1714, 1774	Ramazzini (Italy), Priestley (England)

Table 1.3. (continued)

Molecule	Chemical Formula	Mineral Name	Former Name, Alternate Name, or Meaning	Year Discovered	Discoverer
Molecular nitrogen	$N_2(g)$		Mephitic air	1772	Rutherford (England)
Nitric oxide	$NO(g)$		Nitrous air	1772	Priestley (England)
Nitrous oxide	$N_2O(g)$		Diminished nitrous air, laughing gas	1772	Priestley (England)
Hydrochloric acid	$HCl(g)$		Marine acid air, Muriatic gas	1772	Priestley (England)
Hydrofluoric acid	$HF(g)$		Fluor acid	1773	Scheele (Sweden)
Molecular oxygen	$O_2(g)$		Dephlogisticated air	1774, 1772–5	Priestley (England), Scheele (Sweden)
Chlorine gas	$Cl_2(g)$		Dephlogisticated marine (muriatic) acid gas, "green gas"	1774	Scheele (Sweden)
Acetaldehyde	$CH_3CHO(g)$			1774	Scheele (Sweden)
Carbon monoxide	$CO(g)$			1772– 1779	Priestley (England)
Sulfur dioxide	$SO_2(g)$		Vitriolic acid air	1774– 1779	Priestley (England)
Nitric acid	$HNO_3(g)$			1784	Priestley (England), Cavendish (England)
Hypochlorous acid	$HOCl(g)$			1830	Balard (France)
Ozone	$O_3(g)$		Ozien, "to smell"	1840	Schonbein (Germany)

graphite (plumbago), and charcoal all contained carbon. Carbon in diamonds and graphite is in pure crystalline form. In charcoal, coal, and coke, it takes on a variety of shapes and structures. In the Ancient World, diamonds were valued only for their rarity, not for their beauty, because diamonds were not cut (and thus did not shine) until the fifteenth century. In the Ancient World, graphite was used to make black marks on paper and charcoal was used as a fuel. Today, the emission of elemental carbon (also called *black carbon*) in the form of soot particles exacerbates global warming, visibility, and health problems.

1.2.1.5. Sodium Carbonate (Solid)

Sodium carbonate [$Na_2CO_3(s)$] is a crystal mineral first found by the Egyptians in the Lakes of Natron, a group of six lakes to the west of the Nile Delta. The Egyptians called it *nitrum*. Its name was modified to *nator* by the Hebrews, *nitron* by the Greeks, and *natrum* in the fifteenth century. Today, its mineral name is natrite. For centuries, it has been used as an ingredient in soaps. Some chemical industry names for it have been washing soda, soda ash, and salt cake. The manufacture of sodium carbonate for use in soaps caused acid deposition problems in England and France in the nineteenth century (Chapter 10). In the air, sodium carbonate is present in soil-dust particles.

1.2.1.6. Calcium Carbonate (Solid)

Calcium carbonate [$CaCO_3(s)$] is a crystal present in pure form in the minerals **calcite** and **aragonite** and in mixed form in limestone, marble, chalk, and shells and skeletons of invertebrates. **Limestone** is sedimentary rock containing calcite or **dolomite** [$CaMg(CO_3)_2(s)$], **marble** is recrystallized limestone, and **chalk** is fine-grained rock made of skeletons of microorganisms. In the ancient world, chalk was used for writing. In the air, calcium carbonate is a component of soil-dust particles. The name *calcite* originates from the word "calcspar," itself derived from the Greek word for limestone, *khálix*.

1.2.1.7. Sodium Chloride (Solid)

Sodium chloride [$NaCl(s)$], a crystal mineral formed from the evaporation of ocean water, was well-known in the ancient world. It was found mixed with earthy material and mentioned in the Old Testament to "lose its savor" on its exposure. Today, its mineral name is **halite**, from the Greek word *hals* ("salt"). In the air, sodium chloride is present in sea-spray particles.

1.2.1.8. Potassium Nitrate (Solid)

Potassium nitrate [$KNO_3(s)$] is a crystal mineral also called **saltpeter** ("salt of rock") because it was often found as a saltlike crust on rocks. Saltpeter was an ingredient of Greek fires. In the fifteenth century, it was called *nitrum* (the same early name as sodium carbonate). Today, its mineral name is **nitre**. Potassium nitrate forms chemically in soil-dust and sea-spray particles and may be the most abundant nitrogen-containing solid in the air.

1.2.1.9. Sulfurous Acid (Aqueous)

Ancient Egyptian alchemists obtained **sulfurous acid** [$H_2SO_3(aq)$] ("oil of sulfur") by combusting elemental sulfur in the presence of water. Such burning was also carried out in Homer's time for the purpose of fumigation. Sulfurous acid's use in bleaching wool is mentioned by Pliny the Elder. In the air, sulfurous acid, a precursor to acid deposition, forms when sulfur dioxide gas dissolves in water-containing particles.

1.2.1.10. Calcium Sulfate Dihydrate (Solid)

Calcium sulfate dihydrate [$CaSO_4\text{-}2H_2O(s)$] is a crystal mineral, more commonly known as **gypsum** (*gypsos*, "plaster" in Greek). Gypsum was first referred to in 315 B.C. by the Greek botanist and alchemist, **Theophrastus** (371–286 B.C.), born in Lesbos, who wrote 10 books on botany, stones, metals, and minerals. Gypsum is a naturally occurring mineral that appears worldwide in soils and aerosol particles. It forms chemically when aqueous sulfuric acid reacts with the mineral calcite. When aerosol particles containing sulfuric acid deposit onto marble statues (which contain calcite), a gypsum crust also forms. Gypsum soil beds are mined to produce plaster of paris, obtained by heating pure gypsum and adding water. **Plaster of paris** was named such because early Parisians found gypsum in the clays and muds of the Paris basin and used the gypsum to make plaster and cement. Gypsum is possibly the most common sulfur-containing solid in the atmosphere.

1.2.1.11. Ammonium Chloride (Solid)

Geber (or Abu Abdallah Jaber ben-Hayyam al-Kufi, Fig. 1.2) was an Arabian alchemist who lived about 750–800 A.D. Although the writings attributed to him may have been forged in the thirteenth century, it is clear that Geber or the writer was aware of *sal ammoniac* [$NH_4Cl(s)$, **ammonium chloride**], a mineral crystal obtained from the Libyan desert near the temple of Jupiter Ammon (the ultimate source of the name for the gas, **ammonia**). Ammonium chloride can form when ammonia gas enters sea-spray particles, which contain chlorine. It may be the most abundant ammonium-containing solid in the air.

1.2.1.12. Sulfuric Acid (Aqueous)

Vincent de Beauvais (1190–1264), a French philosopher, mentions the solvent power of the liquid acid distilled from the natural crystal, potassium alum [$KAl(SO_4)_2 \cdot 12H_2O(s)$]. The acid was probably dissolved **sulfuric acid** [$H_2SO_4(aq)$], and de Beauvais may have been the first to record its observation. In 1585, **Andreas Libavius** (1540–1616; Fig. 1.3), a German chemist who wrote one of the first noteworthy chemical textbooks, found that sulfuric acid could also be extracted from "green vitriol" (ferrous sulphate, $FeSO_4 \cdot 7H_2O(s)$, a blue-green natural crystal) and obtained by burning elemental sulfur with saltpeter [$KNO_3(s)$] in the presence of liquid water. Sulfuric acid is present in aerosol particles and responsible for most acid deposition problems today.

Figure 1.2. Geber (c. 750–800).

1.2.1.13. Nitric Acid (Aqueous)

Libavius also reacted elemental sulfur with dissolved **nitric acid** [$HNO_3(aq)$], indicating that dissolved nitric acid was known during his time. It was most likely formed from the reaction of $H_2SO_4(aq)$ with $KNO_3(s)$. Today, nitric acid is an abundant component of aerosol particles.

1.2.1.14. Hydrochloric Acid (Aqueous)

Angelus Sala (1575–1640), a German physician, produced ammonium chloride [$NH_4Cl(s)$] by treating **ammonium carbonate** [$(NH_4)_2CO_3(s)$] with dissolved **hydrochloric acid** [$HCl(aq)$]. This may be the first recorded use of HCl(aq). Hydrochloric acid was probably obtained by reacting common salt [$NaCl(s)$] with sulfuric acid [$H_2SO_4(aq)$]. Hydrochloric acid is an abundant component of sea-spray particles.

Figure 1.3. Andreas Libavius (1540–1616).

1.2.1.15. Sodium Sulfate Decahydrate, Ammonium Nitrate, and Ammonium Sulfate (Solids)

Johann Rudolf Glauber (1604–1688; Fig. 1.4), a German chemist, discovered what is now called **Glauber's salt** [$Na_2SO_4 \cdot 10H_2O(s)$, **sodium sulphate decahydrate**]. He called it *sal mirabile* and referred to it as a universal medicine. Glauber was also aware of the mineral **ammonium nitrate** [$NH_4NO_3(s)$], which he called *nitrum* *flammans*, and the mineral **ammonium sulfate** [$(NH_4)_2SO_4(s)$], which he called *secret sal ammoniac*. In his book, *Miraculum Mundi*, he provided a recipe for producing ammonium sulfate and stated that it may have previously been used by two alchemists, Paracelsus and Van Helmont. Ammonium sulfate is also a natural sublimation product of the fumaroles of Mount Vesuvius and Mount Etna. Mascagni first described the natural occurrence of this salt; therefore, its mineral name today is **mascagnite**. Without the hydrated water, sodium sulfate is a mineral called **thenardite**, named after Baron Louis Jacques Thenard (1777–1857), who found it in Espartinas salt lake, near Madrid, Spain. Ammonium nitrate is not a common naturally occurring mineral in soil, although it was found to exist in Nicojack Cavern, Tennessee. Its mineral name is **nitrammite**, named after its composition. All three salts form chemically within aerosol particles.

Figure 1.4. Johann Rudolf Glauber (1604–1688).

1.2.1.16. Potassium Sulfate (Solid)

In 1663, **Christopher Glaser** (1615–1673), a French apothecary to Louis XIV, combined sulfur with melted saltpeter [$KNO_3(s)$] to form the crystal **potassium sulfate** [$K_2SO_4(s)$], which he named *sal polychrestum glaseri*. Its present mineral name is **arcanite**, from the Latin words *arcanum duplicatum*, an early alchemist name for the salt. Arcanite is a chemically produced component of aerosol particles.

1.2.1.17. Calcium Nitrate (Solid)

In 1675, **Christopher Baldwin** (1600–1682) wrote a book in which he discussed a preparation of chalk [made primarily of calcite, $CaCO_3(s)$] with nitric acid [$HNO_3(aq)$], to produce the crystal **calcium nitrate** [$Ca(NO_3)_2(s)$], which is phosphorescent in the dark. Because of its appearance, he named the substance *phosphorus*, meaning "light-bearer." It is now known as **Baldwin's phosphorus** because it differs from elemental phosphorus. Elemental **phosphorus** (P), a nonmetallic substance that also glows in the dark, was discovered in Germany in 1669 by **Hennig Brand** (?–c.1692) of Sweden by distilling a mixture of sand and evaporated urine. The extraction of phosphorus was replicated by Johann Kunckel (1630–1750) of Germany, who knew both Baldwin and Brand, and called phosphorus the "**phosphorus of Brand**." Kunckel published a treatise on phosphorus in 1678. Calcium nitrate forms chemically in aerosol particles. Phosphorous is a component of the Earth's crust and of soil-dust particles.

1.2.1.18. Magnesium Sulfate (Solid)

In 1695, **Nehemiah Grew**, a London physician, evaporated water from the mineral spring at Epsom to obtain the crystal, **magnesium sulfate** [$MgSO_4\text{-}7H_2O(s)$], which was subsequently called **Epsom salt**. Magnesium sulfate can be found in the air as a constituent of soil-dust and sea-spray particles.

1.2.2. Studies of Gases in the Air, 1450–1790

Gases were more difficult to observe and isolate than were liquids or solids, so the study of gases began only after many liquids and solids had been investigated. In this subsection, the history of discovery of gases from the fifteenth through eighteenth centuries is discussed.

1.2.2.1. Water Vapor

Although water vapor was known in the ancient world, changes in its abundance were not detected until the fifteenth century. In 1450, **Nicolas Cryfts** suggested that changes in atmospheric water vapor could be measured with a **hygroscope**, which could be constructed of dried wool placed on a scale. A change in weight of the wool over time would represent a change in the water-vapor content of the air. Capitalizing on Cryfts notes, **Leonardo da Vinci** (1452–1519) built such a hygroscope. Wood and seaweed were later used in place of wool. In the seventeenth century, gut, string, cord, and hair were also used to measure changes in water vapor because the lengths of these materials would change on their absorption of water from the air.

1.2.2.2. Molecular Hydrogen (Gas)

Paracelsus (1493–1541; Fig. 1.5), an alchemist born near Zurich, may have been the first to observe what is now known as **hydrogen gas** or **molecular hydrogen** [$H_2(g)$]. He found that when sulfuric acid was poured over certain metals, it gave off an inflammable vapor. In 1766, Henry Cavendish found the same result, but isolated the vapor's properties and is more well-known for the discovery of molecular hydrogen. Molecular hydrogen is a well-mixed gas in today's lower atmosphere.

Figure 1.5. Paracelsus (1493–1541).

1.2.2.3. Ammonia and Carbon Dioxide (Gases)

John Baptist Van Helmont (1577–1644), born in Belgium, introduced the term **gas** into the chemical vocabulary. He produced what he called *gas silvestre* ("gas that is wild and dwells in out-of-the-way places") by fermenting alcoholic liquor, burning charcoal, and acidifying marble and chalk. The gas he discovered in all three cases, but did not know at the time, was **carbon dioxide** [$CO_2(g)$]. Another gas he produced was an inflammable vapor evolved from dung. He called this gas *gas pingue*, which was probably impure **ammonia** [$NH_3(g)$]. Today, carbon dioxide is thought to be the main cause of global warming. Ammonia, produced naturally and anthropogenically, dissolves and reacts in aerosol particles.

1.2.2.4. Fire–Air

In 1676, **John Mayow** (1643–1679; Fig. 1.6), an English physician, found that air appeared to contain two components, one that allowed fire to burn and animals to breathe (which Mayow called *nitro–aereo*, or "**fire–air**"), and another that did not.

When he placed a lighted candle and a small animal in a closed vessel, the lighted candle went out before the animal died. When he placed only the animal in the vessel, the animal took twice as long to die. Thus, Mayow showed that air was diminished by combustion and breathing. Fire–air later turned out to be molecular oxygen [$O_2(g)$].

Figure 1.6. John Mayow (1643–1679).

1.2.2.5. Phlogisticated Air

In 1669, Johann Joachim Becher (1635–1682), a German physician, took a step backward in the understanding of the composition of air when he wrote *Physica Subterranea*. In this book, he stated that every combustible material contains different amounts of *terra mercurialis* ("fluid or mercurial earth," thought to be mercury), *terra lapidia* ("strong or vitrifiable earth," thought to be salt), and *terra pinguis* ("fatty earth," thought to be sulfur). During combustion, *terra pinguis* was thought to be expelled to the air. The principle that every combustible material releases its "source" of combustion was not new, but it was more specific than were previous theories.

One of Becher's followers was Georg Ernst Stahl (1660–1734). In 1702, Stahl published *Specimen Becherianum*, in which he restated that every material contains a special combustible substance that escapes to the air when the material is burned. Stahl called the combustible substance, previously named *terra pinguis* by Becher, phlogiston after the Greek word *phlogizein*, "to set on fire." Stahl felt that *phlogiston* disappeared either as fire or as soot, which he felt was the purest form of *phlogiston*. Becher's and Stahl's theories of *terra pinguis* and *phlogiston* turned out to be incorrect because combustion occurs when oxygen from the air combines with a substance on heating, and the resulting oxide of the substance is released as a gas, not when a material alone in a substance is released on heating.

Interestingly, in *Specimen Becherianum*, Stahl was the first to point out that sulfurous acid is more volatile (evaporates more readily) than is sulfuric acid. He called the former *acidum volatile* and the latter *acidum fixum*. He also noted that sulfuric acid is the stronger acid.

1.2.2.6. Carbon Dioxide Again – Fixed Air

In 1756, Joseph Black (1728–1799; Fig. 1.7), a Scottish physician and chemist, performed an experiment in which he heated magnesium carbonate [$MgCO_3(s)$], called *magnesia alba* ("white magnesia") at the time. On heating, $MgCO_3(s)$ lost weight, producing a heavy gas that neither sustained a flame nor supported life. When the gas was exposed to quicklime [$CaO(s)$, calcium oxide], a white-gray crystal, the weight was reabsorbed. Black called the gas "fixed air" because of its ability to attach or "fix" to compounds exposed to it. The fixed air turned out to be carbon dioxide [$CO_2(g)$], and when it was reabsorbed on exposure to $CaO(s)$, it was really forming calcium carbonate [$CaCO_3(s)$]. Fixed air was renamed to carbon dioxide in 1781 by the French chemist, **Antoine Laurent Lavoisier** (1743–1794). What Black did not recognize was that fixed air, or $CO_2(g)$, had previously been discovered by Van Helmont more than a century earlier. In 1756, Black also isolated ammonia gas [$NH_3(g)$], previously observed by Van Helmont and later called **alkaline acid air** by Joseph Priestley. Black is separately known for making the first systematic study of a chemical reaction and developing the concepts of latent heat and specific heat.

1.2.2.7. Molecular Hydrogen Again – Inflammable Air

In 1766, **Henry Cavendish** (1731–1810; Fig. 1.8), an English chemist and physicist, followed up Black's work by producing a gas he called "**inflammable air**." This gas was obtained by diluting either sulfuric acid [$H_2SO_4(aq)$] or hydrochloric acid [$HCl(aq)$] with water and pouring the resulting solution on a metal, such as iron, zinc, or tin. This experiment was similar to that of Paracelsus, who also observed an inflammable vapor. Cavendish thought "inflammable air" was *phlogiston*, but this turned out to be incorrect. Nevertheless, Cavendish isolated the properties of the gas. In 1783, he found that exploding a mixture of the gas with air produced water. Subsequently, Lavoisier called the gas **hydrogen**, the "water producer." More specifically, the gas was **molecular hydrogen** [$H_2(g)$].

In other experiments, Cavendish exposed marble, which contains $CaCO_3(s)$, to hydrochloric acid [$HCl(aq)$] to produce $CO_2(g)$, as Van Helmont had done earlier. Cavendish, took the further step of measuring the properties of $CO_2(g)$. Cavendish is also known for studying the weights of gases and the density of the Earth. In 1783, after oxygen had been discovered, Cavendish calculated that air contained 20.83 percent oxygen by volume, close to the more accurate measurement today of 20.95 percent.

1.2.2.8. Molecular Nitrogen (Gas) – Mephitic Air

In 1772, **Daniel Rutherford** (1749–1819; Fig. 1.9) performed an experiment by which he allowed an animal to breathe the air in an enclosed space until the animal died [removing the molecular oxygen, $O_2(g)$, which had not been discovered yet]. He then exposed the remaining air to the crystal **caustic potash**

Figure 1.7. Joseph Black (1728–1799).

Figure 1.8. Henry Cavendish (1731–1810).

[$KOH(s)$, potassium hydroxide or pot ashes], obtained by burning wood in a large iron pot. $CO_2(g)$ in the remaining air reacted with caustic potash, forming **pearl ash** or **potash** [$K_2CO_2(s)$, potassium carbonate]. The residue after $CO_2(g)$ was removed could not sustain life; thus, Rutherford called it "mephitic [poisonous or foul-smelling] air." Mephitic air is now known as molecular nitrogen gas [$N_2(g)$], which makes up nearly 80 percent of air by volume. The name nitrogen, the "nitre maker," was given by Jean-Antoine Chaptal (1756–1832), a French industrial chemist, because nitrogen was found to be a constituent of the crystal nitre [$KNO_3(s)$].

1.2.2.9. Molecular Oxygen (Gas) – Dephlogisticated Air

Molecular oxygen gas [$O_2(g)$] was discovered independently by two chemists, on August 1, 1774, by Joseph Priestley (1733–1804; Fig. 1.10) and sometime between 1772 and 1775 by Karl Wilhelm Scheele (1742–1786; Fig. 1.11), a Swedish chemist. Although both chemists discovered oxygen near the same time, Priestley announced his discovery in 1774, and Scheele published his discovery in 1777.

To obtain oxygen, Priestley burned the element mercury (Hg), a silvery-white liquid metal, in air to form bright red mercuric oxide [$HgO(s)$], a powder. He then heated the mercuric oxide in a container from which all air had been removed. Burning mercuric oxide in a vacuum released oxygen so that the only gas in the container was molecular oxygen. Due to the container's high oxygen content, flammable material burned more readily in the container than in regular air.

Figure 1.9. Daniel Rutherford (1749–1819).

(a)   (b)

Figure 1.10. (a) Joseph Priestley (1733–1804). (b) Reconstruction of Priestley's oxygen apparatus.

(a) (b)

Figure 1.11. (a) Karl Wilhelm Scheele (1742–1786). (b) Scheele's laboratory, with oven in the center.

Priestley called the new gas "**dephlogisticated air**" because he incorrectly believed that burning occurred so brightly because the gas contained no *phlogiston*. He thought that, on the burning of a substance, the substance emitted *phlogiston* into the gas, causing the flame to die out eventually.

Scheele independently isolated molecular oxygen in at least three ways: heating manganic oxide [$Mn_2O_3(s)$] (a black powder), heating red mercuric oxide [$HgO(s)$], and heating a mixture of nitric acid [$HNO_3(aq)$] and potassium nitrate [$KNO_3(s)$].

At the end of 1774, Priestley went to Paris to explain his method of preparing dephlogisticated air to Lavoisier, who subsequently experimented with the gas for 12 years and revised its name to **oxygen**, the "acid maker," because he believed (incorrectly) that all acids contained oxygen. Almost all oxygen in the air is in the form of **molecular oxygen** [$O_2(g)$].

Subsequently, Lavoisier formalized the oxygen theory of combustion and proved the law of conservation of mass, which states that, in a chemical reaction, mass is conserved. For example, when sulfur, phosphorus, carbon, or another solid is burned, its gas-plus-solid-phase mass increases by an amount equal to the loss in mass of oxygen from the air. Lavoisier used the fact that oxygen combines with a solid to form an oxide of the solid that is released to the air during combustion to disprove the theory of *phlogiston*, which was premised on the belief that only material in the original solid was released on combustion. Lavoisier similarly showed that rusting is a mass-conserving process by which oxygen from the air combines with a solid to form an oxide of the solid.

In 1775–6, Lavoisier found that diamonds contain pure carbon and produce carbon dioxide when heated. In 1781, he renamed Black's fixed air to carbon dioxide and determined its elemental composition. Lavoisier also devised the first chemical system of nomenclature and specified that matter exists in three states – gas, liquid, and solid. He found that gases could be reduced to liquids or solids by cooling the air. Lavoisier is said to be the founder of modern chemistry.

Figure 1.12. "The Arrest of Lavoisier," (1876) by L. Langenmantel. Courtesy of the Edgar Fahs Smith Collection, University of Pennsylvania Library.

Unfortunately, Lavoisier was arrested during the French Revolution (Fig. 1.12) because he was a member of an unpopular political group, the Ferme Générale. On May 8, 1794, after a trial of less than a day, he and 27 others were guillotined and his body was thrown into a common grave.

Ironically, Priestley was attacked for his staunch defense of the principles of the French Revolution. On July 14, 1791, he lost his house, library, and laboratory in Birmingham, England, to a fire set by a mob angry at his public support of the revolution (Fig. 1.13). Priestley ultimately fled to the United States, where he lived until 1804.

1.2.2.10. Additional Discoveries by Priestley

During his career, Priestley discovered several additional gases relevant to air pollution. Between 1767 and 1773, while working at Mill Hill Chapel, in Leeds, Yorkshire, he isolated **nitric oxide** [NO(g), "nitrous air"], **nitrogen dioxide** [NO_2(g), "red nitrous vapor"], **nitrous oxide** [N_2O(g), "diminished nitrous air"], and **hydrochloric acid** gas [HCl(g), "marine acid air"]. Nitrogen dioxide may have been observed earlier by **Bernardo Ramazzini** (1633–1714), an Italian medical doctor and early pioneer in industrial medicine. Priestley also discovered **carbon monoxide** [CO(g)], and **sulfur dioxide** [SO_2(g), "vitriolic acid air"]. He formed gas-phase **nitric acid** [HNO_3(g)], although Cavendish uncovered its composition. Priestley is also known for inventing the eraser and carbonated water (soda pop) and for being the first to observe photosynthesis.

Figure 1.13. The destruction of Priestley's house, library, and laboratory, Fair Hill, Birmingham, 1791. Courtesy of the Edgar Fahs Smith Collection, University of Pennsylvania Library.

Today, NO(g), NO_2(g), and CO(g) are emitted during fossil-fuel combustion and biomass burning and are components of urban smog. N_2O(g) is produced from microbial metabolism, fossil-fuel combustion, and biomass burning. HCl(g) is emitted by volcanos, evaporates from sea-spray particles, and is a product of chlorine reactions in the upper atmosphere. SO_2(g) is emitted by volcanos, coal-fired power plants, and vehicles. NO_2(g) and SO_2(g) are precursors of acid deposition.

1.2.2.11. Hydrofluoric Acid (Gas)

A meticulous artist at his craft, Scheele also discovered **hydrofluoric acid** gas [HF(g)] in 1773. Scheele named HF(g) "fluor acid" after the crystal mineral **fluorspar** [CaF_2(s), fluorite], which contains it. The name *fluorspar* was coined in 1529 by Georigius Agricola from the Latin and French word *fluere*, which means "flow" or "flux," because feldspar appeared to flow. Elemental **fluorine** (F) was isolated from HF(g) only in 1886 by French chemist **Henri Moissan** (1852–1907). Prior to that time, at least two chemists died from toxic exposure trying to isolate F from HF(g). Moissan won a Nobel Prize for isolating fluorine and inventing the electric arc furnace. Today, HF(g) is a product of chemical reactions in the upper atmosphere involving anthropogenically emitted fluorine compounds.

1.2.2.12. Chlorine (Gas)

In 1774, Scheele discovered **chlorine gas** [$Cl_2(g)$], and thus the element chlorine (Cl), by reacting dissolved hydrochloric acid [$HCl(aq)$] with pyrolusite [$MnO_2(s)$]. Chlorine gas is a dense, odorous, greenish-yellow, corrosive, toxic gas. He called it *dephlogisticated marine acid gas*. Lavoisier changed the name to *oxymuriatic acid* because he incorrectly thought it contained oxygen and chlorine. The name was eventually changed to *chlorine*, the "green gas," in 1810 by Sir Humphry Davy, who showed that chlorine was an element and did not contain oxygen. Today, $Cl_2(g)$ is a product of chemical reactions, primarily in the upper atmosphere.

1.2.3. Discoveries after 1790

After 1790, the pace at which gas, liquid, and solid chemicals were discovered increased. In the following subsections, more chemicals of atmospheric importance are discussed.

1.2.3.1. Elemental Potassium, Sodium, Calcium, and Chlorine

In 1807–8, **Sir Humphry Davy** (1778–1829; Fig. 1.14), who, along with Priestley, is the most well-known British chemist, developed electrolysis, which led to the discovery of the elements **potassium** (K), **sodium** (Na), **calcium** (Ca), and **barium** (Ba). **Electrolysis** is the passage of an electric current through a solution to break down a compound or cause a reaction. Potassium was isolated by electrolysis from **caustic potash** [potassium hydroxide, $KOH(s)$]. Potassium is the seventh-most abundant element in the Earth's crust and is emitted into the air in soil-dust and sea-spray particles. Sodium was isolated by electrolysis from **caustic soda** [sodium hydroxide, $NaOH(s)$]. The name *sodium* derives from the Italian word *soda*, a term applied to all alkalis in the Middle Ages. Sodium is the sixth-most abundant element in the Earth's crust and is emitted in soil-dust and sea-spray particles. Calcium was isolated by electrolysis from quicklime [$CaO(s)$]. The name *calcium* was derived from the word *calx*, the name the Romans used for lime. Calcium is the fifth-most abundant element in the Earth's crust and is emitted in soil-dust and sea-spray particles.

In 1810, Davy also named the element **chlorine**, previously called *oxymuriatic acid*. He proved that chlorine was an element and that muriatic gas [$HCl(g)$, hydrochloric acid gas] contains chlorine and hydrogen, but no oxygen. He similarly proved that hydrofluoric acid gas [$HF(g)$] contains no oxygen. Both proofs contradicted Lavoisier's theory that all acids contained oxygen.

Figure 1.14. Sir Humphry Davy (1778–1829).

1.2.3.2. Elemental Silicon and Chemical Symbols

A contemporary of Davy, **Jöns Jakob Berzelius** (1779–1848; Fig. 1.15) of Sweden discovered the elements **silicon** (Si) in 1823, selenium (Se) in 1817, and thorium (Th) in 1828. He also spent 10 years determining the atomic or molecular weights of more than 2,000 elements and compounds, publishing the results in 1818 and 1826. Berzelius isolated *silicon*, a name derived from the Latin word *silex*, meaning "flint," by fusing iron, carbon, and the crystal quartz [$SiO_2(s)$]. Silicon is the second-most abundant element in the Earth's crust, after oxygen, and is present in soil-dust particles.

Berzelius's most noticeable achievement was to invent a system of chemical symbols and notation. For elements, he used the first one or two letters of the element's Latin or Greek name. For example, oxygen was denoted with an O, hydrogen with an H, mercury with Hg (hydrargyrum), and lead with Pb (plumbum). For compounds with more than one atom of an element, he identified the number of atoms of the element with a subscript. For example, he identified water with H_2O.

1.2.3.3. Elemental Bromine and Hypochlorous Acid (Gas)

In 1826, Antoine–Jérôme Balard (1802–1876), a French apothecary, accidentally discovered the element bromine (Br) after analyzing the "bittern" (saline liquor) that remained after common salt had crystallized out of concentrated water in a salt marsh near the Mediterannean sea. *Bromine* means "stench" in Greek. It is a heavy, reddish-brown liquid that evaporates at room temperature to a red gas that irritates the throat and eyes and has a strong smell. It is the only nonmetallic element that can be in the liquid phase at room temperature. Balard is also known for his discovery of hypochlorous acid gas [$HOCl(g)$]. Bromine and hypochlorous acid contribute to ozone destruction in the upper atmosphere today.

1.2.3.4. Organic Chemistry

Baron Justus von Liebig (1803–1873; Fig. 1.16) is considered the founder of organic and agricultural chemistry. Not only did he discover numerous organics and identify their properties, but he also introduced a systematic method of determining the empirical composition of organics, discovered several organic radicals, and isolated the atmospheric versus soil sources of plant nutrients, including carbon dioxide, water, and ammonia. He suggested that mineral fertilizers should be added to plants when their soils become depleted in nutrients.

Figure 1.15. Jöns Jakob Berzelius (1779–1848).

Figure 1.16. Baron Justus von Liebig (1803–1873).

1.2.3.5. Elemental Magnesium

Magnesium (Mg) is the eighth-most abundant element in the Earth's crust. It is present in soil-dust and sea-spray particles. Although Sir Humphry Davy isolated an impure form of magnesium in 1808, it was not until 1828 that French chemist Antoine-Alexandre-Brutus Bussy isolated it in a pure state by reacting magnesium chloride with metallic potassium. Magnesium is named from magnesia [MgO(s), magnesium oxide], the crystal that contains it. Magnesia is named after the ancient city of Magnesia in Thessaly, a region of east-central Greece. The Greeks mined magnesia as an ingredient in the philosopher's stone, an elixir and a mineral that was believed to have the ability to convert metal into gold.

1.2.3.6. Elemental Aluminum

In 1761, chemist Louis Bernard Guyton de Morveau (1737–1816) named the base in potassium alum [KAl(SO$_4$)$_2$·12H$_2$O(s)] alumine. In 1807, Davy proposed the name alumium for the metal although it had yet to be isolated. An impure form of aluminum was isolated by Oersted in 1825, but it was not until 1827 that an associate of Liebig, Friedrich Wöhler (1800–1882), a German chemist, isolated a pure form of the metal and renamed it aluminum. Aluminum is the most abundant metal in the Earth's crust. Pure aluminum is silvery-white. Aluminum is present in soil-dust particles.

Figure 1.17. Christian Friederich Schönbein (1799–1868).

1.2.3.7. Ozone (Gas)

In 1839, one of the most important trace gases in the air, ozone [O$_3$(g)], was discovered by German chemist Christian Friederich Schönbein (1799–1868; Fig. 1.17). Schonbein named ozone after the Greek word, *ozien*, which means "to smell," because ozone has a pungent, sweet smell. Schönbein was also known for his discovery of gun-cotton in 1846. This compound is produced by reaction of either nitric acid or nitric plus sulfuric acid with a carbonaceous compound. Gun-cotton was the first of a group of "nitro-compound" explosives invented.

1.2.3.8. Noble Gases

The air contains several inert noble gases in trace quantities, including helium (He), argon (Ar), neon (Ne), krypton (Kr), and xenon (Xe). All were discovered between 1868 and 1898. In 1868, Pierre Janssen (1824–1907), a French astronomer, observed a yellow line in the spectrum of the sun's chromosphere. Because no known element on Earth could account for this line, he thought it was due to an element unique to the sun. Joseph Norman Lockyer (1836–1920), an English astronomer, confirmed Janssen's findings, and named the new element helium (He), after *Helios*, the Greek god of the sun. The element was not discovered on Earth until 1895, when Sir William Ramsay (1852–1916), a Scottish chemist, found it in the mineral clevite. Swedish chemists Per Theodor Cleve (after whom *clevite* is named) and Nils Abraham Langlet found helium in the mineral at about the same time. Helium is the most abundant element in the universe next to hydrogen. On Earth, helium is emitted to the air following the decay of radioactive minerals.

In 1894, Lord Baron Rayleigh, an English physicist born John William Strutt (1842–1919), found that nitrogen gas from the air was 0.5 percent heavier than was that prepared chemically. He and Sir Ramsay found that the difference was due to an additional gas that they called argon (Ar), after the Greek word *argos*, meaning "lazy" in reference to the inert qualities of the gas. The two shared a Nobel Prize for their discovery. Argon forms from the radioactive decay of potassium (K). In his 1898 book *War of the Worlds*, H. G. Wells wrote that Martians used "toxic brown argon gas" to attack London, but were subdued by the common cold. Argon is neither brown nor poisonous at typical atmospheric concentrations. It is colorless and odorless as a gas and liquid. Sir Ramsay, together with M. W. Travers, went on to discover the elements neon (Ne), krypton (Kr), and xenon (Xe), all in 1898. All three are named after Greek words: *neos* ("new"), *kryptos* ("concealed"), and *xenos* ("guest"), respectively. The source of krypton and xenon is the radioactive decay of elements in the Earth's crust, and the source of neon is volcanic outgassing.

1.2.3.9. Radioactive Gases

In the twentieth century, two radioactive elements of atmospheric importance, polonium (Po) and radon (Rn), were discovered. These elements are carcinogenic and are found in the air of many homes overlying uranium-rich soils. In 1898, French chemists Pierre (1859–1906) and Marie Curie (1867–1934; Fig. 1.18) discovered polonium, which was named after Marie Curie's native country, Poland. In 1903, Pierre and Marie Curie, along with French physicist Antoine Henri Becquerel (1852–1908), won a Nobel Prize for their fundamental research on radioactivity. In 1911, Marie Curie won a second prize for her discoveries of polonium and radium (Ra), a radon precursor. Radon, itself, was discovered in 1900 by German physicist Friedrich Ernst Dorn (1848–1916), who called it radium emanation because it is a product of radioactive decay of radium. The name *radium* is from the Latin word *radius*, meaning "ray." Ramsay and Gray, who isolated radon and determined its density, changed its name to niton in 1908. In 1923, niton was renamed radon.

Figure 1.18. Marie Curie (1867–1934).

1.3. CHEMICAL STRUCTURE AND REACTIVITY

In this section, the structure and reactivity of a few compounds identified in earlier sections are discussed. Table 1.4 shows the chemical structure of selected compounds. Single, double, and triple lines between atoms denote single, double, and

Table 1.4. Structures of Some Common Compounds

Compound Name	Structure Showing Bonds and Free Electrons	Formula with Free Electrons	Formula without Free Electrons
Molecular oxygen	O=O	$O_2(g)$	$O_2(g)$
Molecular nitrogen	N≡N	$N_2(g)$	$N_2(g)$
Ozone		$O_3(g)$	$O_3(g)$
Hydroxyl radical	Ȯ—H	$\dot{O}H(g)$	$OH(g)$
Water vapor		$H_2O(g)$	$H_2O(g)$
Nitric oxide	Ṅ=O	$\dot{N}O(g)$	$NO(g)$
Nitrogen dioxide		$\dot{N}O_2(g)$	$NO_2(g)$
Sulfur dioxide		$SO_2(g)$	$SO_2(g)$
Carbon monoxide	⁻C≡O⁺	$CO(g)$	$CO(g)$
Carbon dioxide	O=C=O	$CO_2(g)$	$CO_2(g)$
Methane		$CH_4(g)$	$CH_4(g)$
Sulfate ion		SO_4^{2-}	SO_4^{2-}

triple bonds, respectively. For some compounds [OH(g), NO(g), $NO_2(g)$], a single dot is shown adjacent to an atom. A single dot indicates that the atom has a free electron. Compounds with a free electron are called **free radicals** and are highly reactive. Some nonfree radicals that have a single bond [e.g., $O_3(g)$] are also reactive because single bonds are readily broken. Compounds with triple bonds [$N_2(g)$, CO(g)] are not so reactive because triple bonds are difficult to break. **Noble elements** (He, Ar, Ne, Kr, Xe) have no free electrons and no potential to form bonds with other elements; thus, they are chemically unreactive (inert).

For some compounds in Table 1.4 [$NO_2(g)$, $O_3(g)$, CO(g)], positive and negative charges are shown. Such a charge distribution arises when one atom transfers charge to another atom during molecular formation. During $NO_2(g)$ formation, for example, a net negative charge is transferred to an oxygen atom from the nitrogen atom, resulting in the charge distribution shown. Compounds with both positive and negative charges have zero net charge and are not ions, but the positive (negative) end of the compound is likely to attract negative (positive) charges from other compounds, enhancing the reactivity of the compound. For SO_4^{2-}, a net negative charge is shown, indicating that it is an ion.

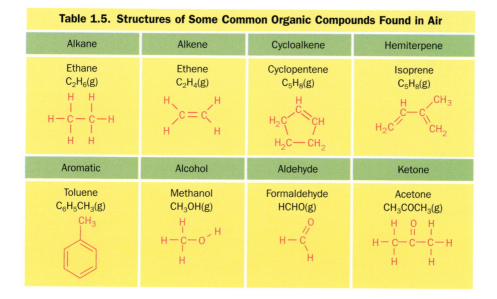

Table 1.5. Structures of Some Common Organic Compounds Found in Air

Alkane	Alkene	Cycloalkene	Hemiterpene
Ethane $C_2H_6(g)$	Ethene $C_2H_4(g)$	Cyclopentene $C_5H_8(g)$	Isoprene $C_5H_8(g)$
Aromatic	Alcohol	Aldehyde	Ketone
Toluene $C_6H_5CH_3(g)$	Methanol $CH_3OH(g)$	Formaldehyde $HCHO(g)$	Acetone $CH_3COCH_3(g)$

When oxygen combines with an element or compound during a chemical reaction, the process is called **oxidation**, and the resulting substance is said to be **oxidized**. The substances $O_2(g)$, $O_3(g)$, $OH(g)$, $H_2O(g)$, $NO(g)$, $NO_2(g)$, $SO_2(g)$, $CO(g)$, and $CO_2(g)$ are oxidized. When oxygen is removed from a substance during a reaction, the process is called **reduction**, and the resulting element or compound is said to be **reduced**. The substances $H_2(g)$, $N_2(g)$, $NH_3(g)$, and $CH_4(g)$ are reduced.

Table 1.4 shows structures of inorganic compounds and methane, an organic compound. Table 1.5 shows structures of additional organic compounds. **Inorganic compounds** are compounds that contain any element, including hydrogen (H) or carbon (C), but not both H and C. **Organic compounds** are compounds that contain both H and C, but may also contain other elements. Methane is the simplest organic compound.

Organic compounds that contain only H and C are **hydrocarbons**. Hydrocarbons include alkanes, cycloalkanes, alkenes, cycloalkenes, alkynes, aromatics, and terpenes. Examples of some of these groups are given in Table 1.5. **Alkanes** (paraffins) are open-chain (noncyclical) hydrocarbons with a single bond between each pair of carbon atoms and have the molecular formula C_nH_{2n+2}. **Cycloalkanes** (not shown) are like alkanes, but with a cyclical structure. **Alkenes** (olefins) are open-chain hydrocarbons with a double bond between one pair of carbon atoms and have the molecular formula C_nH_{2n}. **Cycloalkenes** are similar to alkenes, but with a cyclical structure. **Alkynes** (acetylenes, not shown) are open-chain hydrocarbons with a triple bond between at least one pair of carbon atoms. **Terpenes** are a class of naturally occurring hydrocarbons that include hemiterpenes (C_5H_8) **monoterpenes** ($C_{10}H_{16}$), **sesquiterpenes** ($C_{15}H_{24}$), **diterpenes** ($C_{20}H_{32}$), and so on. **Aromatic hydrocarbons** are hydrocarbons with a benzene ring and possibly other carbon and hydrogen atoms attached to the ring. Two representations of a benzene ring are shown in Fig. 1.19.

Aromatics are so named because the first aromatics isolated were obtained from substances that had a pleasant fragrance, or *aroma*. Around 1868, Austrian chemist

Joseph Loschmidt (1821–1895) found that such aromatic compounds could be obtained by replacing one or more hydrogen atoms on a benzene ring with another atom or group. The name *aromatic* was subsequently applied to any compound that had a benzene ring in its structure. Loschmidt was the first to explain the structure of benzene, toluene, and ozone. He is also the first to quantify accurately Avogadro's number (Section 3.4).

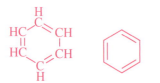

Figure 1.19. Two representations of a benzene ring.

When methane, a fairly unreactive hydrocarbon, is excluded from the list of hydrocarbons, the remaining hydrocarbons are called **nonmethane hydrocarbons (NMHCs)**. When oxygenated functional groups, such as aldehydes, ketones, alcohols, acids, and nitrates, are added to hydrocarbons, the resulting compounds are called **oxygenated hydrocarbons**. In Table 1.5, the alcohol, aldehyde, and ketone are oxygenated hydrocarbons. Nonmethane hydrocarbons and oxygenated hydrocarbons are **reactive organic gases (ROGs)**. **Total organic gas (TOG)** is the sum of ROGs and methane. **Volatile organic compounds (VOCs)** are organic compounds with relatively low boiling points that, therefore, readily evaporate. Although all VOCs are not necessarily ROGs, these terms are often interchanged. Finally, aldehydes and ketones are called **carbonyls**. The sum of nonmethane hydrocarbons and carbonyls is **nonmethane organic carbon (NMOC)**.

1.4. CHEMICAL REACTIONS AND PHOTOPROCESSES

Many of the pollution problems today are exacerbated by atmospheric chemical reactions. Reactions are initiated by sunlight, lightning, changes in temperature, or molecular collisions. In this section, chemical reactions are briefly discussed.

Gas-phase chemical reactions are conveniently divided into *photolysis reactions* (also called *photoprocesses*, *photodissociation reactions*, or *photolytic reactions*) and *chemical kinetic reactions*. Photolysis reactions are **unimolecular** (involving one reactant) and are initiated when solar radiation strikes a molecule and breaks it into two or more products. An example of a **photolysis reaction** is

$$\overset{\bullet}{N}O_2(g) + h\nu \longrightarrow \overset{\bullet}{N}O(g) + \overset{\bullet}{\cdot}O(g) \qquad \lambda < 420 \text{ nm}$$

Nitrogen dioxide   Nitric oxide   Atomic oxygen   (1.1)

where $h\nu$ implies a photon of solar radiation and λ is the wavelength of the radiation (defined in Chapter 2).

Chemical kinetic reactions are usually **bimolecular** (involving two reactants). Types of kinetic reactions include thermal decomposition, isomerization, and standard collision reactions. Thermal decomposition and isomerization reactions occur when a reactant molecule collides with an air molecule. The kinetic energy of the collision elevates the reactant to an energy state high enough that it can thermally decompose or isomerize. **Thermal decomposition** occurs when the excited reactant dissociates into two or more products. **Isomerization** occurs when the excited reactant changes chemical structure, but not composition or molecular weight.

An example of a bimolecular **thermal decomposition reaction** is

$$N_2O_5\,(g) + M \longrightarrow \overset{\bullet}{N}O_2\,(g) + N\overset{\bullet}{O}_3\,(g) + M \tag{1.2}$$

Dinitrogen   Nitrogen   Nitrate
pentoxide   dioxide   radical

where M is the molecule that provides the collisional energy. M can be any molecule. Because molecular oxygen [$O_2(g)$] and nitrogen [$N_2(g)$] together make up more than 99 percent of the gas molecules in the air today, M is most likely to be $O_2(g)$ or $N_2(g)$.

Because M in Reaction 1.2 does not change concentration, the reaction can also be written as

$$N_2O_5\,(g) \xrightarrow{\;M\;} \overset{\bullet}{N}O_2\,(g) + N\overset{\bullet}{O}_3\,(g) \tag{1.3}$$

Dinitrogen   Nitrogen   Nitrate
pentoxide   dioxide   radical

Thermal decomposition reactions are temperature dependent. At high temperatures, they proceed faster than at low temperatures. Isomerization reactions are similar to Reaction 1.3, except that an isomerization reaction has one product, which is another form of the reactant. An example of an isomerization reaction is

(1.4)

Excited Criegee   Excited formic
biradical   acid

The bimolecular **collision reaction** is the most common type of kinetic reaction and may occur between any two chemically active reactants that collide. A prototypical collision reaction is

$$CH_4\,(g) + \overset{\bullet}{O}H\,(g) \longrightarrow \overset{\bullet}{C}H_3\,(g) + H_2O\,(g) \tag{1.5}$$

Methane   Hydroxyl   Methyl   Water
          radical   radical   vapor

In some cases, bimolecular reactions result in **collision complexes** that ultimately break into products. Such reactions have the form $A + B \rightleftharpoons AB^* \rightarrow D + F$, where AB^* is a molecule that has weak bonds and is relatively unstable, and the double arrow indicates that the reaction is **reversible**.

Termolecular (involving three reactants) collision reactions are rare because the probability that three trace gases collide simultaneously and change form is not large. For descriptive purposes, however, pairs of reactions, can be written as termolecular **combination reactions**. For example, the combination of the bimolecular kinetic reaction $NO_2(g) + NO_3(g) \rightleftharpoons N_2O_5(g)^*$ with the isomerization reaction $N_2O_5(g)^* + M \rightleftharpoons N_2O_5(g) + M$ gives

$$\overset{\bullet}{N}O_2\,(g) + N\overset{\bullet}{O}_3\,(g) + M \rightleftharpoons N_2O_5\,(g) + M \tag{1.6}$$

Nitrogen   Nitrate   Dinitrogen
dioxide   radical   pentoxide

In this case, M is any molecule, whose purpose is to carry away energy released during the reaction. The purpose of M in Reaction 1.6 differs from its purpose in Reaction 1.2, where it provided collisional energy for the reaction. In both cases, M is usually either $N_2(g)$ or $O_2(g)$. Reactions 1.2 and 1.6 are pressure dependent because the concentration of M is proportional to the air pressure. Because M in Reaction 1.6 does not change concentration, Reaction 1.6 can also be written as

$$\overset{\bullet}{N}O_2(g) + \overset{\bullet}{N}O_3(g) \overset{M}{\rightleftharpoons} N_2O_5(g)$$

Nitrogen   Nitrate   Dinitrogen
dioxide   radical   pentoxide   (1.7)

1.5. LIFETIMES OF CHEMICALS

Some gases are important because their concentrations are high, suggesting that these gases do not degrade quickly. Others are important because they react quickly to form one or more products that are harmful or otherwise important. Gases that do not react away quickly include $N_2(g)$, $O_2(g)$, and $CO_2(g)$. Some that do react away quickly (but may also reform quickly) include $OH(g)$, $NO(g)$, $NO_2(g)$, and $O_3(g)$, most of which are free radicals. The time required for the concentration of a gas to decrease to $1/e$ its original concentration as a result of chemical reaction is called an **e-folding lifetime**. This parameter is similar to the **half-lifetime**, which is the time required for a gas concentration to decrease to one-half its original concentration. Throughout this text, these terms are used to evaluate the importance of different chemicals.

1.6. SUMMARY

In this chapter, atoms, molecules, elements, and compounds were defined and a history of the discovery of elements and compounds of atmospheric importance was given. Only a few elements, including carbon, sulfur, and certain metals, and a few solid compounds, including calcite, halite, and nitre, were known in ancient times. An acceleration of the discovery of elements and compounds, particularly of gases, occurred near the end of the eighteenth century. Several types of chemical reactions occur in the air, including photolysis, kinetic, thermal decomposition, isomerization, and combination reactions. The rate of reaction depends on the reactivity and concentration of molecules. The chemical *e*-folding lifetime of a substance is the time required for its concentration to decrease to $1/e$ its original value and gives an indication of the reactivity of the substance. Molecules with free electrons are called *free radicals* and are highly reactive.

1.7. PROBLEMS

1.1.   What are the main differences between gases and aerosol particles?

1.2.   What compound might you expect to form on the surface of a statue made of marble or limestone (both of which contain calcite – calcium carbonate) if aqueous sulfuric acid deposits onto the statue?

1.3.   Describe one experiment you could devise to isolate molecular oxygen.

1.4. What was the fundamental flaw with the theory of phlogiston?

1.5. Why did Lavoisier name oxygen as he did? Was his definition correct? Why or why not?

1.6. Is a termolecular combination reaction the result of the collision of three molecules simultaneously? Why or why not?

1.7. If the chemical e-folding lifetimes of the harmless substances A, B, and C are 1 hour, 1 week, and 1 year, respectively, and all three substances produce harmful products when they break down, which substance would you prefer to eliminate from urban air first? Why?

1.8. Match each person below with a surrogate name or description of a chemical s/he discovered.

(a) Priestly	(1) "gas that is wild and dwells in out-of-the-way places"
(b) Schönbein	(2) "Poland"
(c) M. Curie	(3) "foul-smelling air"
(d) Baldwin	(4) "stench"
(e) Theophrastus	(5) "plaster"
(f) Paracelsus	(6) "water maker"
(g) Van Helmont	(7) "lazy gas"
(h) Balard	(8) "acid maker"
(i) Rayleigh	(9) "light bearer"
(j) D. Rutherford	(10) "to smell"

THE SUN, THE EARTH, AND THE EVOLUTION OF THE EARTH'S ATMOSPHERE

Anthropogenic pollution problems result from the enhancement of gas and aerosol-particle concentrations above background concentrations. In this chapter, the evolution of the background atmosphere is discussed. The discussion requires a description of the sun and its origins because sunlight has affected much of the evolution of the Earth's atmosphere. The description also requires a discussion of the Earth's composition and structure because the inner Earth affects atmospheric composition through outgassing, and the crust affects atmospheric composition through exchange processes, including soil-dust emission. Earth's earliest atmosphere contained mostly hydrogen and helium. Carbon dioxide replaced these gases during the onset of the Earth's second atmosphere. Today, nitrogen and oxygen are the prevalent gases. Processes controlling the changes in atmospheric composition over time include outgassing from the Earth's interior, microbial metabolism, and atmospheric chemistry. These processes still affect the natural composition of the air today.

2.1. THE SUN AND ITS ORIGIN

The sun provides the energy to power the Earth. Most of that energy originates from the sun's surface, not from its interior. The reason for this is discussed as follows.

About 15 billion years ago (b.y.a.), all mass in the known universe may have been compressed into a single point, estimated to have a density of 10^9 kg m^{-3} and a temperature of 10^{12} K (kelvin). With the "**Big Bang**," this point of mass exploded, ejecting material in all directions. Aggregates of ejected material collapsed gravitationally to form the earliest stars. When temperatures in the cores of early stars reached 10 million K, nuclear fusion of hydrogen (H) into helium (He) and higher elements began, releasing energy that powered the stars. As early stars aged, they ultimately exploded, ejecting stellar material into space. Table 2.1 gives the abundance of hydrogen in the universe today relative to the abundances of other interstellar elements.

About 4.6 b.y.a., some interstellar material aggregated to form a cloudy mass, the **solar nebula**. The composition of the solar nebula was the same as that of 95 percent of the other stars in the universe. Gravitational collapse of the solar nebula resulted in the formation of the sun.

Table 2.1. Cosmic Abundance of Hydrogen Relative to Those of Other Elements

Element	Atomic Mass	Abundance of H Relative to Element	Element	Atomic Mass	Abundance of H Relative to Element
Hydrogen (H)	1.01	1:1	Silicon (Si)[a]	28.1	26,000:1
Helium (He)	4.00	14:1	Iron (Fe)[a]	55.8	29,000:1
Oxygen (O)	16.0	1,400:1	Sulfur (S)	32.1	53,000:1
Carbon (C)	12.0	2,300:1	Argon (Ar)	39.9	260,000:1
Neon (Ne)	20.2	10,000:1	Aluminum (Al)[a]	27.0	306,000:1
Nitrogen (N)	14.0	11,000:1	Calcium (Ca)[a]	40.1	413,000:1
Magnesium (Mg)[a]	24.3	24,000:1	Sodium (Na)[a]	23.0	433,000:1

[a]Rock-forming elements. All other elements vaporize more readily.

Adapted from Goody (1995).

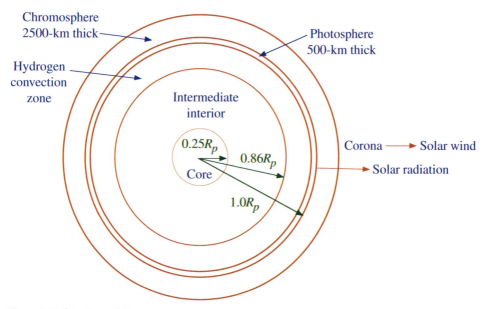

Figure 2.1. Structure of the sun.

Today, the sun is divided into concentric layers, including interior and atmospheric layers. About 90 percent of atoms in the sun are hydrogen and 9.9 percent are helium. The remainder are the other natural elements of the periodic table.

The sun's interior consists of liquid and gas, and its atmosphere consists predominately of gas. The interior consists of the core, the intermediate interior, and the hydrogen convection zone (Fig. 2.1). The photosphere, which is primarily gaseous, is a transition region between the sun's atmosphere and its interior. Beyond the photosphere, the sun's atmosphere consists of the chromosphere, the corona, and solar wind discharge.

The sun has an effective radius (R_p) of 696,000 km, or 109 times the radius of the Earth (6378 km). The sun's effective radius is the distance between the center of the sun and the top of the photosphere. The mass of the sun is about 1.99×10^{30} kg, or 333,000 times the mass of the Earth (5.98×10^{24} kg). The sun's surface gravity at the top of the photosphere is about 274 m s^{-2}, or 28 times that of the Earth (9.8 m s^{-2}).

The **sun's core** lies between its center and about 0.25 R_p. Temperatures in the core reach 15 million K. At these temperatures, electrons are stripped from hydrogen atoms. The remaining nuclei (single protons) collide in intense thermonuclear fusion reactions, producing helium and releasing energy in the form of photons of radiation that power the sun. Energy from the core radiates to the **intermediate interior**, which has a thickness of about 0.61 R_p and a temperature ranging from 8 million K at its base to about 5 million K at its top. Energy transfer through the intermediate interior is also radiative. The next layer is the **hydrogen convection zone (HCZ)**, a region in which convection of hydrogen atoms due to buoyancy takes over from radiation as the predominant mechanism of transferring energy toward the sun's surface. Temperatures at the base of the HCZ are around 5 million K; temperatures at its top are near 6,400 K. In the HCZ, a strong temperature gradient exists and the **mean free path** (average distance between collisions) of photons with hydrogen and helium atoms decreases with increasing distance from the core. The HCZ is a region that prevents much of the sun's internal radiation from escaping to its surface. In fact, a photon of radiation emitted from the core of the sun takes about 10 million years to reach the top of the HCZ. The HCZ thickness is difficult to determine and may range from about 0.14 to 0.3 R_p.

Above the HCZ lies the **photosphere** ("light sphere"), which is a relatively thin (500 km thick) transition region between the sun's interior and its atmosphere. Temperatures in the photosphere range from 6,400 K at its base to 4,000 K at its top and average 5,785 K. The photosphere is the source of most solar energy that reaches the planets, including the Earth. Although the sun's interior is much hotter than is its photosphere, most energy produced in its interior is confined by the HCZ.

Above the photosphere lies the **chromosphere** ("color sphere"), which is a 2,500 km thick region of hot gases. Temperatures at the base of the chromosphere are around 4,000 K. Those at the top are up to 1 million K. The name *chromosphere* arises because at the high temperatures found in this region, hydrogen is energized and decays back to its ground state, emitting wavelengths of radiation in the visible part of the solar spectrum. For example, hydrogen decay results in radiation emission at 0.6563 μm, which is in the red part of the spectrum, giving the chromosphere a characteristic red coloration observed during solar eclipses.

The **corona** is the outer shell of the solar atmosphere and has an average temperature of about 1 to 2 million K. Because of the high temperatures, all gases in the corona, particularly hydrogen and helium, are ionized. A low-concentration, steady stream of these ions escapes the corona and the sun's gravitational field and propagates through space, intercepting the planets with speeds ranging from 300 to 1000 km s^{-1}. This stream is called the **solar wind**. The solar wind is the outer boundary of the corona and extends from the chromosphere to the outermost reaches of the solar system.

The **Earth–sun distance** (R_{es}) is about 150 million km. At the Earth, the solar wind temperature is about 200,000 K, and the number concentration of solar-wind

Figure 2.2. The Aurora Australis as seen from Kangaroo Island, southern Australia. Photo by David Miller, National Geophysical Data Center, available from NOAA Central Library.

ions is a few to tens per cubic centimeter of space. As the solar wind approaches the Earth, the Earth's magnetic fields bend the path of the wind toward the North and South Poles. In the atmosphere above these regions the ionized gases collide with air molecules, creating luminous bands of streaming, colored lights. In the Northern Hemisphere, these lights are called the **Northern Lights** or **Aurora Borealis** ("northern dawn" in Latin), and in the Southern Hemisphere, they are called the **Southern Lights** or **Aurora Australis** ("southern dawn"). These lights, one of the seven natural wonders of the Earth, can be seen at high latitudes, such as in northern Scotland, Scandinavia, and parts of Canada in the Northern Hemisphere and in southern Australia and Argentina in the Southern Hemisphere. Figure 2.2 shows a photograph of the Aurora Australis.

2.2. SPECTRA OF THE RADIATION OF THE SUN AND THE EARTH

Life on Earth would not have evolved to its present state without heating by solar radiation. Next, the sun's radiation spectrum is described.

Radiation is the emission or propagation of energy in the form of a photon or an electromagnetic wave. Whether radiation is considered a photon or a wave is still debated. A **photon** is a particle or quantum of energy that has no mass, no electric charge, and an indefinite lifetime. An **electromagnetic wave** is a disturbance traveling through a medium, such as air or space, that transfers energy from one object to another without permanently displacing the medium itself.

Because radiative energy can be transferred even in a vacuum, it is not necessary for gas molecules to be present for radiative energy transfer to occur. Thus, such transfer can occur through space, where few gas molecules exist, or through the Earth's atmosphere, where many molecules exist.

Radiation is emitted by all bodies in the universe that have a temperature above absolute zero (0 K). During emission, a body releases electromagnetic energy at different wavelengths, where a **wavelength** is the difference in distance between peaks or troughs in a wave. The intensity of emission from a body varies with wavelength, temperature, and efficiency of emission. Bodies that emit radiation with perfect efficiency are termed *blackbodies*. A **blackbody** is a body that absorbs all radiation incident on it. No incident radiation is reflected by a blackbody. No bodies are true blackbodies, although the Earth and the sun are close, as are black carbon, platinum black, and black gold. The term *blackbody* was coined because good absorbers of visible radiation generally appear black. However, good absorbers of infrared radiation are not necessarily black. For example, one such absorber is white oil-based paint.

Bodies that absorb radiation incident upon them with perfect efficiency also emit radiation with perfect efficiency. The wavelength of peak intensity of emission of a blackbody is inversely proportional to the absolute temperature of the body. This law, called **Wien's displacement law**, was derived in 1893 by German physicist Wilhelm Wien (1864–1928). Wien's law states

$$\lambda_p(\mu m) \approx \frac{2,897}{T(K)} \tag{2.1}$$

where λ_p is the wavelength (in micrometers, μm) of peak blackbody emission, and T is the temperature (K) of the body. Wien won a Nobel prize in 1911 for his discovery.

EXAMPLE 2.1.

Calculate the peak wavelength of radiative emission for each the sun and the Earth.

Solution

The effective temperature of the sun's photosphere is 5,785 K, giving the sun a blackbody emission peak wavelength of 0.5 μm from Equation 2.1. The average surface temperature of the Earth is 288 K, giving the Earth a blackbody emission peak wavelength of 10 μm.

At any wavelength, the intensity of radiative emission from an object increases with increasing temperature. Thus, hotter bodies (such as the sun) emit radiation more intensely than do colder bodies (such as the Earth). Figure 2.3 shows the radiation intensity versus wavelength for blackbodies at four temperatures. The figure shows that at 15 million K, a temperature at which nuclear fusion reactions occur in the sun's center, **gamma radiation** wavelengths (10^{-8} to 10^{-4} μm) and **X radiation** wavelengths (10^{-4} to 0.01 μm) are the most intensely emitted wavelengths. At 6,000 K, **visible** wavelengths (0.38–0.75 μm) are the most intensely emitted wavelengths, although shorter **ultraviolet (UV)** wavelengths (0.01–0.38 μm), and longer **infrared (IR)** wavelengths (0.75–100 μm) are also emitted. At 300 K, infrared wavelengths are the most intensely emitted wavelengths.

Figure 2.4 focuses on the 6,000 and the 300 K spectra in Fig. 2.3. These are the radiation spectra for the sun's photosphere and the Earth, respectively. The **solar spectrum** is divided into the UV, visible, and IR spectra. The UV spectrum is divided into the **far UV** (0.01–0.25 μm) and **near UV** (0.25–0.38 μm) spectra, as shown in Fig. 2.5. The near UV spectrum is further divided into **UV-A** (0.32–0.38 μm), **UV-B** (0.29–0.32 μm), and **UV-C** (0.25–0.29 μm) wavelengths. The visible spectrum

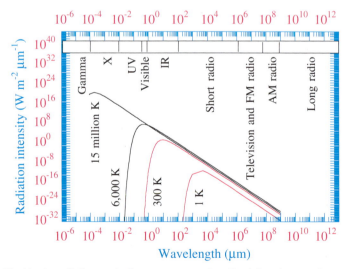

Figure 2.3. Blackbody radiation emission versus wavelength at four temperatures. Units are watts (joules of energy per second) per square meter of area per micrometer wavelength. The 15 million K spectrum represents emission from the center of the sun (most of which does not penetrate to the sun's exterior) The 6,000 K spectrum represents emission from the sun's surface and received at the top of the Earth's atmosphere (not at its surface). The 300 K spectrum represents emission from the Earth's surface. The 1 K spectrum is close to the coldest temperature possible (0 K).

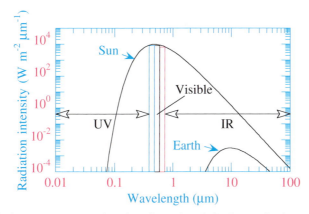

Figure 2.4. Radiation spectrum as a function of wavelength for the sun's photosphere and the Earth when both are considered blackbodies. The sun's spectrum is received at the top of the Earth's atmosphere.

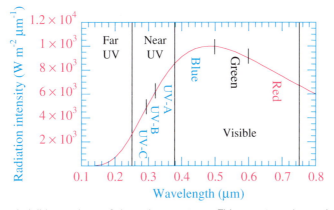

Figure 2.5. UV and visible portions of the solar spectrum. This spectrum is received at the top of the Earth's atmosphere.

contains the colors of the rainbow. For convenience, visible light is divided into **blue** (0.38–0.5 μm), **green** (0.5–0.6 μm), and **red** (0.6–0.75 μm) wavelengths. Infrared wavelengths are divided into **solar (near)-infrared** (0.75–4 μm) and **thermal (far)-infrared** (4–100 μm) wavelengths. The intensity of the sun's emission is strongest in the visible spectrum. That of the Earth is strongest in the thermal-IR spectrum.

Figures 2.3 to 2.5 show wavelength dependencies of the intensity of radiation emission at a given temperature. Integrating the intensity over all wavelengths (summing the area under any of the curves) gives the total intensity of emission of a body at a given temperature. This intensity is proportional to the fourth power of the object's kelvin temperature (T) and is given by the **Stefan–Boltzmann law**, derived empirically in 1879 by Austrian physicist **Josef Stefan** (1835–1893) and theoretically in 1889 by Austrian physicist **Ludwig Boltzmann** (1844–1906). The law states

$$F_b = \varepsilon \sigma_B T^4 \tag{2.2}$$

where F_b is the radiation intensity (W m^{-2}), summed over all wavelengths, emitted by a body at a given temperature, ε is the emissivity of the body, and $\sigma_B = 5.67 \times 10^{-8}$ W m^{-2} K^{-4} is the Stefan–Boltzmann constant. The **emissivity**, which ranges from 0 to 1, is the efficiency at which a body emits radiation in comparison with the emissivity of a blackbody, which is unity. Soil has an emissivity of 0.9 to 0.98, and water has an emissivity of 0.92 to 0.97. All the curves in Figs. 2.3 to 2.5 show emission spectra for blackbodies ($\varepsilon = 1$).

EXAMPLE 2.2.

How does doubling the kelvin temperature of a blackbody change the intensity of radiative emission of the body? What is the ratio of intensity of the sun's radiation compared with that of the Earth's?

Solution

From Equation 2.2, the doubling of the kelvin temperature of a body increases its intensity of radiative emission by a factor of 16. The temperature of the sun's photosphere (5,785 K) is about 20 times that of the Earth (288 K). Assuming both are blackbodies ($\varepsilon = 1$), the intensity of the sun's radiation (63.5 million W m^{-2}) is 163,000 times that of the Earth's (390 W m^{-2}).

2.3. PRIMORDIAL EVOLUTION OF THE EARTH AND ITS ATMOSPHERE

Earth formed when rock-forming elements (identified in Table 2.1), present as gases at high temperatures in the solar nebula, condensed into small solid grains as the nebula cooled. The grains grew by collision to centimeter-sized particles. Additional grains accreted onto the particles, resulting in **planetesimals**, which are small-body precursors to planet formation. Accretion of grains and particles onto planetesimals resulted in the formation of **asteroids** (Fig. 2.6), which are rocky bodies 1 to 1,000 km in size that orbit the sun. Asteroids collided to form the planets. The growth of planets was aided by the bombardment of **meteorites**, which are solid minerals or rocks that reach the planet's surface without vaporizing. Meteorite bombardment was intense for about 500 million years. Although the solar nebula has since cooled and most of it has been converted to solar or planetary material or has been swept away from the solar system, some planetary growth still continues today, as leftover asteroids and meteorites occasionally strike the planets. Table 2.2 shows the average composition of stony meteorites, the total Earth, and the Earth's continental and oceanic crusts. The table indicates that meteorite composition is relatively similar to that of the total Earth, supporting the theory that meteorites played a role in the Earth's formation.

Meteorites and asteroids consist of **rock-forming elements** (e.g., Mg, Si, Fe, Al, Ca, Na, Ni, Cr, Mn) that condensed from the gas phase in the cooling solar nebula, and **noncondensable elements** (e.g., H, He, O, C, Ne, N, S, Ar, P). How did noncondensable elements enter meteorites and asteroids, particularly as they were too light to attract to these bodies gravitationally? One theory is that noncondensable elements may have chemically reacted as gases in the solar nebula to form high molecular weight compounds that were condensable, although less condensable (more volatile) than were rock-forming elements. When meteorites and asteroids collided with the Earth, they brought with them volatile compounds and rock-forming elements. Whereas some of the volatiles vaporized on impact, others have taken longer to vaporize and have been outgassed ever since through volcanos, fumaroles, steam wells, and geysers.

Earth's first atmosphere likely contained hydrogen (H) and helium (He), the most abundant elements in the solar nebula. During the formation of the Earth, the sun was also forming. Early stars are known to blast off a large amount of gas into space. This outgassed solar material, the **solar wind**, was previously introduced as an extension of

Figure 2.6. The asteroid Ida and its moon, Dactyl, taken by the Galileo spacecraft as it passed within 10,878 km of the asteroid on August 28, 1993. Available from National Space Science Data Center.

Table 2.2. Mass Percentages of Major Elements in Stony Meteorites, the Total Earth, and the Earth's Continental and Oceanic Crusts

Element	Stony Meteorites	Total Earth	Earth's Continental Crust	Earth's Oceanic Crust
Oxygen (O)	33.24	29.50	46.6	45.4
Iron (Fe)	27.24	34.60	5.0	6.4
Silicon (Si)	17.10	15.20	27.2	22.8
Magnesium (Mg)	14.29	12.70	2.1	4.1
Sulfur (S)	1.93	1.93	0.026	0.026
Nickel (Ni)	1.64	2.39	0.075	0.075
Calcium (Ca)	1.27	1.13	3.6	8.8
Aluminum (Al)	1.22	1.09	8.1	8.7
Sodium (Na)	0.64	0.57	2.8	1.9

Adapted from Cattermole and Moore (1985).

the sun's corona. During the birth of the sun, nuclear reactions in the sun were enhanced, and solar wind speeds and densities were much larger than they are today. This early stage of the sun is called the T-Tauri stage after the first star observed at this point in its evolution. The enhanced solar wind is thought to have stripped away the first atmosphere of not only the Earth, but also all other planets in the solar system. Additional H and He were lost from the Earth's first atmosphere after escaping the

Earth's gravitational field. As a result of these two loss processes (solar wind stripping and gravitational escape), the ratios of H and He to other elements in the Earth's atmosphere today are less than are the corresponding ratios in the sun.

2.3.1. Solid-Earth Formation

The rock-forming elements that reached the Earth reacted to form compounds with different melting points, densities, and chemical reactivities. Dense compounds and compounds with high melting points, including many iron- and nickel-containing compounds, settled to the center of the Earth, called the **Earth's core**. Table 2.2 shows that the total Earth contains more than 34 percent iron and 2 percent nickel by mass, but the Earth's **crust** (its top layer) contains less than 7 percent iron and 0.1 percent nickel by mass, supporting the contention that iron and nickel settled to the core. Low-density compounds and compounds with low melting points, including silicates of aluminum, sodium, and calcium, rose to the surface and are the most common compounds in the Earth's crust. Table 2.2 supports this contention. Some moderately dense and moderately high-melting-point silicates, such as those containing magnesium or iron, settled to the Earth's **mantle**, which is a layer of Earth's interior between its crust and its core.

During the formation of the Earth's core, between 4.5 and 4.0 b.y.a., temperatures in the core were hotter than they are today, and the only mechanism of heat escape to the surface was **conduction**, the transfer of energy from molecule to molecule. Because conduction is a slow process, the Earth's internal energy could not dissipate easily, and its temperature increased until the entire body became molten. At that time, the Earth's surface consisted of **magma oceans**, a hot mixture of melted rock and suspended crystals. When the Earth was molten, **convection**, the mass movement of molecules, became the predominant form of energy transfer between the core and surface. Convection occurred because temperatures in the core were hot enough for core material to expand and float to the crust, where it cooled and sank down again. This process enhanced energy dissipation from the Earth's center to space. After sufficient energy dissipation (cooling), the magma oceans solidified, creating the Earth's crust. The crust is estimated to have formed 3.8 to 4.0 b.y.a., but possibly as early as 4.2 to 4.3 b.y.a. (Crowley and North, 1991). The core cooled as well, but its outer part, now called the **outer core**, remains molten. Its inner part, now called the **inner core**, is solid.

Figure 2.7 shows temperature, density, and pressure profiles inside the Earth today. The Earth's crust extends from the topographical surface to about 10 to 75 km below continents and to about 8 km below the ocean floor. The crust itself contains low-density, low-melting-point silicates. The continental crust contains primarily granite, whereas the ocean crust contains primarily basalt. **Granite** is a type of rock composed mainly of quartz [$SiO_2(s)$] and potassium feldspar [$KAlSi_3O_8(s)$]. **Basalt** is a type of rock composed primarily of plagioclase feldspar [[$NaAlSi_3O_3$-$CaAl_2Si_2O_8(s)$]] and pyroxene (multiple compositions). The densities of both granite and basalt are about 2,800 kg m^{-3}.

Below the Earth's crust is its **mantle**, which consists of an upper and lower part, both made of iron–magnesium–silicate minerals. The upper mantle extends from the crust down to about 700 km. At that depth, a density gradation occurs due to a change in crystal packing. This gradation roughly defines the base of the upper mantle and the top of the lower mantle. Below 700 km, the density gradually increases to the mantle–core boundary at 2,900 km.

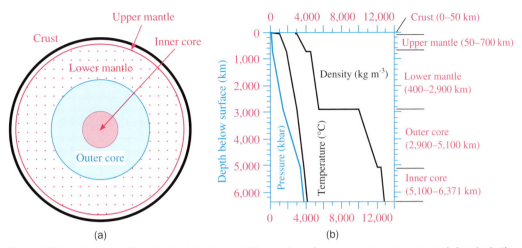

Figure 2.7. (a) Diagram of the Earth's interior and (b) variation of pressure, temperature, and density in the Earth's interior.

The outer core extends from 2,900 km down to about 5,100 km. This region consists of liquid iron and nickel, although the top few hundred kilometers contain liquids and crystals. The inner core extends from 5,100 km down to the Earth's center and is solid, also consisting of iron and nickel, but packed at a higher density. Temperatures, densities, and pressures at the center of the Earth are estimated to be 4,300°C, 13,000 kg m^{-3}, and 3,850 kbar, respectively (1 kilobar = 1,000 bar = 10^8 N m^{-2} = 10^8 Pa – for comparison, surface air pressures are about 1 bar).

2.3.2. Prebiotic Atmosphere

Earth's second atmosphere evolved as a result of outgassing from the Earth's mantle. As temperatures increased during the molten stage, hydrogen and oxygen, bound in crustal minerals as hydroxyl molecules (OH), became detached, forming the gas-phase hydroxyl radical. The hydroxyl radical then reacted with reduced gases, such as molecular hydrogen [$H_2(g)$], methane [$CH_4(g)$], ammonia [$NH_3(g)$], molecular nitrogen [$N_2(g)$], and hydrogen sulfide [$H_2S(g)$], to form oxidized gases, such as water [$H_2O(g)$], carbon monoxide [$CO(g)$], carbon dioxide [$CO_2(g)$], nitrogen dioxide [$NO_2(g)$], and sulfur dioxide [$SO_2(g)$]. As the molten rock rose to the Earth's surface during convection, oxidized and reduced gases were ejected into the air by volcanos, fumaroles, steam wells, and geysers.

After the crust and mantle solidified, outgassing continued. The resulting secondary atmosphere contained no free elemental oxygen. All oxygen was tied up in oxidized molecules. Indeed, if any free oxygen did exist, it would have been removed by chemical reaction.

Most outgassed water vapor in the air condensed to form the oceans. The oceans are a critical part of today's hydrologic cycle. In this cycle, ocean water evaporates, the vapor is transported and recondenses to clouds, rain precipitates back to land or ocean surfaces, and water on land flows gravitationally back to the oceans. Sizable oceans have been present during almost all of the Earth's history (Pollack and Yung, 1980). In 1955, geochemist William W. Rubey (1898–1974) calculated that all the water in the Earth's oceans and atmosphere could be accounted for by the release of water vapor from volcanos that erupted throughout the Earth's history.

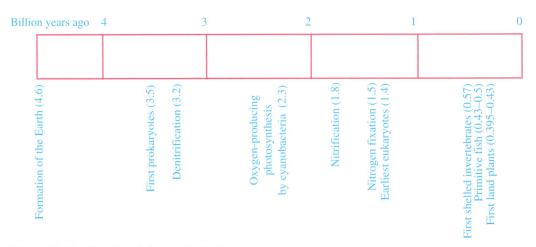

Figure 2.8. Timeline of evolution on the Earth.

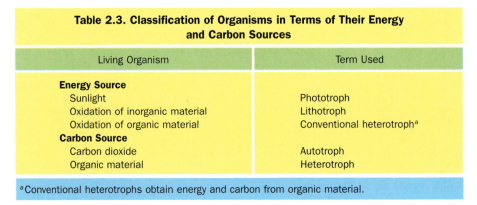

Table 2.3. Classification of Organisms in Terms of Their Energy and Carbon Sources

Living Organism	Term Used
Energy Source	
Sunlight	Phototroph
Oxidation of inorganic material	Lithotroph
Oxidation of organic material	Conventional heterotroph[a]
Carbon Source	
Carbon dioxide	Autotroph
Organic material	Heterotroph

[a] Conventional heterotrophs obtain energy and carbon from organic material.

2.3.3. Biotic Atmosphere Before Oxygen

Figure 2.8 shows an approximate timeline of important steps during the evolution of the Earth's atmosphere. Living organisms have been responsible for most of the changes.

Living organisms can be classified according to their energy and carbon sources. Table 2.3 shows that energy sources for organisms include sunlight and both inorganic and organic compounds. Organisms that obtain their energy from sunlight are **phototrophs**. Organisms that obtain their energy from oxidation of inorganic compounds (e.g., carbon dioxide [$CO_2(g)$], molecular hydrogen [$H_2(g)$], hydrogen sulfide [$H_2S(g)$], the ammonium ion [NH_4^+], the nitrite ion [NO_2^-]) are **lithotrophs**. Organisms that obtain their energy from oxidation of organic compounds are **conventional heterotrophs**.

The two sources of carbon for organisms include carbon dioxide and organic compounds. Organisms that obtain their carbon from $CO_2(g)$ are **autotrophs**. Organisms that obtain their carbon from organic compounds are **heterotrophs**.

Table 2.4 classifies organisms according to their energy and carbon sources. **Photoautotrophs** derive their energy from sunlight and their carbon from carbon dioxide. **Photoheterotrophs** derive their energy from sunlight and their carbon from organic material. **Lithotrophic autotrophs** derive their energy and carbon from

Table 2.4. Examples of Organisms Classified According to Their Energy and Carbon Sources

Organism Classification	Examples
Photoautotrophs	Green plants, most algae, cyanobacteria, some purple and green bacteria
Photoheterotrophs	Some algae, most purple and green bacteria, some cyanobacteria
Lithotrophic autotrophs	Hydrogen bacteria, colorless sulfur bacteria, methanogenic bacteria, nitrifying bacteria, iron bacteria
Lithotrophic heterotrophs	Some colorless sulfur bacteria
Conventional heterotrophs	All animals, all fungi, all protozoa, most bacteria

Figure 2.9. Hot sulfur springs in Lassen National Park, California. Boiling water of geothermal origin is rich in hydrogen sulfide. Lithotrophic autotrophic bacteria thrive in the springs and oxidize the hydrogen sulfide to sulfuric acid, which dissolves the surrounding mineral and converts part of the spring into a "mud pot." The steam contains mostly water vapor, but hydrogen sulfide, sulfuric acid, and other gases are also present. Courtesy of Alfred Spormann, Stanford University.

inorganic material (Fig. 2.9). **Lithotrophic heterotrophs** derive their energy from inorganic material and their carbon from organic material. **Conventional heterotrophs** derive their energy and carbon from organic material. Today, all animals, fungi, protozoa, and most bacteria are conventional heterotrophs.

About 3.5 b.y.a., the first microorganisms appeared on the Earth after a period during which the building blocks of life, *amino acids*, were produced by **abiotic synthesis**. Abiotic synthesis is the process by which life is created from chemical reactions and electrical discharges. It was demonstrated in 1953 by American chemist Stanley Miller (b. 1930), who was working in the laboratory of Harold Urey (1893–1981). Miller discharged electricity (simulating lightning) through a flask containing $H_2(g)$, $H_2O(g)$, $CH_4(g)$, and $NH_3(g)$ and liquid water. He let the bubbling mixture sit for a week and, after analyzing the results, found that he had produced complex organic molecules, including amino acids. Later experiments showed that the same results could be obtained with different gases and with UV radiation as opposed to with an electrical discharge. In all cases, much of the initial gas had to be highly reduced (Miller and Orgel, 1974).

In the prebiotic atmosphere, the amino acids required for the production of deoxyribonucleic acid (DNA) were first developed. About 3.5 b.y.a., the first microscopic cells containing DNA appeared. These cells, termed **prokaryotic cells**, contained a single strand of DNA, but had no nucleus. Many early prokaryotes were conventional heterotrophs because they obtained their energy and carbon from organic molecules produced during abiotic synthesis. Today, prokaryotic microorganisms include certain bacteria and blue-green algae.

2.3.3.1. Early Carbon Dioxide

During early biotic evolution, the major energy-producing process, carried out by prokaryotic conventional heterotrophs, was **fermentation**, which produced **carbon dioxide** gas. One fermentation reaction is

$$C_6H_{12}O_6 \text{ (aq)} \longrightarrow 2C_2H_5OH(aq) + 2CO_2(g) \tag{2.3}$$

Glucose   Ethanol   Carbon dioxide

which is exothermic (energy releasing). The energy source, glucose in this case, is only partially oxidized to carbon dioxide; thus, the reaction is inefficient.

2.3.3.2. Early Methane

A source of **methane** gas in the Earth's early atmosphere (and today) was metabolism by methanogenic bacteria. Such bacteria obtain their carbon from carbon dioxide and their energy from oxidation of molecular hydrogen; thus, methanogenic bacteria are lithotrophic autotrophs. Their methane-producing reaction is

$$4H_2(g) + CO_2(g) \longrightarrow CH_4(g) + 2H_2O(aq) \tag{2.4}$$

Molecular hydrogen   Carbon dioxide   Methane   Liquid water

Methanogenic bacteria use about 90 to 95 percent of carbon dioxide available to them for this process. The rest is used for synthesis of cell carbon. In Reaction 2.4, $CO_2(g)$ is reduced to $CH_4(g)$. A reaction such as this, in which cells produce energy by breaking down compounds in the absence of molecular oxygen, is called an **anaerobic respiration** reaction, where *anaerobic* means in the absence of oxygen. Anaerobic respiration produces energy more efficiently than does fermentation.

2.3.3.3. Early Molecular Nitrogen

In the Earth's early atmosphere, the most important source of **molecular nitrogen** was ammonia photolysis by the two step process,

$$NH_3(g) + h\nu \longrightarrow \cdot \dot{\ddot{N}} \ (g) + 3 \ \dot{H}(g) \tag{2.5}$$

Ammonia   Atomic nitrogen   Atomic hydrogen

$$\cdot \dot{\ddot{N}} \ (g) + \cdot \dot{\ddot{N}}(g) \xrightarrow{\ M\ } N_2(g) \tag{2.6}$$

Atomic nitrogen   Molecular nitrogen

Once oxygen levels built up in the air sufficiently (about 0.4 b.y.a.), ammonia photolysis became obsolete because oxygen absorbs the sun's wavelengths capable of photolyzing ammonia. About 3.2 b.y.a., some anaerobic conventional heterotrophs developed a new mechanism of producing molecular nitrogen. In this two step process, called **denitrification**, organic compounds react with the **nitrate ion** [NO_3^-] and the resulting **nitrite ion** [NO_2^-] by

$$\text{Organic compound} + NO_3^- \longrightarrow CO_2(g) + NO_2^- + .. \tag{2.7}$$

	Nitrate ion	Carbon dioxide	Nitrite ion

$$\text{Organic compound} + NO_2^- \longrightarrow CO_2(g) + N_2(g) + .. \tag{2.8}$$

	Nitrite ion	Carbon dioxide	Molecular nitrogen

Denitrification is the source of most molecular nitrogen in the air today.

2.3.3.4. Anoxygenic Photosynthesis

Most organisms in the Earth's early atmosphere relied on the conversion of organic or inorganic material to obtain their energy. During the microbial era, such organisms most likely lived underground or in water to avoid exposure to harmful UV radiation. At some point, certain bacteria, called *phototrophs*, developed the ability to obtain their energy from sunlight by a new process, called **photosynthesis**. An example of a photosynthetic reaction by blue, green, and yellow sulfur bacteria, which are photoautotrophs, is

$$CO_2(g) + 2H_2S(g) + h\nu \longrightarrow CH_2O(aq) + H_2O(aq) + 2\overset{\bullet}{S}(g) \tag{2.9}$$

Carbon dioxide	Hydrogen sulfide		Carbohydrate	Liquid water	Atomic sulfur

where $CH_2O(aq)$ represents a generic carbohydrate dissolved in water. Because reduced compounds, such as $H_2S(g)$, were not abundant on the surface of the Earth, photoautotrophs flourished only in limited environments. Early **sulfur-producing photosynthesis** did not result in the production of oxygen; thus, it is referred to as **anoxygenic photosynthesis**.

2.3.4. The Oxygen Age

The earth's atmosphere lacked molecular oxygen and ozone until the onset of **oxygen producing photosynthesis** about 2.3 b.y.a, halfway into the present lifetime of the earth. For the next 1.9 billion years, oxygen-producing photosynthesis was carried out primarily by cyanobacteria (Figure 2.10). During this period, oxygen buildup was slow. About 1.4 b.y.a., oxygen levels were still only 1 percent of those today. Land plants evolved from blue-green algae, descendents of cyanobacteria, about 395–430 million years ago (m.y.a.). Like cyanobacteria, plants photosynthesized to produce oxygen. Subsequent to land-plant evolution, oxygen levels began to increase rapidly. Today, 21 out of every 100 molecules in the air are molecular oxygen.

Oxygen-producing photosynthesis in plants is similar to that in bacteria. In both cases, $CO_2(g)$ and sunlight are required, and reactions occur in **chlorophylls**. Chlorophylls reside in photosynthetic membranes. In bacteria, the membranes are cell membranes; in plants and algae, photosynthetic membranes are found in **chloroplasts**.

Figure 2.10. Hot spring in Yellowstone National Park, Wyoming. The hot, mineral-rich water provides ideal conditions for colored photosynthetic cyanobacteria to grow at the perimeter of the spring, where the temperatures drop to about 70 °C. The colors identify different photosynthetic bacteria with different temperature optima. Courtesy of Alfred Spormann, Stanford University.

Chlorophylls are made of **pigments**, which are organic molecules that absorb visible light. Plant and tree leaves generally contain two pigments, chlorophyll *a* and *b*, both of which absorb blue wavelengths (shorter than 500 nm) and red wavelengths (longer than 600 nm) of visible light. Chlorophyll *a* absorbs red wavelengths more efficiently than does chlorophyll *b*, and chlorophyll *b* absorbs blue wavelengths more efficiently than does chlorophyll *a*. Because neither chlorophyll absorbs between 500 and 600 nm, the green part of the visible spectrum, green wavelengths are reflected by chlorophyll, giving leaves a green color. Photosynthetic bacteria generally appear purple, blue, green, or yellow, indicating that their pigments absorb blue, green, and/or red wavelengths to different degrees.

The oxygen-producing photosynthesis process in green plants is

$$6CO_2(g) + 6H_2O(aq) + h\nu \longrightarrow C_6H_{12}O_6(aq) + 6O_2(g) \qquad (2.10)$$

Carbon dioxide / Liquid water / Glucose / Molecular oxygen

where the result, glucose, is dissolved in water in the photosynthetic membrane of the plant. The source of molecular oxygen during photosynthesis in green plants is not carbon dioxide, but water. This can be seen by first dividing Reaction 2.10 by 6, then adding water to each side of the equation. The result is

$$CO_2(g) + 2H_2O(aq) + h\nu \longrightarrow CH_2O(aq) + H_2O(aq) + O_2(g) \qquad (2.11)$$

Carbon dioxide / Liquid water / Carbohydrate / Liquid water / Molecular oxygen

A comparison of Reaction 2.11 with Reaction 2.9 indicates that because the source of atomic sulfur in Reaction 2.9 is hydrogen sulfide, the analogous source of oxygen in Reaction 2.11 should be water. This was first hypothesized in 1931 by Cornelius B. Van Niel, a Dutch microbiologist working at Stanford University, and later proved to be correct experimentally with the use of isotopically labeled water.

With photosynthesis in cyanobacteria and green plants came the production of molecular oxygen and ozone, which helped to shield the Earth's surface from UV radiation. Molecular oxygen absorbs far-UV radiation, and ozone absorbs far-UV, UV-C and a large portion of UV-B radiation. With the slow production of these gases, the surface of the Earth became protected from such radiation. The increase in ozone allowed organisms to migrate to the top of the oceans and land plants to develop 430–395 m.y.a.

2.3.5. Aerobic Respiration and the Oxygen Cycle

The introduction of molecular oxygen and ozone 2.3 b.y.a. also resulted in biological changes in organisms that shaped our present atmosphere. Most important was the development of **aerobic respiration**, which is the process by which molecular oxygen reacts with organic cell material to produce energy during cellular respiration. **Cellular respiration** is the oxidation of organic molecules in living cells.

Whereas aerobic respiration may have developed first in prokaryotes (bacteria and blue-green algae), its spread coincided with the development of another type of organism, the **eukaryote**, about 1.4 b.y.a. A eukaryotic cell contains DNA surrounded by a true membrane-enclosed nucleus. This differs from a prokaryotic cell, which contains a single strand of DNA but not a nucleus. Unlike prokaryotes, many eukaryotes became multicellular. Today, the cells of all higher animals, plants, fungi, protozoa, and most algae are eukaryotic. Prokaryotic cells never evolved past the microbial stage.

Almost all eukaryotic cells respire aerobically. In fact, such cells usually switch from fermentation to aerobic respiration when oxygen concentrations reach about 1 percent of the present oxygen level (Pollack and Yung, 1980). Thus, eukaryotic cells probably developed only 1.4 b.y.a., after the oxygen level increased to above 1 percent of its present level.

The products of aerobic respiration are carbon dioxide and water. Aerobic respiration of glucose, a typical cell component, occurs by

$$C_6H_{12}O_6(aq) + 6O_2(g) \longrightarrow 6CO_2(g) + 6H_2O(aq) \tag{2.12}$$

Glucose   Molecular   Carbon   Liquid
            oxygen   dioxide   water

This process produces energy more efficiently than does fermentation or anaerobic respiration. Thus, Reaction 2.12 was an evolutionary improvement.

Table 2.5 summarizes the current sources and sinks of $O_2(g)$. The primary source is photosynthesis. The major sinks are photolysis in and above the stratosphere and aerobic respiration.

2.3.6. The Nitrogen Cycle

With the development of aerobic respiration came the evolution of organisms that affect the nitrogen cycle. This cycle centers around molecular nitrogen $[N_2(g)]$, the most abundant gas in the air today (making up about 78 percent of it by volume).

Table 2.5. Sources and Sinks of Atmospheric Molecular Oxygen	
Sources	Sinks
Photosynthesis by green plants Chemical production in the stratosphere and above	Photolysis and kinetic reaction Aerobic respiration Dissolution into ocean water Rusting Chemical reaction on soil surfaces Fuel combustion and biomass burning

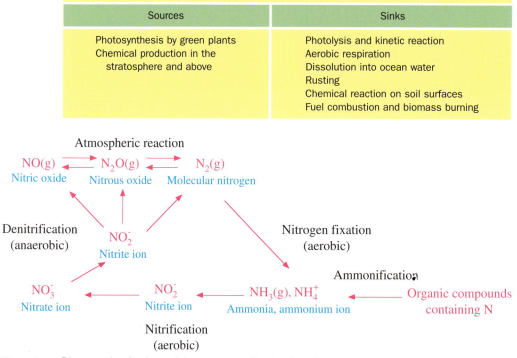

Figure 2.11. Diagram showing bacterial processes affecting the nitrogen cycle.

Figure 2.11 summarizes the major steps in the nitrogen cycle. Four of the five steps are carried out by bacteria in soils. The fifth involves nonbiological chemical reactions occurring in the air.

The direct source of molecular nitrogen in the air is **denitrification**, the two step process carried out by anaerobic bacteria in soils that was described by Reactions 2.7 and 2.8. The second step of denitrification produces nitric oxide [$NO(g)$], nitrous oxide [$N_2O(g)$], or $N_2(g)$, with $N_2(g)$ being the dominant product. $N_2(g)$ is also produced chemically from $N_2O(g)$, which can form from $NO(g)$.

$N_2(g)$ is slowly removed from the air by **nitrogen fixation**. During this process, nitrogen-fixing bacteria, such as *Rhizobium*, *Azotobacter*, and *Beijerinckia*, convert $N_2(g)$ to ammonium [NH_4^+], some of which evaporates back to the air as ammonia gas [$NH_3(g)$]. Another source of ammonium in soils is **ammonification**, a process by which bacteria decompose organic compounds to ammonium. Today, an anthropogenic source of ammonium is fertilizer.

Ammonium is converted to nitrate in soils during a two step process called **nitrification**. This process occurs only in aerobic environments. In the first step, nitrosofying (nitrite-forming) bacteria produce nitrite from ammonium. In the second step, nitrifying (nitrate-forming) bacteria produce nitrate from nitrite. Once nitrate is formed, the nitrogen cycle continues through the denitrification process.

$N_2(g)$ has few chemical sinks. Because its chemical loss is slow and because its removal by nitrogen fixation is slower than is its production by denitrification, $N_2(g)$'s concentration has built up over time. Table 2.6 summarizes the sources and sinks of $N_2(g)$.

Table 2.6. Sources and Sinks of Atmospheric Molecular Nitrogen

Sources	Sinks
Bacterial denitrification Atmospheric reaction and photolysis of $N_2O(g)$	Bacterial nitrogen fixation Atmospheric reaction High-temperature combustion

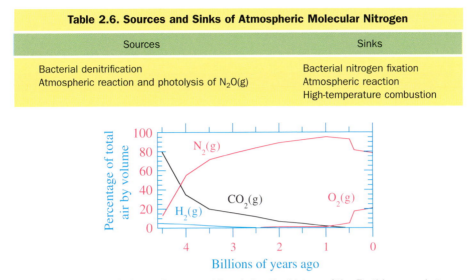

Figure 2.12. Estimated change in composition during the history of the Earth's second atmosphere.

Modified from Cattermole and Moore (1985).

2.3.7. Summary of Atmospheric Evolution

Figure 2.12 summarizes the estimated variations in $N_2(g)$, $O_2(g)$, $CO_2(g)$, and $H_2(g)$ during the evolution of the Earth's second atmosphere. The figure shows that the atmosphere of the early Earth may have been dominated by carbon dioxide. Nitrogen gradually increased due to denitrification. The oxygen concentration increased following the onset of oxygen-producing photosynthesis 2.3 b.y.a. It reached 1 percent of its present level 1.4 b.y.a., but did not approach its present level until after the evolution of green plants around 400 m.y.a.

2.4. SUMMARY

The sun formed from the condensation of the solar nebula about 4.6 b.y.a. Solar radiation incident on the Earth originates from the sun's photosphere. The photosphere emits radiation with an effective temperature near 6,000 K. The solar spectrum consists of UV, visible, and near-IR wavelength regimes. The Earth formed from the same nebula as the sun. Most of the Earth's growth was due to asteroid and meteorite bombardment. The composition of the Earth, percentage-wise, is similar to that of stony meteorites. Dense compounds and compounds with high melting points settled to the center of the Earth. Light compounds and those with low melting points became concentrated in the crust. The first atmosphere of the Earth, which consisted of hydrogen and helium, may have been swept away by an enhanced solar wind during early nuclear explosions in the sun. The second atmosphere, which resulted from outgassing, initially consisted of carbon dioxide, water vapor, and assorted gases. When microbes first evolved, they converted carbon dioxide, ammonia, hydrogen sulfide, and organic material to methane, molecular nitrogen, sulfur dioxide, and carbon dioxide, respectively. Oxygen-producing photosynthesis led to the production of oxygen and ozone. The presence of oxygen resulted in the evolution of aerobic respiration, which led to a more efficient means of producing molecular nitrogen, the major constituent in today's atmosphere.

2.5. PROBLEMS

2.1.   Explain why the Earth's core consists primarily of iron whereas its crust consists primarily of oxygen and silicon.

2.2.   Why does the Earth receive radiation from the sun as if the sun is an emitter with an effective temperature of near 6,000 K, when in fact, the hottest temperatures of the sun are more than 15 million K?

2.3.   Even though the moon is effectively the same distance from the sun as the Earth, the moon has no effective atmosphere. Why?

2.4.   What is the intensity of radiation emitted by a hot desert (330 K) relative to that emitted by the stratosphere over the South Pole during July (190 K)?

2.5.   What prevents the Earth from having magma oceans on its surface today, even though temperatures in the interior of the Earth exceed 4,000 K?

2.6.   What peak wavelength of radiation is emitted in the center of the Earth, where the temperature is near 4,000 K?

2.7.   What elements do you expect to be most abundant in soil-dust particles lifted by the wind? Why?

2.8.   Describe the nitrogen cycle in today's atmosphere. What would happen to molecular nitrogen production if nitrification were eliminated from the cycle? (*Hint:* Consider the processes occurring in the preoxygen atmosphere.)

2.9.   Identify at least three ways in which oxygen improved the opportunity for higher life forms to develop on the Earth.

2.10.   What is the advantage of aerobic respiration over fermentation?

2.11.   What is the source of oxygen during photosynthesis in green plants? Explain.

STRUCTURE AND COMPOSITION OF THE PRESENT-DAY ATMOSPHERE

I n this chapter, the structure and composition of the present-day atmosphere are described. The structure is defined in terms of the variation of pressure, density, and temperature with height. Pressure and density are controlled by gases in the air, the most abundant of which are molecular nitrogen and oxygen. The temperature structure is controlled by the distribution of gases that absorb ultraviolet (UV) and thermal-infrared (IR) radiation. Pressure, density, and temperature are inter-related by the equation of state. Gases in the air include fixed and variable gases. In the following, the structure of the atmosphere in terms of pressure, density, and tempera-ture variations with height is discussed. The main constituents of the air are then examined in terms of their sources and sinks, abundances, health effects, and impor-tance with respect to different air pollution issues.

3.1. AIR PRESSURE AND DENSITY STRUCTURE

Figure 3.1. Experiment with mercury barome-ter, conducted by Evangelista Torricelli (1608–1647) and directed by Blaise Pascal (1623–1662). The third person is the artist, Ernest Board. Courtesy of the Edgar Fahs Smith Collection, University of Pennsylvania Library.

Air consists of gases and particles, but the mass of air is dominated by gases. Of all gas molecules in the air, more than 99 percent are molecular nitrogen or oxygen. Thus, oxygen and nitrogen are responsi-ble for the current pressure, temperature, and density structure of the Earth's atmosphere. **Air pressure**, the force of air per unit area, can be cal-culated as the summed weight of all gas molecules between a horizontal plane and the top of the atmosphere and divided by the area of the plane. Thus, the more oxygen and nitrogen present, the greater the air pressure. Because the weight of air per unit area above a given altitude is always greater than that above any higher altitude, air pres-sure decreases with increasing altitude. In fact, pressure decreases exponentially with increasing altitude. **Standard sea-level pressure** is 1013 mil-libars [mb, in which 1 mb = 100 N m^{-2} = 100 kg m^{-1} s^{-2} = 100 Pascal (Pa) = 1 hectaPascal (hPa)]. The sea-level pressure at a given location and time typically differs by $+10$ to -20 mb from standard sea-level pressure. In a strong low-pressure system, such as at the center of a hurricane, the actual sea-level pressure may be more than 50 mb lower than standard sea-level pressure.

Atmospheric pressure was first measured in 1643 by Italian physicist **Evangelista Torricelli** (1608–1647; Fig. 3.1), an associate of **Galileo Galilei** (1564–1642). Torricelli filled a glass tube 1.2 m long with mercury and inverted it onto a dish. He found that only a portion of the mercury flowed from the tube into the dish, and the resulting space above the mercury in the tube was devoid of air (a **vacuum**). Torricelli was the first person to record a sustained vacuum. After further observations, he suggested that the change in height of the mercury in the tube each day was caused by a change in atmospheric pressure. Air pressure balanced the pressure exerted by the column of mercury in the tube, preventing the mercury from flowing freely from the tube. Decreases in air pressure caused more mercury to flow out of the tube and the mercury level to drop. Increases in air pressure had the opposite effect. The inverted tube Torricelli used to derive this conclusion was called a **mercury barometer**, where a barometer is a device for measuring atmospheric pressure.

EXAMPLE 3.1.

To what height must mercury in a barometer rise to balance an atmospheric pressure of 1,000 mb, assuming the density of mercury is 13,558 kg m^{-3}?

Solution

The pressure (kg m^{-1} s^{-2}) exerted by mercury equals the product of its density (kg m^{-3}), gravity (9.81 m s^{-2}), and the height (h, in meters) of the column of mercury. Equating this pressure with the pressure exerted by air and solving for the height gives

$$h = \frac{1{,}000 \text{ mb}}{13{,}558\frac{\text{kg}}{\text{m}^3} \times 9.81\frac{\text{m}}{\text{s}^2}} \times \frac{100 \text{ kg m}^{-1}\text{s}^{-2}}{1 \text{ mb}} = 0.752 \text{ m} = 29.6 \text{ inches}$$

indicating that a column of mercury 29.6 inches (75.2 cm) high exerts as much pressure at the Earth's surface as a 1,000-mb column of air extending between the surface and the top of the atmosphere.

Soon after Torricelli's discovery, French mathematician **Blaise Pascal** (1623-1662) confirmed Torricelli's theory. He and his brother-in-law, Florin Périer, each carried a glass tube of mercury, inverted in a bath of mercury, up the hill of Puy-de-Dôme, France. Each recorded the level of mercury at the same instant at different altitudes on the hill, confirming that atmospheric pressure decreases with increasing altitude.

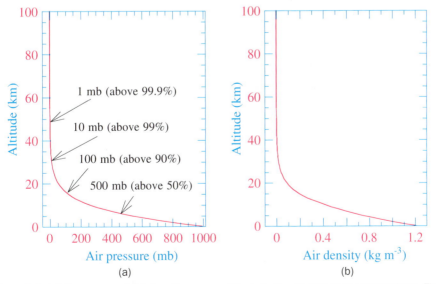

Figure 3.2. (a) Pressure and (b) density versus altitude in the Earth's lower atmosphere. The pressure diagram shows that 99.9 percent of the atmosphere lies below an altitude of about 48 km (1 mb), and 50 percent lies below about 5.5 km (500 mb).

In 1663, the Royal Society of London built its own mercury barometer based on Torricelli's model. A more advanced **aneroid barometer** was developed in 1843. The aneroid barometer measured pressure by gauging the expansion and contraction of a metal tightly sealed in a case containing no air.

Air density is the mass of air per unit volume of air. Because oxygen and nitrogen are concentrated near the Earth's surface, air density peaks near the surface. Air density decreases exponentially with increasing altitude.

Figure 3.2 shows standard profiles of air pressure and density. Figure 3.2(a) shows that 50 percent of atmospheric mass lies between sea level and 5.5 km. About 99.9 percent of mass lies below about 48 km. The Earth's radius is about 6,378 km. Thus, almost all of the Earth's atmosphere lies in a layer thinner than 1 percent of the radius of the Earth.

3.2. PROCESSES AFFECTING TEMPERATURE

Temperature is proportional to the average kinetic energy of an air molecule. From gas kinetic theory, the absolute temperature (K) of air is obtained from

$$\frac{4}{\pi} k_B T = \frac{1}{2} \overline{M} \, \overline{v}_a^2 \tag{3.1}$$

where k_B is **Boltzmann's constant** (1.3807×10^{-23} kg m^2 s^{-2} K^{-1} molecule^{-1}), $\overline{M}$ is the average mass of one air molecule (4.8096×10^{-26} kg molecule^{-1}), and $\overline{v}_a$ is the average **thermal speed** of an air molecule (m s^{-1}). The right side of Equation 3.1 is the kinetic energy of an air molecule at its average thermal speed.

EXAMPLE 3.2.

What is the thermal speed of an air molecule at 200 K? At 300 K?

Solution

From Equation 3.1, the thermal speed of an air molecule is 382 m s^{-1} at 200 K and 468 m s^{-1} at 300 K.

Measurements of relative air temperature changes were first attempted in 1593 by Galileo, who devised a thermoscope to measure the expansion and contraction of air upon its heating and cooling, respectively. However, the instrument did not have a scale and was unreliable. In the mid-seventeenth century, the thermoscope was replaced by the liquid-in-glass thermometer developed in Florence, Italy. In the early eighteenth century, useful thermometer scales were developed by **Gabriel Daniel Fahrenheit** (1686–1736) of Germany and **Anders Celsius** (1701–1744) of Sweden.

The temperature at a given location and time is affected by energy transfer processes, including conduction, convection, advection, and radiation. These processes are discussed briefly in the following subsections.

3.2.1. Conduction

Conduction is the transfer of energy in a medium (the conductor) from one molecule to the next in the presence of a temperature gradient. The medium, as a whole, experiences no molecular movement. Conduction occurs through soil, air, and particles. Conduction

Table 3.1. Thermal Conductivities of Four Media

Substance	Thermal Conductivity (κ) at 298.15 K (J m^{-1} s^{-1} K^{-1})
Dry air at constant pressure	0.0256
Liquid water	0.6
Clay	0.920
Dry sand	0.298

affects air temperature by transferring energy between the soil surface and the bottom molecular layers of the air. The rate of a material's conduction is determined by its **thermal conductivity** (κ, J m^{-1} s^{-1} K^{-1}), which quantifies the rate of flow of thermal energy through a material in the presence of a temperature gradient. Table 3.1 gives thermal conductivities of a few substances. It shows that liquid water, clay, and dry sand are more conductive than is dry air. Thus, energy passes through air slower than it passes through other materials given the same temperature gradient. Clay is more conductive and dry sand is less conductive than is liquid water.

The flux of energy due to conduction (W m^{-2}) can be approximated with the **conductive heat flux equation**,

$$H_c = -\kappa \frac{\Delta T}{\Delta z} \tag{3.2}$$

where ΔT (K) is the change in temperature over a distance Δz (m). At the ground, molecules of soil and water transfer energy by conduction to overlying molecules of air. Because the temperature gradient ($\Delta T/\Delta z$) between the surface and a thin (e.g., 1 mm) layer of air just above the surface is large, the conductive heat flux at the ground is large. Above the ground, temperature gradients are smaller and conductive heat fluxes through the air are smaller than they are at the ground.

EXAMPLE 3.3.

Compare the conductive heat flux through a 1-mm thin layer of air touching the surface if T = 298 K and ΔT = −12 K with that through the free troposphere, where T = 273 K and $\Delta T/\Delta z$ = −6.5 K km^{-1}.

Solution

For air, κ = 0.0256 J m^{-1} s^{-1} K^{-1}; thus, the conductive heat flux at the surface is H_c = 307 W m^{-2}. In the free troposphere, H_c = 1.5 x 10^{-4} W m^{-2}, which is much smaller than is the value at the surface. Thus, heat conduction through the air is important only adjacent to the ground.

3.2.2. Convection

Convection is the transfer of energy, gases, and particles by the mass movement of air, predominantly in the vertical direction. It differs from conduction in that during conduction, energy is transferred from one molecule to another, whereas during convection, energy is transferred as the molecules themselves move. Two important types of convection are *forced* and *free*.

Forced convection is an upward or downward vertical movement of air caused by mechanical means. Forced convection occurs, for example, when (1) horizontal near-surface winds converge (diverge), forcing air to rise (sink); (2) horizontal winds encounter a topographic barrier, forcing air to rise; or (3) winds blow over objects protruding from the ground, creating swirling motions of air, or **eddies**, which mix air vertically and horizontally. Objects of different size create eddies of different size. **Turbulence** is the effect of groups of eddies of different size. Turbulence from wind-generated eddies is called **mechanical turbulence**.

Free convection (thermal turbulence) is a predominantly vertical motion produced by buoyancy, which occurs when the sun heats different areas of the ground differential-ly. Differential heating occurs because clouds or hills block the sun in some areas but not in others, or different surfaces lie at different angles relative to the sun. Over a warm, sunlit surface, conduction transfers energy from the ground to molecules of air adjacent to the ground. The warmed air above the ground rises buoyantly, producing a **thermal**. Cool air from nearby is drawn down to replace the rising air. Near the surface, the cool air heats by conduction then rises, feeding the thermal. Free convection occurs most readily over land when the sky is cloud-free and winds are light.

3.2.3. Advection

Advection is the horizontal movement of energy, gases, and particles by the wind. Like convection, advection results in the mass movement of molecules.

3.2.4. Radiation

Radiation, first defined in Section 2.2, is the transfer of energy by electromagnetic waves or photons, which do not require a medium, such as air, for their transmission. Thus, radiative energy transfer can occur even when no atmosphere exists, such as above the moon's surface. Conduction cannot occur above the moon's surface because too few molecules are present in the moon's atmosphere to transfer energy away from its surface. Similarly, neither convection nor advection can occur in the moon's atmosphere.

3.3. TEMPERATURE STRUCTURE OF THE ATMOSPHERE

The bottom 100 km of the Earth's atmosphere is called the **homosphere**, which is a region in which major gases are well mixed. The homosphere is divided into four lay-ers in which temperatures change with altitude. These are, from bottom to top, the **troposphere**, **stratosphere**, **mesosphere**, and **thermosphere**. Figure 3.3 shows an average profile of the temperature structure of the homosphere.

3.3.1. Troposphere

The troposphere is divided into the boundary layer (ignored in Fig. 3.3) and the free troposphere. These regions are briefly discussed as follows.

3.3.1.1. Boundary Layer

The **boundary layer** extends from the surface to between 500 and 3,000 m alti-tude. All humans live in the boundary layer, so it is this region of the atmosphere in which pollution buildup is of the most concern. Pollutants emitted near the ground accumulate in the boundary layer. When pollutants escape the boundary layer, they can travel horizontally long distances before they are removed from the air. The boundary

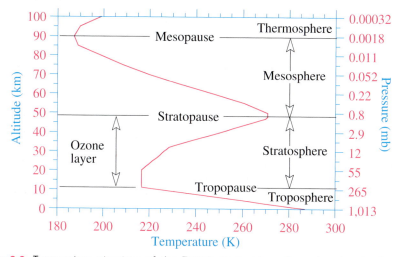

Figure 3.3. Temperature structure of the Earth's lower atmosphere, ignoring the boundary layer.

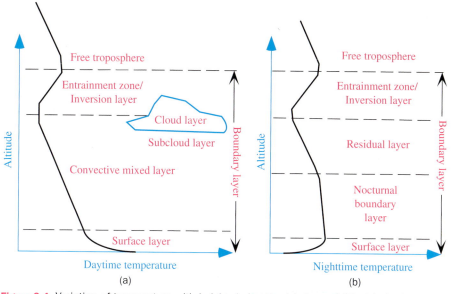

Figure 3.4. Variation of temperature with height during the (a) day and (b) night in the atmospheric boundary layer over land under a high-pressure system. Adapted from Stull (1988).

layer differs from the free troposphere in that the temperature profile in the boundary layer responds to changes in ground temperatures over a period of less than an hour, whereas the temperature profile in the free troposphere responds to changes in ground temperatures over a longer period (Stull, 1988).

Figure 3.4 shows a typical temperature variation in the boundary layer over land during the day and night. During the day [Fig. 3.4(a)], the boundary layer is characterized by a surface layer, a convective mixed layer, and an entrainment zone. The **surface layer**, which comprises the bottom 10 percent of the boundary layer, is a region of strong change of wind speed with height (wind shear). Because the boundary-layer depth ranges from 500 to 3000 m, the surface layer is about 50 to

300 m thick. Wind shear occurs in the surface layer simply because wind speeds at the ground are zero and those above the ground are not.

The **convective mixed layer** is the region of air just above the surface layer. When sunlight warms the ground during the day, some of the energy is transferred from the ground to the air just above the ground by conduction. Because the air above the ground is now warm, it rises buoyantly as a thermal. Thermals originating from the surface layer rise and gain their maximum acceleration in the convective mixed layer. As thermals rise, they displace cooler air aloft downward; thus, upward and downward motions occur, allowing air and pollutants to mix in this layer.

The top of the mixed layer is often bounded by a **temperature inversion**, which is an increase in temperature with increasing height. The inversion inhibits the rise of thermals originating from the surface layer or the mixed layer. Some mixing (entrainment) between the inversion and mixed layer does occur; thus, the inversion layer is called an **entrainment zone**. Pollutants are generally trapped beneath or within an inversion; thus, the closer the inversion is to the ground, the higher pollutant concentrations become.

Other features of the daytime boundary layer are the cloud and subcloud layers. A region in which clouds appear in the boundary layer is the **cloud layer**, and the region underneath is the **subcloud layer**.

During the night [Fig. 3.4(b)], the ground cools radiatively, causing air temperatures to increase with increasing height from the ground, creating a surface inversion. Once the nighttime surface inversion forms, pollutants, when emitted, are confined to the surface layer.

Cooling at the top of the surface layer at night cools the bottom of the mixed layer, reducing the buoyancy and associated mixing at the base of the mixed layer. The portion of the daytime mixed layer that loses its buoyancy at night is the **nocturnal boundary layer**. The remaining portion of the mixed layer is the **residual layer**. Because thermals do not form at night, the residual layer does not undergo much change at night, except at its base. At night, the nocturnal boundary layer thickens, eroding the residual-layer base. Above the residual layer, the inversion remains.

3.3.1.2. Free Troposphere

The free **troposphere** lies between the boundary layer and the tropopause. It is a region in which, on average, the temperature decreases with increasing altitude. The average rate of temperature decrease in the free troposphere is about 6.5 K km^{-1}. The temperature decreases with increasing altitude in the free troposphere for the following reason: The ground surface receives energy from the sun daily, heating the ground, but the top of the troposphere continuously radiates energy upward, cooling the upper troposphere. The troposphere, itself, has relatively little capacity to absorb solar energy; thus, it relies on energy-transfer processes from the ground to maintain its temperature. Convective thermals from the surface transfer energy upward, but as these thermals rise into regions of lower pressure, they expand and cool, resulting in a decrease of temperature with increasing height in the troposphere.

The **tropopause** is the upper boundary of the troposphere. Above the tropopause base, temperatures are relatively constant with increasing height before increasing with increasing height in the stratosphere.

Figure 3.5(a) and (b) show zonally averaged temperatures for a generic January and July, respectively. A **zonally averaged** temperature is found by averaging temperatures over all longitudes at a given latitude and altitude. The figure indicates that tropopause heights are higher (15 to 18 km) over the equator than over the poles (8 to 10 km). Strong

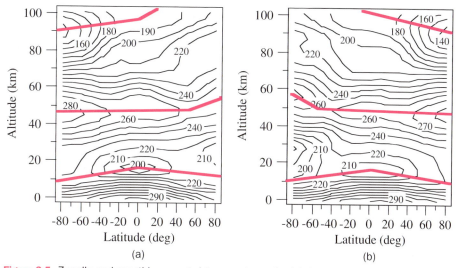

Figure 3.5. Zonally and monthly averaged temperatures for (a) January and (b) July. The thick solid lines are, from bottom to top, the tropopause, stratopause, and mesopause, respectively. Data for the plots were compiled by Fleming et al. (1988).

vertical motions over the equator raise the base of the ozone layer there. Because ozone is responsible for warming above the tropopause, pushing ozone to greater heights over the equator increases the altitude at which warming begins. Near the poles, downward motions push stratospheric ozone downward, lowering the tropopause height over the poles.

Temperatures at the tropopause over the equator are colder than they are over the poles. One reason is that the higher base of the ozone layer over the equator allows tropospheric temperatures to cool to a greater altitude over the equator than over the poles. A second reason is that lower- and mid-tropospheric water vapor contents are much higher over the equator than they are over the poles. Water vapor absorbs thermal-IR radiation emitted from the Earth's surface, preventing that radiation from reaching the upper troposphere.

3.3.2. Stratosphere

The stratosphere is a large temperature inversion. The inversion is caused by ozone, which absorbs the sun's UV radiation and reemits thermal-IR radiation, heating the stratosphere. Peak stratospheric temperatures occur at the top of the stratosphere because this is the altitude at which ozone absorbs the shortest UV wavelengths reaching the stratosphere (about 0.175 μm). Although the ozone content at the top of the stratosphere is low, each ozone molecule can absorb short wavelengths, increasing the average kinetic energy and, thus, temperature (through Equation 3.1) of all molecules. Short UV wavelengths do not penetrate to the lower stratosphere. Ozone densities in the stratosphere peak at 25 to 32 km.

3.3.3. Mesosphere

Temperatures decrease with increasing altitude in the **mesosphere** just as they do in the free troposphere. Ozone densities are too low, in comparison with those of oxygen and nitrogen, for ozone absorption of UV radiation to affect the average temperature of all molecules in the mesosphere.

3.3.4. Thermosphere

In the thermosphere, temperatures increase with increasing altitude because $O_2(g)$ and $N_2(g)$ absorb very short far-UV wavelengths in this region. Peak temperatures in the thermosphere range from 1,200 to 2,000 K, depending on solar activity. Air in the thermosphere does not *feel* hot to the skin because the thermosphere contains so few gas molecules. Because each gas molecule in the thermosphere is highly energized, the average temperature is high. Because molecular oxygen and nitrogen absorb very short wavelengths in the thermosphere, such wavelengths do not penetrate to the mesosphere.

3.4. EQUATION OF STATE

Pressure, density, and temperature in the atmosphere are related by the equation of state. The equation of state describes the relationship among pressure, volume, and absolute temperature for a real gas. The ideal gas law describes this relationship for an ideal gas. An ideal gas is a gas for which the product of the pressure and volume is proportional to the absolute temperature. A real gas is ideal only to the extent that intermolecular forces are small, which occurs when pressures are low enough or temperatures are high enough for the gas to be sufficiently dilute. Under typical atmospheric temperature and pressure conditions, the ideal gas law can reasonably approximate the equation of state.

The ideal gas law is expressed as a combination of Boyle's law, Charles's law, and Avogadro's law. In 1661, Robert Boyle (1627–1691; Fig. 3.6), an English natural philosopher and chemist, found that doubling the pressure exerted on a gas at constant temperature reduced the volume of the gas by one half. This relationship is embodied in Boyle's law,

$$p \propto \frac{1}{V} \quad \text{at constant temperature} \tag{3.3}$$

The Honble: Robert Boyle.

London Printed for Tho. Cockerill at ye 3 Legqs in ye Poultry.

Figure 3.6. Robert Boyle (1627–1691).

where p is the pressure exerted on the gas (mb) and V is the volume enclosed by the gas (m³ or cm³). Boyle's law describes the compressibility of a gas. When high pressure is exerted on a gas, such as in the lower atmosphere, the gas compresses until it exerts an equal pressure on its surroundings. When a gas is subject to low pressure, such as in the upper atmosphere, the gas expands until it exerts an equal pressure on its surroundings.

In 1787, French chemist Jacques Charles (1746–1823; Fig. 3.7) found that increasing the absolute temperature of a gas at constant pressure increased the volume of the gas. This relationship is embodied in Charles's law,

$$V \propto T \quad \text{at constant pressure} \tag{3.4}$$

where T is the temperature of the gas (K). Charles's law states that, at constant pressure, the volume of a gas must decrease when its temperature decreases. Because gases change phase to liquids or solids before 0 K, Charles's law cannot be extrapolated to 0 K.

Charles is also known for his development of a balloon filled with hydrogen gas [$H_2(g)$]. On June 4, 1783, Joseph Michel (1740–1810) and Jacques-Étienne (1745–1799) Montgolfier launched the first untethered hot-air balloon in a marketplace in Annonay, Southern France. The balloon was filled with air heated by burning straw and wool under the opening of a light paper or fabric bag. Prompted by this discovery, the French Academy of Sciences asked Charles to replicate the feat. Instead of filling his balloon with hot air, Charles filled it with $H_2(g)$, a gas 14 times lighter than air that was first observed by Paracelsus and isolated by Cavendish. The balloon was launched on August 27, 1783, and flew for 45 minutes. When it landed, the balloon was hacked to pieces by frightened farmers.

Figure 3.7. Jacques Alexandre Cesar Charles (1746–1823).

In 1811, Amedeo Avogadro (1776–1856; Fig. 3.8), an Italian natural philosopher and chemist, discerned the difference between atoms and molecules. He found that molecular oxygen and nitrogen gas, thought to be single atoms at the time, were really molecules, each consisting of two atoms. He went on to hypothesize that equal volumes of all gases at the same temperature and pressure contained the same number of molecules. In other words, the volume of a gas is proportional to the number of molecules of gas present and independent of the type of gas. This relationship today is Avogadro's law,

$$V \propto n \quad \text{at constant pressure and temperature} \quad (3.5)$$

where n is the number of gas moles. The number of molecules in a mole is constant for all gases and given by Avogadro's number, $A = 6.02252 \times 10^{23}$ molecules mole^{-1}. Avogadro did not devise this number, nor was the term *mole* in the chemical vocabulary during his lifetime. In 1865, Austrian chemist Joseph Loschmidt (1821–1895), who also isolated the first aromatic compounds, estimated the size of an air molecule and the number of molecules in a cubic centimeter of gas. Avogadro's number was devised soon after.

Figure 3.8. Amedeo Avogadro (1776–1856).

Combining Boyle's law, Charles's law, and Avogadro's law gives the ideal gas law or simplified equation of state as

$$p = \frac{nR^*T}{V} = \frac{nA}{V}\left(\frac{R^*}{A}\right)T = Nk_BT \tag{3.6}$$

where

$$N = \frac{nA}{V} \tag{3.7}$$

is the number concentration of gas molecules (molecules of gas per cubic meter or cubic centimeter of air), R^* is the universal gas constant (0.083145 m³ mb mole⁻¹ K⁻¹ or 8.314×10^4 cm³ mb mole⁻¹ K⁻¹), and

$$k_B = \frac{R^*}{A} \tag{3.8}$$

is Boltzmann's constant (1.3807×10^{-25} m³ mb K⁻¹ or 1.3807×10^{-19} cm³ mb K⁻¹). The Appendix contains alternative units for R^* and k_B.

EXAMPLE 3.4.

Calculate the number concentration of air molecules in the atmosphere at standard sea-level pressure and temperature and at a pressure of 1 mb.

Solution

At standard sea level, p = 1013 mb and T = 288 K. Thus, from Equation 3.6, $N = 2.55 \times 10^{19}$ molecules cm⁻³. From Fig. 3.2(a), p = 1 mb occurs at 48 km. At this altitude and pressure, T = 270 K, as shown in Fig. 3.3. Under such conditions, $N = 2.68 \times 10^{16}$ molecules cm⁻³.

Figure 3.9. John Dalton (1766–1844).

Equation 3.6 can be used to relate the partial pressure exerted by a gas to its number concentration. In 1803, **John Dalton** (1766–1844; Fig. 3.9), an English chemist and physicist, stated that total atmospheric pressure equals the sum of the partial pressures of the individual gases in the air. This is **Dalton's law of partial pressure**. The **partial pressure** exerted by a gas in a mixture is the pressure the gas exerts if it alone occupies the same volume as the mixture. Mathematically, the partial pressure of gas q is

$$p_q = N_q k_B T \tag{3.9}$$

where N_q is the number concentration of the gas (molecules cm⁻³). Total atmospheric pressure is

$$p_a = \sum_q p_q \tag{3.10}$$

Dalton is also known for proposing the atomic theory of matter and studying color blindness (Daltonism).

Total atmospheric pressure can also be written as

$$p_a = p_d + p_v \tag{3.11}$$

where p_d is the partial pressure exerted by dry air and p_v is the partial pressure exerted by water vapor. Dry air consists of all gases in the air, except water vapor. Together, $N_2(g)$, $O_2(g)$, $Ar(g)$, and $CO_2(g)$ constitute 99.996 percent of dry air by volume. The partial pressures of all gases aside from these four can be ignored, without much loss in accuracy, when dry-air pressure is calculated. This assumption is convenient because the concentrations of most trace gases vary in time and space.

The partial pressure of dry air is related to the mass density and number concentration of dry air through the **equation of state for dry air**,

$$p_d = \rho_d R'T = N_d k_B T \tag{3.12}$$

where ρ_d is the mass density of dry air (kg m^{-3} or g cm^{-3}), R' is the gas constant for dry air (2.8704 m^3 mb kg^{-1} K^{-1} or 2870.3 cm^3 mb g^{-1} K^{-1} – alternative units are given in the Appendix), and N_d is the number concentration of dry air molecules (molecules cm^{-3}). The dry-air mass density, number concentration, and gas constant are further defined as

$$\rho_d = \frac{n_d m_d}{V} \qquad N_d = \frac{n_d A}{V} \qquad R' = \frac{R^*}{m_d} \tag{3.13}$$

respectively, where n_d is the number of moles of dry air and m_d is the molecular weight of dry air, which is a volume-weighted average of the molecular weights of $N_2(g)$, $O_2(g)$, $Ar(g)$, and $CO_2(g)$. The standard value of m_d is 28.966 g mole^{-1}. The equation of state for water vapor is analogous to that for dry air.

EXAMPLE 3.5.

When $p_d = 1013$ mb and $T = 288$ K, what is the density of dry air?

Solution

From Equation 3.12, $\rho_d = 1.23$ kg m^{-3}.

The number concentration of a gas (molecules per unit volume of air) is an absolute quantity. The abundance of a gas may also be expressed in terms of a relative quantity, **volume mixing ratio**, defined as the number of gas molecules per molecule of dry air, and expressed for gas q as

$$\chi_q = \frac{N_q}{N_d} = \frac{p_q}{p_d} \tag{3.14}$$

where N_q and p_q are the number concentration and partial pressure, respectively, of gas q. Volume mixing ratios may be multiplied by 100 and expressed as a **percentage of dry air volume**, multiplied by 10^6 and expressed in **parts per million volume** (ppmv), multiplied by 10^9 and expressed in **parts per billion volume** (ppbv), or multiplied by 10^{12} and expressed in **parts per trillion volume** (pptv).

EXAMPLE 3.6.

Find the number concentration and partial pressure of ozone if its volume mixing ratio is $\chi_q = 0.10$ ppmv. Assume $T = 288$ K and $p_d = 1{,}013$ mb.

Solution

From Example 3.4, $N_d = 2.55 \times 10^{19}$ molecules cm^{-3}. Thus, from Equation 3.14, the number concentration of ozone is $N_q = 0.10$ ppmv $\times 10^{-6} \times 2.55 \times 10^{19}$ molecules cm$^{-3} = 2.55 \times 10^{12}$ molecules cm^{-3}. From Equation 3.9, the partial pressure exerted by ozone is $p_q = 0.000101$ mb.

3.5. COMPOSITION OF THE PRESENT-DAY ATMOSPHERE

The present-day atmosphere below 100 km (the homosphere) contains only a few well-mixed gases that, together, make up more than 99 percent of all gas molecules in this region. These well-mixed gases are called **fixed gases** because their mixing ratios do not vary much in time or space. Nevertheless, it is the **variable gases**, whose mixing ratios are small but vary in time and space, that are the most important gases with respect to air pollution issues. Fixed and variable gases are discussed in the following subsections.

3.5.1. Fixed Gases

Table 3.2 gives the volume mixing ratios of fixed gases in the homosphere. At any altitude, $O_2(g)$ makes up about 20.95 percent and $N_2(g)$ makes up about 78.08 percent of all non-water gas molecules by volume. Although the mixing ratios of these gases are constant with increasing altitudes, their partial pressures decrease with increasing altitude because air pressure decreases with increasing altitude [Fig. 3.2(a)], and $O_2(g)$ and $N_2(g)$ partial pressures are constant fractions of air pressure.

Together, $N_2(g)$ and $O_2(g)$ make up 99.03 percent of all gases in the atmosphere by volume. Argon (Ar) makes up most of the remaining 0.97 percent. Argon, the "lazy gas," is colorless and odorless. Like other noble gases, it is inert and does not react chemically. Other fixed but inert gases present in trace concentrations include neon, helium, krypton, and xenon.

3.5.2. Variable Gases

Gases whose volume mixing ratios change in time and space are variable gases. Table 3.3 summarizes the volume mixing ratios of some variable gases in the clean

Table 3.2. Volume Mixing Ratios of Fixed Gases in the Lowest 100 km of the Earth's Atmosphere

Gas	Chemical Formula	Volume Mixing Ratio	
		Percent	ppmv
Molecular nitrogen	$N_2(g)$	78.08	780,000
Molecular oxygen	$O_2(g)$	20.95	209,500
Argon	$Ar(g)$	0.93	9,300
Neon	$Ne(g)$	0.0015	15
Helium	$He(g)$	0.0005	5
Krypton	$Kr(g)$	0.0001	1
Xenon	$Xe(g)$	0.000005	0.05

Table 3.3. Volume Mixing Ratios of Some Variable Gases in Three Atmospheric Regions

Gas Name	Chemical Formula	Volume Mixing Ratio (ppbv)		
		Clean Troposphere	Polluted Troposphere	Stratosphere
Inorganic				
Water vapor	$H_2O(g)$	3,000–4.0(+7)[a]	5.0(+6)–4.0(+7)	3,000–6,000
Carbon dioxide	$CO_2(g)$	365,000	365,000	365,000
Carbon monoxide	$CO(g)$	40–200	2,000–10,000	10–60
Ozone	$O_3(g)$	10–100	10–350	1,000–12,000
Sulfur dioxide	$SO_2(g)$	0.02–1	1–30	0.01–1
Nitric oxide	$NO(g)$	0.005–0.1	0.05–300	0.005–10
Nitrogen dioxide	$NO_2(g)$	0.01–0.3	0.2–200	0.005–10
CFC-12	$CF_2Cl_2(g)$	0.55	0.55	0.22
Organic				
Methane	$CH_4(g)$	1,800	1,800–2,500	150–1,700
Ethane	$C_2H_6(g)$	0–2.5	1–50	—
Ethene	$C_2H_4(g)$	0–1	1–30	—
Formaldehyde	$HCHO(g)$	0.1–1	1–200	—
Toluene	$C_6H_5CH_3$	—	1–30	—
Xylene	$C_6H_4(CH_3)_2(g)$	—	1–30	—
Methyl chloride	$CH_3Cl(g)$	0.61	0.61	0.36

[a]4.0(+7) means 4.0×10^7. —, indicates that the volume mixing ratio is negligible, on average.

troposphere, the polluted troposphere (e.g., urban areas), and the stratosphere. Many organic gases degrade chemically before they reach the stratosphere, so their mixing ratios are low in the stratosphere.

3.6. CHARACTERISTICS OF SELECTED GASES AND AEROSOL PARTICLE COMPONENTS

Table 3.4 lists gases and aerosol particle components relevant to each of five air pollution problems. The table indicates that each air pollution problem involves a different set of pollutants, although some pollutants are common to two or more problems.

Next, a few gases and aerosol particle components listed in Table 3.4 are discussed in terms of their relevance, abundance, sources, sinks, and health effects.

3.6.1. Water Vapor

Water vapor [$H_2O(g)$] is the most important variable gas in the air. It is a greenhouse gas in that it readily absorbs thermal-IR radiation, but it is also a vital link in the hydrologic cycle on Earth. As a natural greenhouse gas, it is much more important than is carbon dioxide for maintaining a climate suitable for life on Earth. Water vapor is not considered an air pollutant; thus, no regulations control its concentration or emission.

Sources and Sinks

Table 3.5 summarizes the sources and sinks of water vapor. The main source is evaporation from the oceans. Approximately 85 percent of water vapor originates from ocean-water evaporation.

Table 3.4. Some Gases and Aerosol Particle Components Important for Specified Air Pollution Topics

Indoor Air Pollution	Outdoor Urban Air Pollution	Acid Deposition	Stratospheric Ozone Reduction	Global Climate Change
Gases				
Nitrogen dioxide	Ozone	Sulfur dioxide	Ozone	Water vapor
Carbon monoxide	Nitric oxide	Sulfuric acid	Nitric oxide	Carbon dioxide
Formaldehyde	Nitrogen dioxide	Nitrogen dioxide	Nitric acid	Methane
Sulfur dioxide	Carbon monoxide	Nitric acid	Hydrochloric acid	Nitrous oxide
Organic gases	Ethene	Hydrochloric acid	Chlorine nitrate	Ozone
Radon	Toluene	Carbon dioxide	CFC-11	CFC-11
	Xylene		CFC-12	CFC-12
	PAN			
Aerosol Particle Components				
Black carbon	Black carbon	Sulfate	Chloride	Black carbon
Organic matter	Organic matter	Nitrate	Sulfate	Organic matter
Sulfate	Sulfate	Chloride	Nitrate	Sulfate
Nitrate	Nitrate			Nitrate
Ammonium	Ammonium			Ammonium
Allergens	Soil dust			Soil dust
Asbestos	Sea spray			Sea spray
Fungal spores	Tire particles			
Pollens	Lead			
Tobacco smoke				

Mixing Ratios

The mixing ratio of water vapor varies with location and time but is physically limited to no more than about 4 to 5 percent of total air by its **saturation mixing ratio**, the maximum water vapor the air can hold at a given temperature before the water vapor condenses as a liquid. When temperatures are low, such as over the North and South Poles and in the stratosphere, saturation mixing ratios are low, and water vapor readily deposits as ice or condenses as liquid water. When temperatures are high, such as over the equator, saturation mixing ratios are high, and liquid water may evaporate to the gas phase.

Health Effects

Water vapor has no harmful effects on humans. Liquid water in aerosol particles indirectly causes health problems when it comes in contact with pollutants because many gases dissolve in liquid water. Small drops can subsequently be inhaled, causing health problems in some cases.

Table 3.5. Sources and Sinks of Atmospheric Water Vapor

Sources	Sinks
Evaporation from the oceans, lakes, rivers, and soil	Condensation to liquid water in clouds
Sublimation from sea ice and snow	Vapor deposition to ice crystals in clouds
Transpiration from plant leaves	Transfer to oceans, ice caps, and soils
Kinetic reaction	Kinetic reaction

3.6.2. Carbon Dioxide

Carbon dioxide $[CO_2(g)]$ is a colorless, odorless, natural greenhouse gas, that is also responsible for much of the global warming that has occurred to date. It is a by-product of chemical reactions, but it is not an important outdoor air pollutant in the classic sense because it does not chemically react to form further products nor is it harmful to health at typical mixing ratios. $CO_2(g)$ plays a background role in acid deposition problems because it is responsible for the natural acidity of rainwater, but such natural acidity does not cause environmental damage. $CO_2(g)$ plays a subtle role in stratospheric ozone depletion because global warming near the Earth's surface due to $CO_2(g)$ enhances global cooling of the stratosphere, and such cooling feeds back to the ozone layer. Mixing ratios of carbon dioxide are not regulated in any country. $CO_2(g)$ emission controls are the subject of an ongoing effort by the international community to reduce global warming.

Carbon Reservoirs

The present-day atmosphere contains about 700 gigatons (700 GT, or 700×10^9 tons) of total carbon, primarily in its most oxidized form, $CO_2(g)$. Carbon in the air also appears in its most reduced form, methane $[CH_4(g)]$, and in the form of a variety of other gas and particle components. The mass of carbon as $CO_2(g)$ in the air is more than 200 times that of its nearest carbon-containing rival, $CH_4(g)$. Although 700 GT of carbon sounds like a lot, it pales in comparison with the amount of carbon stored in other carbon reservoirs, particularly the deep oceans, ocean sediments, and carbonate rocks. Table 3.6 shows the relative abundance of carbon in each of these reservoirs

Exchanges of carbon among the reservoirs include exchanges between the surface ocean (0 to 60 m below sea level) and deep ocean (below the surface ocean) by up- and downwelling of water, the deep ocean and sediments by sedimentation and burial of dead organic matter and shell material, the sediments and atmosphere by volcanism, the land and atmosphere by green-plant photosynthesis and bacterial metabolism, and the surface ocean and atmosphere by evaporation and dissolution.

Table 3.6. Storage Reservoirs of Carbon in the Earth's Atmosphere, Oceans, Sediments, and Land

Location and Form of Carbon	GT-C
Atmosphere	
Gas and particulate carbon	700
Surface oceans	
Live organic carbon	5
Dead organic carbon	30
Bicarbonate ion	500
Deep oceans	
Dead organic carbon	3,000
Bicarbonate ion	40,000
Ocean sediments	
Dead organic carbon	10,000,000
Land/ocean sediments	
Carbonate rock	60,000,000
Land	
Live organic carbon	800
Dead organic carbon	2,000

Sources and Sinks

Table 3.7 lists the major sources and sinks of $CO_2(g)$. $CO_2(g)$ is produced during many of the biological processes discussed in Chapter 2, including fermentation (Reaction 2.3), denitritication (Reactions 2.7 and 2.8), and aerobic respiration (Reaction 2.12). These processes are carried out by heterotrophic bacteria. Aerobic respiration is also carried out by plant and animal cells. Other sources of $CO_2(g)$ include evaporation from the oceans, chemical oxidation of carbon monoxide and organic gases, volcanic outgassing, natural and anthropogenic biomass burning (Fig. 3.10), and fossil-fuel combustion. The single largest source of $CO_2(g)$ is bacterial

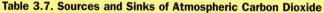

Table 3.7. Sources and Sinks of Atmospheric Carbon Dioxide

Sources	Sinks
Heterotrophic bacterial fermentation	Oxygen-producing photosynthesis
Heterotrophic bacterial anaerobic respiration	Autotrophic bacterial photosynthesis
Heterotrophic bacterial aerobic respiration	Autotrophic bacterial anaerobic respiration
Plant, animal, fungus, protozoa aerobic respiration	Dissolution into the oceans
Evaporation of dissolved CO_2 from the oceans	Transfer to soils and ice caps
Photolysis and kinetic reaction	Chemical weathering of carbonate rocks
Volcanic outgassing	Photolysis
Biomass burning	
Fossil-fuel combustion	
Cement production	

Figure 3.10. Natural forest fire in Yellowstone National Park on August 1, 1988. Emissions from the fire include gases (e.g., carbon dioxide, carbon monoxide, nitric oxide, organics) and aerosol particles (e.g., soot, organic matter). Photo by U.S. Forest Service, available from National Renewable Energy Laboratory.

decomposition of dead organic matter. Indoor sources of $CO_2(g)$ include human exhalation and complete combustion of gas, kerosene, wood, and coal.

$CO_2(g)$ is removed from the air by oxygen-producing photosynthesis (Reaction 2.10), dissolution into ocean water, transfer to soils and ice caps, and chemical weathering. Its *e*-folding lifetime, from emission to removal, due to all loss processes ranges from 50 to 200 years. Like water vapor, carbon dioxide is a greenhouse gas. Unlike water vapor, $CO_2(g)$ has few chemical loss processes in the gas phase. Its main chemical loss is photolysis to $CO(g)$ in the upper stratosphere and mesosphere. An important removal mechanism of $CO_2(g)$ is its dissolution in ocean water. Dissolution occurs by the reversible (denoted by the double arrows) reaction,

$$CO_2(g) \rightleftharpoons CO_2(aq)$$

Gaseous   Dissolved
carbon   carbon   (3.16)
dioxide   dioxide

followed by the rapid combination of $CO_2(aq)$ with water to form carbonic acid [$H_2CO_3(aq)$] and the dissociation of carbonic acid to a **hydrogen ion** [H^+], the **bicarbonate ion** [HCO_3^-], or the **carbonate ion** [CO_3^{2-}] by the reversible reactions

$$CO_2(aq) + H_2O(aq) \rightleftharpoons H_2CO_3(aq) \rightleftharpoons H^+ + HCO_3^- \rightleftharpoons 2H^+ + CO_3^{2-}$$

Dissolved   Liquid   Dissolved   Hydrogen Bicarbonate Hydrogen Carbonate
carbon dioxide water   carbonic acid   ion   ion   ion   ion
(3.17)

Under typical ocean conditions, nearly all dissolved $CO_2(g)$ dissociates to the bicarbonate ion and a small fraction dissociates to the carbonate ion. Certain organisms in the ocean are able to synthesize CO_3^{2-} with the calcium ion [Ca^{2+}] to form calcium carbonate [$CaCO_3(s)$, calcite] shells by

$$Ca^{2+} + CO_3^{2-} \longrightarrow CaCO_3(s)$$

Calcium   Cabonate   Calcium   (3.18)
  ion   ion   carbonate

When shelled organisms die, they sink to the bottom of the ocean, where they are ultimately buried and their shells are turned into calcite rock.

Another removal process of $CO_2(g)$ from the air is **chemical weathering**, which is the breakdown and reformation of rocks and minerals at the atomic and molecular level by chemical reaction. One chemical weathering reaction is

$$CaSiO_3(s) + CO_2(g) \rightleftharpoons CaCO_3(s) + SiO_2(s)$$

Generic   Carbon   Calcium   Silicon
calcium   dioxide   carbonate   dioxide   (3.19)
silicate   (calcite)   (quartz)

in which calcium-bearing silicate rocks react with $CO_2(g)$ to form calcium carbonate rock and quartz rock [$SiO_2(s)$]. At high temperatures, such as in the Earth's mantle, the reverse reaction also occurs, releasing $CO_2(g)$, which is expelled to the air by volcanic eruptions.

Another chemical weathering reaction involves carbon dioxide and calcite rock. During this process, $CO_2(g)$ enters surface water or groundwater by Reaction 3.16 and forms carbonic acid [$H_2CO_3(aq)$] by Reaction 3.17. The acid reacts with calcite, producing Ca^{2+} and HCO_3^- by

$$CaCO_3(s) + CO_2(g) + H_2O(aq) \rightleftharpoons CaCO_3(s) + H_2CO_3(aq) \rightleftharpoons Ca^{2+} + 2HCO_3^-$$

Calcium   Gaseous   Liquid   Calcium   Carbonic   Calcium   Bicarbonate
carbonate   carbon   water   carbonate   acid   ion   ion
             dioxide   (3.20)

Because Reaction 3.20 is reversible, it can proceed either to the right or left. When the partial pressure of $CO_2(g)$ is high, the reaction proceeds to the right, breaking down calcite, removing $CO_2(g)$ and producing Ca^{2+}. Within soils, root and microorganism respiration and organic matter decomposition cause the partial pressure of $CO_2(g)$ to be about 10 to 100 times that in the atmosphere (Brook et al., 1983). Thus, calcite is broken down and $CO_2(g)$ is removed more readily within soils than at soil surfaces. Dissolved calcium ultimately flows with runoff back to the oceans, where some of it is stored and the rest of it is converted to shell material.

Mixing Ratios

Figure 3.11 shows how outdoor $CO_2(g)$ mixing ratios have increased steadily since 1958 at the Mauna Loa Observatory, Hawaii. Average global $CO_2(g)$ mixing ratios have increased from approximately 280 ppmv in the mid-1800s to approximately 370 ppmv today. The yearly increases are due to increased $CO_2(g)$ emission from fossil-fuel combustion. The seasonal fluctuation in $CO_2(g)$ mixing ratios is due to photosynthesis and bacterial decomposition. When annual plants grow in the spring and summer, photosynthesis removes $CO_2(g)$ from the air. When such plants die in the fall and winter, their decomposition by bacteria adds $CO_2(g)$ to the air. Typical indoor mixing ratios of $CO_2(g)$ are 700 to 2,000 ppmv, but can exceed 3,000 ppmv when unvented appliances are used (Arashidani et al., 1996).

Health Effects

Outdoor mixing ratios of $CO_2(g)$ are too low to cause noticeable health problems. In indoor air, $CO_2(g)$ mixing ratios may build up enough to cause some discomfort, but those higher than 15,000 ppmv are necessary to affect human respiration. Mixing ratios higher than 30,000 ppmv are necessary to cause headaches, dizziness, or nausea (Schwarzberg, 1993). Such mixing ratios do not generally occur.

3.6.3. Carbon Monoxide

Carbon monoxide [$CO(g)$] is a tasteless, colorless, and odorless gas. Although $CO(g)$ is the most abundantly emitted variable gas aside from $CO_2(g)$ and $H_2O(g)$, it plays a small role in ozone formation in urban areas. In the background troposphere, it plays a

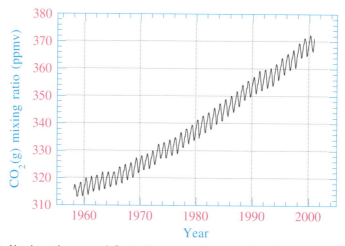

Figure 3.11. Yearly and seasonal fluctuations in carbon dioxide mixing ratio at Mauna Loa Observatory, Hawaii, since 1958. Data for 1958–1999 from Keeling and Whorf (2000) and for 2000 from Mauna Loa Data Center (2001).

Table 3.8. Sources and Sinks of Atmospheric Carbon Monoxide

Sources	Sinks
Fossil-fuel combustion	Kinetic reaction to carbon dioxide
Biomass burning	Transfer to soils and ice caps
Photolysis and kinetic reaction	Dissolution in ocean water
Plants and biological activity in oceans	

larger role in ozone formation. $CO(g)$ is not a greenhouse gas, but its emission and oxidation to $CO_2(g)$ affect global climate. $CO(g)$ is not important with respect to stratospheric ozone reduction or acid deposition. $CO(g)$ is an important component of urban and indoor air pollution because it has harmful short-term health effects. $CO(g)$ is one of six pollutants, called **criteria air pollutants** (Section 8.1.5), for which U.S. National Ambient Air Quality Standards (NAAQS) were set by the U.S. Environmental Protection Agency (U.S. EPA) under the 1970 U.S. Clean Air Act Amendments (CAAA70). $CO(g)$ is now regulated in many other countries as well (Section 8.2).

Sources and Sinks

Table 3.8 summarizes the sources and sinks of $CO(g)$. A major source of $CO(g)$ is incomplete combustion in automobiles, trucks, and airplanes. $CO(g)$ emission sources include wildfires, biomass burning, nontransportation combustion, some industrial processes, and biological activity. Indoor sources of $CO(g)$ include water heaters, coal and gas heaters, and gas stoves. The major sink of $CO(g)$ is chemical conversion to $CO_2(g)$. It is also lost by deposition to soils and ice caps and dissolution in ocean water. Because it is relatively insoluble, its dissolution rate is slow.

Table 3.9 shows that, in 1997, total emissions of $CO(g)$ in the United States were 90 million short tons (1 metric ton = 1.1023 short tons). The largest source was transportation. $CO(g)$ emissions decreased in the United States between 1988 and 1997 by about 25 percent, primarily due to the increased use of the catalytic converter in vehicles (Chapter 8).

Mixing Ratios

Mixing ratios of $CO(g)$ in urban air are typically 2 to 10 ppmv. On freeways and in traffic tunnels, they can rise to more than 100 ppmv. Typical $CO(g)$ mixing ratios inside automobiles in urban areas range from 9 to 56 ppmv (Finlayson-Pitts and Pitts, 1999). In indoor air, hourly average mixing ratios can reach 6–12 ppmv when a gas stove is turned on (Samet et al., 1987). In the absence of indoor sources, $CO(g)$ indoor mixing ratios are usually less than are those outdoors (Jones, 1999). In the free troposphere, $CO(g)$ mixing ratios vary from 50 to 150 ppbv.

Health Effects

Exposure to 300 ppmv of $CO(g)$ for one hour causes headaches; exposure to 700 ppmv of $CO(g)$ for one hour causes death, $CO(g)$ poisoning occurs when it dissolves in blood and replaces oxygen as an attachment to hemoglobin [Hb(aq)], an iron-containing compound. The conversion of $O_2Hb(aq)$ to $COHb(aq)$ (carboxyhemoglobin) causes suffocation. $CO(g)$ can also interfere with $O_2(g)$ diffusion in cellular mitochondria and with intracellular oxidation (Gold, 1992). For the most part, the effects of $CO(g)$ are reversible once exposure to $CO(g)$ is reduced. Following acute exposure,

Table 3.9. Estimated Total Emissions and Percentage of Total Emissions by Source Category in the United States in 1997

Substance	Chemical Formula or Acronym	Total Emissions (10^6 short tons per year)	Industrial Processes (area sources; percentage of total)	Fuel Combustion (point sources; percentage of total)	Transportation (mobile sources; percentage of total)	Miscellaneous (percentage of total)
Carbon monoxide	$CO(g)$	90	6.9	5.5	76.6	11.0
Nitrogen oxides	$NO_x(g)$	24	3.9	45.4	49.2	1.5
Sulfur dioxide	$SO_2(g)$	20	8.4	84.7	6.6	0.3
Particulate matter ≤10 μm[a]	$PM_{10}(aq,s)$	37	3.9	3.2	2.2	90.7
Lead	$Pb(s)$	0.004	74.1	12.6	13.3	0
Reactive organic gases	ROGs	20	51.2	4.5	39.9	4.4

[a]PM_{10} is particulate matter with diameter ≤10 μm. Miscellaneous PM_{10} sources include fugitive dust (57.9 percent of total PM_{10} emissions), agricultural and forest emissions (14.0 percent), wind erosion (15.8 percent), and other combustion sources (3.0 percent).

Source: U.S. EPA (1997).

however, individuals may still express neurologic or psychologic symptoms for weeks or months, especially if they become unconscious temporarily (Choi, 1983).

3.6.4. Methane

Methane [$CH_4(g)$] is the most reduced form of carbon in the air. It is also the simplest and most abundant hydrocarbon and organic gas. Methane is a greenhouse gas that absorbs thermal-IR radiation 25 times more efficiently, molecule for molecule, than does $CO_2(g)$, but mixing ratios of carbon dioxide, are much larger than are those of methane. Methane slightly enhances ozone formation in photochemical smog, but because the incremental ozone produced from methane is small in comparison with ozone produced from other hydrocarbons, methane is a relatively unimportant component of photochemical smog. In the stratosphere, methane has little effect on the ozone layer, but its chemical decomposition provides one of the few sources of stratospheric water vapor. Neither the emission nor ambient concentration of methane is regulated in any country.

Sources and Sinks

Table 3.10 summarizes the sources and sinks of methane. Methane is produced in anaerobic environments, where methanogenic bacteria consume organic material and excrete methane (Equation 2.4). Ripe anaerobic environments include rice paddies (Fig. 3.12), landfills, wetlands, and the digestive tracts of cattle, sheep, and termites. Methane is also produced in the ground from the decomposition of fossilized carbon. The resulting natural gas, which contains more than 90 percent methane, often leaks to the air or is harnessed and used for energy. Methane is also produced during

Table 3.10. Sources and Sinks of Atmospheric Methane

Sources	Sinks

Methanogenic bacteria (lithotrophic autotrophs)
Natural gas leaks during fossil-fuel mining
   and transport
Biomass burning
Fossil-fuel combustion
Kinetic reaction
	Kinetic reaction
	Transfer to soils and ice caps
	Methanotrophic bacteria (conventional
	   heterotrophs)

Figure 3.12. Rice paddies, such as this one in Sundarbans, West Bengal, India, produce not only an important source of food, but also methane gas. Photo by Jim Welch, available from the National Renewable Energy Laboratory.

biomass burning, fossil-fuel combustion, and atmospheric chemical reactions. Its sinks include chemical reactions, transfer to soils, ice caps, the oceans, and consumption by methanotrophic bacteria. The *e*-folding lifetime of methane due to chemical reaction is about 8 to 12 years, which is slow in comparison with the lifetimes of other organic gases. Because methane is relatively insoluble, its dissolution rate into ocean water is slow. Approximately 80 percent of the methane in the air today is biogenic in origin; the rest originates from fuel combustion and natural gas leaks

Mixing Ratios

Methane's average mixing ratio in the troposphere is near 1.8 ppmv, which is an increase from about 0.8 ppmv in the mid-1800s (Ethridge et al., 1992). Its tropospheric mixing ratio has increased steadily due to increased biomass burning, fossil-fuel combustion, fertilizer use, and landfill development. Mixing ratios of methane are relatively constant with height in the troposphere, but decrease in the stratosphere due to chemical loss. At 25 km, methane's mixing ratio is about half that in the troposphere.

Health Effects

Methane has no harmful human health effects at typical outdoor or indoor mixing ratios.

3.6.5. Ozone

Ozone [$O_3(g)$] is a relatively colorless gas at typical mixing ratios. It appears faintly purple when its mixing ratios are high because it weakly absorbs green wavelengths of visible light and transmits red and blue, which combine to form purple. Ozone exhibits an odor when its mixing ratios exceed 0.02 ppmv. In urban smog or indoors, it is considered an air pollutant because of the harm that it does to humans, animals, plants, and materials. In the United States, it is one of the six criteria air pollutants that requires control under CAAA70. It is also regulated in many other countries. In the stratosphere, ozone's absorption of UV radiation provides a protective shield for life on Earth. Although ozone is considered to be "good" in the stratosphere and "bad" in the boundary layer, ozone molecules are the same in both cases.

Sources and Sinks

Table 3.11 summarizes the sources and sinks of ozone. Ozone is not emitted. Its only source into the air is chemical reaction. Sinks of ozone include reaction, transfer to soils and ice caps, and dissolution in ocean waters. Because ozone is relatively insoluble, its dissolution rate is relatively slow.

Mixing Ratios

In the free troposphere, ozone mixing ratios are 20 to 40 ppbv near sea level and 30 to 70 ppbv at higher altitudes. In urban air, ozone mixing ratios range from less than 0.01 ppmv at night to 0.50 ppmv (during afternoons in the most polluted cities world wide), with typical values of 0.15 ppmv during moderately polluted afternoons. Indoor ozone mixing ratios are almost always less than are those outdoors. In the stratosphere, peak ozone mixing ratios are around 10 ppmv.

Health Effects

Ozone causes headaches at mixing ratios greater than 0.15 ppmv, chest pains at mixing ratios greater than 0.25 ppmv, and sore throat and cough at mixing ratios greater than 0.30 ppmv. Ozone decreases lung function for people who exercise steadily for more than an hour while exposed to concentrations greater than 0.30 ppmv. Symptoms of respiratory problems include coughing and breathing discomfort. Small decreases in lung function affect people with asthma, chronic bronchitis, and emphysema. Ozone may also accelerate the aging of lung tissue. At levels greater than 0.1 ppmv, ozone affects animals by increasing their susceptibility to bacterial infection. It also interferes with the growth of plants and trees and deteriorates organic materials, such as rubber,

Table 3.11. Sources and Sinks of Atmospheric Ozone

Sources	Sinks
Chemical reaction of O(g) with O_2(g)	Photolysis Kinetic reaction Transfer to soils and ice caps Dissolution in ocean water

Table 3.12. Sources and Sinks of Atmospheric Sulfur Dioxide

Sources	Sinks
Oxidation of DMS(g)	Kinetic reaction to H_2SO_4(g)
Volcanic emission	Dissolution in cloud drops and ocean water
Fossil-fuel combustion	Transfer to soils and ice caps
Mineral ore processing	
Chemical manufacturing	

textile dyes and fibers, and some paints and coatings (U.S. EPA, 1978). Ozone increases plant and tree stress and their susceptibility to disease, infestation, and death.

3.6.6. Sulfur Dioxide

Sulfur dioxide [SO_2(g)] is a colorless gas that exhibits a taste at levels greater than 0.3 ppmv and a strong odor at levels greater than 0.5 ppmv. SO_2(g) is a precursor to sulfuric acid [H_2SO_4(aq)], an aerosol particle component that affects acid deposition, global climate, and the global ozone layer. SO_2(g) is one of the six air pollutants for which NAAQS standards are set by the U.S. EPA under CAAA70. SO_2(g) is now regulated in many countries.

Sources and Sinks

Table 3.12 summarizes the major sources and sinks of SO_2(g). Some sources include coal-fired power plants, automobile tailpipes, and volcanos. SO_2(g) is also produced chemically in the air from biologically produced dimethylsulfide [DMS(g)] and hydrogen sulfide [H_2S(g)]. SO_2(g) is removed by chemical reaction, dissolution in water, and transfer to soils and ice caps. SO_2(g) is relatively soluble. SO_2(g) emissions decreased in the U.S. between 1988 and 1997 by about 12 percent. Table 3.9 shows that, in 1997, the total mass of emitted SO_2(g) in the United States was 20 million short tons. Between 1988 and 1997, SO_2(g) emissions decreased in the United States by 17 percent.

Mixing Ratios

In the background troposphere, SO_2(g) mixing ratios range from 10 pptv to 1 ppbv. In polluted air, they range from 1 to 30 ppbv. SO_2(g) levels are usually lower indoors than outdoors. The indoor to outdoor ratio of SO_2(g) is typically between 0.1:1 to 0.6:1 in buildings without indoor sources (Jones, 1999). In one study, indoor mixing ratios were found to be 30 to 57 ppbv in homes equipped with kerosene heaters or gas stoves (Leaderer et al., 1984, 1993).

Health Effects

Because SO_2(g) is soluble, it is absorbed in the mucous membranes of the nose and respiratory tract. Sulfuric acid [H_2SO_4(aq)] is also soluble, but its deposition rate into the respiratory tract depends on the size of the particle in which it dissolves (Maroni et al., 1995). High concentrations of SO_2(g) and H_2SO_4(aq) can harm the lungs (Islam and Ulmer, 1979). Bronchiolar constrictions and respiratory infections can occur at mixing ratios greater than 1.5 ppmv. Long-term exposure to SO_2(g) from coal burning is associated with impaired lung function and other respiratory ailments (Qin et al., 1993). People exposed to open coal fires emitting SO_2(g) are likely to suffer from breathlessness and wheezing more than are those not exposed to such fires (Burr et al., 1981).

3.6.7. Nitric Oxide

Nitric oxide [NO(g)] is a colorless gas and a free radical. It is important because it is a precursor to tropospheric ozone, nitric acid [$HNO_3(g)$], and particulate nitrate [NO_3^-]. Whereas NO(g) does not directly affect acid deposition, nitric acid does. Whereas NO(g) does not affect climate, ozone and particulate nitrate do. Natural NO(g) reduces ozone in the upper stratosphere. Emissions of NO(g) from jets that fly in the stratosphere also reduce stratospheric ozone. Outdoor levels of NO(g) are not regulated in any country.

Sources and Sinks

Table 3.13 summarizes the sources and sinks of NO(g). NO(g) is emitted by microbes in soils and plants during denitrification, and it is produced by lightning, combustion, and chemical reaction. Combustion sources include aircraft, automobiles, oil refineries, and biomass burning. The primary sink of NO(g) is chemical reaction.

Mixing Ratios

A typical sea-level mixing ratio of NO(g) in the background troposphere is 5 pptv. In the upper troposphere, NO(g) mixing ratios are 20 to 60 pptv. In urban regions, NO(g) mixing ratios reach 0.1 ppmv in the early morning, but may decrease to zero by midmorning due to reaction with ozone.

Health Effects

Nitric oxide has no harmful human health effects at typical outdoor or indoor mixing ratios.

3.6.8. Nitrogen Dioxide

Nitrogen dioxide [$NO_2(g)$] is a brown gas with a strong odor. It absorbs short (blue and green) wavelengths of visible radiation, transmitting the remaining green and all red wavelengths, causing $NO_2(g)$ to appear brown. $NO_2(g)$ is an intermediary between NO(g) emission and $O_3(g)$ formation. It is also a precursor to nitric acid, a component of acid deposition. Natural $NO_2(g)$, like natural NO(g), reduces ozone in the upper stratosphere. $NO_2(g)$ is one of the six criteria air pollutants for which ambient standards are set by the U.S. EPA under CAAA70. It is now regulated in many countries.

Sources and Sinks

Table 3.14 summarizes sources and sinks of $NO_2(g)$. Its major source is oxidation of NO(g). Minor sources are fossil fuel combustion and biomass burning. During

Table 3.13. Sources and Sinks of Atmospheric Nitric Oxide

Sources	Sinks
Denitrification in soils and plants	Kinetic reaction
Lightning	Dissolution in ocean water
Fossil-fuel combustion	Transfer to soils and ice caps
Biomass burning	
Photolysis and kinetic reaction	

Table 3.14. Sources and Sinks of Atmospheric Nitrogen Dioxide

Sources	Sinks
Photolysis and kinetic reaction	Photolysis and kinetic reaction
Fossil-fuel combustion	Dissolution in ocean water
Biomass burning	Transfer to soils and ice caps

combustion or burning, $NO_2(g)$ emissions are about 5 to 15 percent those of $NO(g)$. Indoor sources of $NO_2(g)$ include gas appliances, kerosene heaters, woodburning stoves, and cigarettes. Sinks of $NO_2(g)$ include photolysis, chemical reaction, dissolution into ocean water, and transfer to soils and ice caps. $NO_2(g)$ is relatively insoluble in water. Table 3.9 shows that in 1997, 24 million short tons of $NO_2(g)$ were emitted in the United States.

Mixing Ratios

Mixing ratios of $NO_2(g)$ near sea level in the free troposphere range from 20 to 50 pptv. In the upper troposphere, mixing ratios are 30 to 70 pptv. In urban regions, they range from 0.1 to 0.25 ppmv. Outdoors, $NO_2(g)$ is more prevalent during midmorning than during midday or afternoon because sunlight breaks down most $NO_2(g)$ past midmorning. In homes with gas-cooking stoves or unvented gas space heaters, weekly average $NO_2(g)$ mixing ratios can range from 21 to 50 ppbv, although peak mixing ratios may reach 400–1,000 ppbv (Spengler, 1993; Jones et al., 1999).

Health Effects

Although exposure to high mixing ratios of $NO_2(g)$ harms the lungs and increases respiratory infections (Frampton et al., 1991), epidemiologic evidence indicates that exposure to typical mixing ratios of $NO_2(g)$ has little effect on the general population. Children and asthmatics are more susceptible to illness associated with high NO_2 mixing ratios than are adults (Li et al., 1994). Pilotto et al. (1997) found that levels of $NO_2(g)$ greater than 80 ppbv resulted in increased reports of sore throats, colds, and absences from school. Goldstein et al. (1988) found that exposure to 300 to 800 ppbv $NO_2(g)$ in kitchens reduced lung capacity by about 10 percent. $NO_2(g)$ may trigger asthma by damaging or irritating and sensitizing the lungs, making people more susceptible to allergic response to indoor allergens (Jones, 1999). At mixing ratios unrealistic under normal indoor or outdoor conditions, $NO_2(g)$ can result in acute bronchitis (25 to 100 ppmv) or death (150 ppmv).

3.6.9. Lead

Lead [Pb(s)] is a gray-white, solid heavy metal with a low melting point that is present in air pollution as an aerosol particle component. It is soft, malleable, a poor conductor of electricity, and resistant to corrosion. It was first regulated as a criteria air pollutant in the United States in 1976. Many countries now regulate the emission and outdoor concentration of lead.

Sources and Sinks

Table 3.15 summarizes the sources and sinks of atmospheric lead. Lead is emitted during combustion of leaded fuel, manufacture of lead-acid batteries, crushing of lead

Table 3.15. Sources and Sinks of Atmospheric Lead

Sources	Sinks
Leaded-fuel combustion Lead-acid battery manufacturing Lead-ore crushing and smelting Dust from soils contaminated with lead-based paint Solid-waste disposal Crustal physical weathering	Deposition to soils, ice caps, and oceans Inhalation

ore, condensation of lead fumes from lead-ore smelting, solid-waste disposal, uplift of lead-containing soils, and crustal weathering of lead ore. Between the 1920s and the 1970s, the largest source of atmospheric lead was automobile combustion.

In December 1921, General Motors researcher **Thomas J. Midgley Jr.** (1889–1944; Fig. 3.13) discovered that tetraethyl lead was a useful fuel additive for reducing engine knock, increasing octane levels, and increasing engine power and efficiency in automobiles. (Midgley later discovered chlorofluorocarbons, CFCs, the precursors to stratospheric ozone destruction–Section 11.5.1.)

Although Midgley also found that ethanol/benzene blends reduced knock in engines, he chose to push tetraethyl lead, and it was first marketed in 1923 under the name "Ethyl gasoline". That same year, Midgley and three other General Motors laboratory employees experienced lead poisoning. Despite his personal experience and warnings sent to him from leading experts on the poisonous effects of lead, Midgley countered, "The exhaust does not contain enough lead to worry about, but no one knows what legislation might come into existence fostered by competition and fanatical health cranks" (Kovarik, 1999,). Between September 1923 and April 1925, 17 workers at du Pont, General Motors, and Standard Oil died and 149 were injured due to lead poisoning during the processing of leaded gasoline. Five of the workers died in October 1924 at a Standard Oil of New Jersey refinery after they became suddenly insane from the cumulative exposure to high concentrations of tetraethyl lead. Despite the deaths and public outcry, Midgley continued to defend his additive. In a paper presented at the American Chemical Society conference in April 1925, he stated

…[T] etraethyl lead is the only material available which can bring about these [antiknock] results, which are of vital importance to the continued economic use by the general public of all automotive equipment, and unless a grave and inescapable hazard exists in the manufacture of tetraethyl lead, its abandonment cannot be justified. (Midgley, 1925)

Midgley's claim about the lack of antiknock alternatives contradicted his own work with ethanol/benzene blends, iron carbonyl, and other mixes that prevented knock.

In May 1925, the U.S. Surgeon General (head of the Public Health Service) put together a committee to study the health effects of tetraethyl lead. The Surgeon General argued that because no regulatory precedent existed, the committee would have to find striking evidence of serious and immediate harm for action to be taken against lead (Kovarik, 1999). Based on measurements that showed lead contents in fecal pellets of typical drivers and garage workers lower than those of lead-industry workers, and based on the observations that drivers and garage workers had not experienced direct lead poisoning, the Surgeon General concluded that there were "no grounds for prohibiting the use of ethyl gasoline" (U.S. Public Health Service, 1925). He did caution that further studies should be carried out (U.S. Public Health Service, 1925). Despite the caution, more studies were not carried out for thirty years, and effective opposition to the use of leaded gasoline ended.

By the mid-1930s, 90 percent of U.S. gasoline was leaded. Industrial backing of lead became so strong that in 1936, the U.S. Federal Trade Commission issued a restraining order forbidding commercial criticism of tetraethyl lead, stating that it is

> entirely safe to the health of (motorists) and to the public in general when used as a motor fuel, and is not a narcotic in its effect, a poisonous dope, or dangerous to the life or health of a customer, purchaser, user or the general public. (Federal Trade Commission, 1936)

Only in 1959 did the Public Health Service reinvestigate the issue of tetraethyl lead. At that time, they found it "regrettable that the investigations recommended by the Surgeon General's Committee in 1926 were not carried out by the Public Health Service" (U.S. Public Health Service, 1959). Despite the concern, tetraethyl lead was not regulated as a pollutant in the United States until 1976. In 1975, the catalytic converter, which reduced emission of carbon monoxide, hydrocarbons, and eventually oxides of nitrogen from cars, was invented. Because lead deactivates the catalyst in the catalytic converter, cars using catalytic converters could run only on unleaded fuels. Thus, the required use of the catalytic converter in new cars inadvertently provided a convenient method to phase out the use of lead. The regulation of lead as a criteria air pollutant in the United States in 1976 due to its health effects also hastened the phase out of lead as a gasoline additive. Between 1970 and 1997, total lead emissions in the United States decreased from 219,000 to 4,000 short tons per year. Table 3.9 shows that, in 1997, only 13.3 percent of total lead emissions originated from transportation. Today, the largest sources of atmospheric lead in the United States are lead-ore crushing, lead-ore smelting, and battery manufacturing. Since the 1980s, leaded gasoline has been phased out in many countries, although it is still an additive to gasoline in several others.

Figure 3.13. Thomas J. Midgley, Jr. (1889–1944). Inventor of leaded gasoline and chlorofluorocarbons (CFCs), Midgley was born in Beaver Falls, Pennsylvania in 1889. He grew up in Dayton and Columbus, Ohio, and graduated from Cornell University with a degree in Mechanical Engineering in 1911. In 1916, he joined the Dayton Engineering Laboratories Company (DELCO) as a researcher. Delco became the main research laboratory for General Motors in 1919. In 1921, Midgley invented leaded gasoline, which he named Ethyl. In 1923, he became vice president of the Ethyl Gasoline Corporation, a subsidiary of General Motors and Standard Oil. In 1924, he was forced to step down due to management problems. He returned to research on synthetic rubber at the Thomas and Hochwalt Laboratory in Dayton, Ohio, with funding from General Motors. In 1928, Midgley and two assistants invented chlorofluorocarbons (CFCs) as a substitute refrigerant for ammonia. Midgley moved on to became vice president of Kinetic Chemicals, Inc. (1930), director and vice president of the Ethyl-Dow Chemical Company (1933) and director and vice president of the Ohio State University Research Foundation (1940–1944). In 1940, he became afflicted with polio, which became so severe that he lost a leg and designed a system of ropes to pull himself out of bed. On November 2, 1944, he died of strangulation in the rope system. Some consider his death a suicide.

Table 3.16. Selected Toxic Compounds, Their Major Sources, and Their Major Effects

Compound	Source	Effect
Benzene	Gasoline combustion, solvents, tobacco smoke	Respiratory irritation, dizziness, headache, nausea, chromosome aberrations, leukemia, produces ozone
Styrene	Plastic and resin production, clothing, building materials	Eye and throat irritation, carcinogenic, produces ozone
Toluene	Gasoline combustion, biomass burning, petroleum refining, detergent production, painting, building materials	Skin and eye irritation, fatigue, nausea, confusion, fetal toxicity, anemia, liver damage, dysfunction of central nervous system, coma, death, produces ozone
Xylene	Gasoline combustion, lacquers, glues	Eye, nose, and throat irritation, liver and nerve damage, produces ozone
1,3-Butadiene	Manufacture of synthetic rubber, combustion of fossil fuels, tobacco smoke	Irritation of eyes, nose, throat, central nervous system damage, cancer, produces ozone
Acetone	Nail polish and paint remover, cleaning solvent	Nose and throat irritation, dizziness, produces ozone
Methyethylketone	Solvent in paints, adhesives, and cosmetics	Headaches, vision reduction, memory loss
Methylene chloride	Solvent, paint stripper, degreaser	Skin irritation, heart and nervous system disorders, carcinogenic
Vinyl chloride	Polyvinylchloride (PVC) plastics, building materials	Liver, brain and lung cancer, mutagenic

Sources: U.S. EPA (1998), Rushton and Cameron (1999), Turco (1997).

Concentrations

The U.S. National Ambient Air Quality Standard for lead is 1.5 μg m^{-3}, averaged over a calendar quarter. Ambient concentrations of lead between 1988 and 1997 decreased from about 0.17 to 0.06 μg m^{-3}, or by 67 percent (U.S. EPA, 1998). The highest concentrations of lead are now found near lead-ore smelters and battery manufacturing plants.

Health Effects

Health effects of lead were known by the early Romans. Marcus Vitruvius Pollio, a Roman engineer, stated in the first-century B.C.,

> We can take example by the workers in lead who have complexions affected by pallor. For when, in casting, the lead receives the current of air, the fumes from it occupy the members of the body, and burning them thereon, rob the limbs of the virtues of the blood. Therefore it seems that water should not be brought in lead pipes if we desire to have it wholesome. (Kovarik, 1998)

Lead accumulates in bones, soft tissue, and blood. It can affect the kidneys, liver, and nervous system. Severe effects of lead poisoning include mental retardation, behavior disorders, and neurologic impairment. A disease associated with lead accumulation is **plumbism**. Symptoms at various stages include abdominal pains, a black line near the base of the gums, paralysis, loss of nerve function, dizziness, blindness, deafness, coma, and death. Low doses of lead have been linked to nervous system damage in fetuses and young children, resulting in learning deficits and low IQs. Lead may also contribute to high blood pressure and heart disease (U.S. EPA, 1998).

3.6.10. Hazardous Organic Compounds

The 1990 Clean Air Act Amendments (CAAA90) required that the U.S. EPA develop emission standards for each of 189 hazardous air pollutants that were thought to pose a risk of cancer or birth defects or environmental or ecological damage. About 8.1 million tons of toxic compounds are released into the air in the United States each year. Of the toxic emissions, about 18 percent originate from area sources (e.g., buildings, industrial complexes), 61 percent from stationary point sources (e.g., smoke stacks), and 21 percent from mobile sources (U.S. EPA, 1998). Whereas most toxics do not affect concentrations of ozone, the main component of photochemical smog, the U.S. EPA has identified several that do. Table 3.16 identifies selected hazardous organics, their sources, and their effects on health and ozone levels.

3.7. SUMMARY

In this chapter, the structure and composition of the present-day atmosphere were discussed. Pressure, density, and temperature are interrelated by the equation of state. Pressure and density decrease exponentially with increasing altitude throughout the atmosphere. Temperature decreases with increasing altitude in the troposphere and mesosphere but increases with increasing altitude in the stratosphere and thermosphere. Temperatures depend on energy transfer processes, including conduction, convection, advection, and radiation. The troposphere is divided into the boundary layer and free troposphere. The daytime boundary layer is often characterized by a surface layer, a convective mixed layer, and an elevated inversion layer. The nighttime boundary layer is often characterized by a surface layer, a nocturnal boundary layer, a residual layer, and an elevated inversion layer. Pollutants emitted from the surface are often confined to the boundary layer. Total air pressure consists of the partial pressures of all gases in the air, which consist of fixed gases – namely molecular nitrogen, molecular oxygen, and argon – and variable gases, including water vapor, carbon dioxide, methane, ozone, and nitric oxide. Some variable gases have adverse health effects; others affect radiation transfer through the air. Most variable gases are chemically reactive.

3.8. PROBLEMS

3.1   How tall must a column of liquid water, with a density of 1,000 kg m^{-3}, be to balance an atmospheric pressure of 1,000 mb?

3.2.   What do the balloon of Charles and an air parcel experiencing free convection have in common?

3.3.   Calculate the conductive heat flux through 1 cm of clay soil if the temperature at the top of the soil is 283 K and that at the bottom is 284 K. How does this flux compare with the fluxes from Example 3.3?

HW 3.4.   If $T = 295$ K at 1 mm above the ground and the conductive heat flux is $H_c = 250$ W m^{-2}, estimate the air temperature at the ground.

Quiz 3.5.   If oxygen, nitrogen, and ozone did not absorb UV radiation, what would you expect the temperature profile in the atmosphere to look like between the surface and top of the thermosphere?

3.6.   If temperatures in the middle of the sun are 15 million kelvin and those on Earth are near 300 K, what is the relative ratio of the thermal speed of a hydrogen atom in the middle of the sun to that on the Earth? (Assume the mass of one hydrogen atom is 1.66×10^{-27} kg molecule^{-1}.) What are the actual speeds in both cases? How long would a hydrogen atom take to travel from the center of the sun to the Earth if it could escape the sun and if no energy losses occurred during its journey?

3.7.   Do convection, conduction, or advection occur in the moon's atmosphere? Why or why not?

3.8.   In the absence of an elevated inversion layer in Fig. 3.4a, do you expect pollutant concentrations to build up or decrease? Why?

3.9.   Why does the lower part of the daytime convective mixed layer lose its buoyancy at night? How does the loss of buoyancy in this region affect concentrations of pollutants emitted at night?

3.10.   Why are the coldest temperatures in the lowest 50 km on Earth generally at the tropical tropopause?

HW 3.11.   Why does the tropopause height decrease with increasing latitude?

HW 3.12.   According to the equation of state, if temperature increases with increasing height and pressure decreases with increasing height in the stratosphere, how must density change with increasing height?

HW 3.13.   If $N_q = 1.5 \times 10^{12}$ molecules cm^{-3} for ozone gas, $T = 285$ K, and $p_d = 980$ mb, find the volume mixing ratio and partial pressure of ozone.

3.14.   If carbon dioxide were not removed from the oceans by shell production and sedimentation, what would happen to its atmospheric mixing ratios?

3.15.   When the carbon dioxide mixing ratio in the atmosphere increases, what happens to the concentration of dissolved carbon dioxide in the ocean?

3.16.   Show and explain a possible chemical weathering process between carbon dioxide and magnesite [$MgCO_3$(s)].

3.17.   Explain why the volume mixing ratio of oxygen is constant but its number concentration decreases exponentially with altitude in the bottom 100 km of the atmosphere.

New book
3.18
3.19

URBAN AIR POLLUTION

4

U rban air pollution problems have existed for centuries and result from the burning of wood, vegetation, coal, natural gas, oil, gasoline, kerosene, diesel, waste, and chemicals. Two general types of urban-scale pollution were identified in the twentieth century: *London-type smog* and *photochemical smog*. The former results from the burning of coal and other raw materials in the presence of a fog or strong inversion, and the latter results from the emission of hydrocarbons and oxides of nitrogen in the presence of sunlight. In most places, urban pollution consists of a combination of the two. In this chapter, gas-phase urban air pollution is discussed in terms of its early history, early regulation, and chemistry.

4.1. HISTORY AND EARLY REGULATION OF URBAN AIR POLLUTION

Before the twentieth century, air pollution was not treated as a science but as a regulatory or legal problem. Because regulations were often weak or not enforced and health problems associated with air pollution were not well understood, pollution problems were rarely mitigated. In this section, a brief history of air pollution and its regulation until the 1940s is discussed.

4.1.1. Before 1200

In ancient Greece, town leaders were responsible for keeping sources of odors outside of town. In ancient Rome, air pollution resulted in civil lawsuits. The Roman poet Horace noted thousands of wood-burning fires (Hughes, 1994) and the blackening of buildings (Brimblecombe, 1999). Air pollution events caused by emissions under strong inversions in Rome were called *heavy heavens* (Hughes, 1994).

Another ancient source of pollution was copper smelting. The smelting of copper to produce coins near the Mediterranean Sea during Roman times and in China during the Song dynasty (960–1279) caused airborne copper concentrations to increase, as detected by Greenland ice-core measurements (Hong et al., 1996).

4.1.2. 1200–1700

In London in the Middle Ages, a major source of pollution was the heating of limestone [which contains $CaCO_3(s)$, calcium carbonate] in kilns, using oak brushwood as an energy source, to produce quicklime [$CaO(s)$, calcium oxide]. Quicklime was then mixed with water to produce a cement, slaked lime [$Ca(OH)_2(s)$, calcium hydroxide], a building material. This process released organic gases, nitric oxide, carbon dioxide, and organic particulate matter into the air.

Sea coal was introduced into London as early as 1228 and gradually replaced the use of wood as a source of energy in lime kilns (and forges). Wood shortages may have led to a surge in the use of sea coal by the mid-1200s. The burning of sea coal resulted in the release of sulfur dioxide, carbon dioxide, nitric oxide, soot, and particulate organic matter. Coal merchants in London lived in Sea Coal Lane, and they would sell their coal to limeburners on nearby Limeburner's Lane (Brimblecombe, 1987). The quantity of coal burned per forge may have been only one-thousandth of that burned per lime kiln.

The pollution in London due to the burning of sea coal became so severe that, starting in 1285, a commission was set up to remedy the situation. The commission met for several years, and by 1306, King Edward I banned the use of coal in lime kilns. The punishment was "grievous ransom," which may have meant fines and furnace confiscation (Brimblecombe, 1987). By 1329, the ban had either been lifted or lost its effect.

Between the thirteenth and eighteenth centuries, the use of sea coal and charcoal increased in England. Coal was used not only in lime kilns and forges, but also in glass furnaces, brick furnaces, breweries, and home heating. One of the early writers on air pollution was **John Evelyn** (1620–1706), who wrote *Fumifugium, or The Inconveniencie of the Aer and the Smoke of London Dissipated* in 1661. He explained how smoke in London was responsible for the fouling of churches, palaces, clothes, furnishings, paintings, rain, dew, water, and plants. He blamed "Brewers, Diers, Limeburners, Salt and Sope-boylers" for the problems.

4.1.3. 1700–1840 – The Steam Engine

Air quality in Great Britain (the union of England, Scotland, and Wales) worsened in the eighteenth century due to the invention of the steam engine, a machine that burned coal to produce mechanical energy. The idea for the steam engine originated with the French-born English physicist **Denis Papin** (1647–1712), who invented the pressure cooker (1679) while working with Robert Boyle. In this device, water was boiled under a closed lid. The addition of steam (water vapor at high temperature) to the air in the cooker increased the total air pressure exerted on the cooker's lid. Papin noticed that the high pressure pushed the lid up. The phenomenon gave him the idea that steam could be used to push up a piston in a cylinder, and the movement of the cylinder could be used to do work. Although he designed such a cylinder-and-piston steam engine, Papin never built one.

Capitalizing on the idea of Papin, **Thomas Savery** (1650–1715), an English engineer, patented the first practical steam engine in 1698. The engine replaced horses as a source of energy for pumping water out of coal mines. Its main limitation was that it did not work well under high pressure. Following Savery, **Thomas Newcomen** (1663–1729), an English engineer, developed a modified steam engine in 1712 that overcame some of the problems in Savery's engine. Because Savery had a patent on the steam engine, Newcomen was forced to enter into partnership with Savery to market the Newcomen engine. Newcomen's engine was used to pump water out of mines and power waterwheels. Steam engines in the early eighteenth century were inefficient, capturing only 1 percent of their maximum possible energy (McNeill, 2000). Because coal mines were not located in cities, early steam engines did not contribute greatly to urban pollution.

Figure 4.1. James Watt (1736–1819).

In 1763, Scottish engineer and inventor **James Watt** (1736–1819; Fig. 4.1) was given a Newcomen steam engine to repair. While fixing the engine, he realized that its efficiency could be improved. After years of work on the idea, he developed an engine that contained a separate chamber for condensing steam. In 1769, he received a patent for the revised steam engine. Watt made further modifications until 1800, including an engine in which the steam was supplied to both sides of the piston and an engine in which motions were circular instead of up and down. Watt's engines were used not only to pump water out of mines, but also to provide energy for paper mills, iron mills, flour mills, cotton mills, steel mills, distilleries, canals, waterworks, and loco- motives. For many of these uses, steam engines were located in urban areas, increasing air pollution. Pollution became particularly severe because, although Watt had improved the steam engine, it still captured only 5 percent of the energy it used by 1800 (McNeill, 2000). Because the steam engine was a large, centralized source of energy, it was responsible for the shift from the artisan shop to the factory system of industrial production during the Industrial Revolution of 1750–1880 (Rosenberg and Birdzell, 1986).

In the nineteenth century, the steam engine was used not only in Great Britain, but also in many other countries, providing a new source of energy and pollution in those countries. The steam engine played a large part in a hundred-fold global increase in coal combustion between 1800 and 1900. Industries centered around coal combustion arose in the United States, Belgium, Germany, Russia, Japan, India, South Africa, and Australia, among other nations.

Pollution problems in Great Britain worsened not only because of steam-engine emissions, but also because of coal combustion in furnaces and boilers and chemical combustion in factories. Between 1800 and 1900, air pollution may have killed people in Great Britain at a rate four to seven times the rate it killed people worldwide (Clapp, 1994).

4.1.4. Regulation in the United Kingdom, 1840–1930

The severity of pollution in the United Kingdom (the 1803 union of Great Britain and Northern Ireland) was sufficient that, in 1843, a committee was set up in London to obtain information about pollution from furnaces and heated steam boilers. Bills were brought before Parliament in 1843 and 1845 to limit emissions, but they were defeat- ed. A third bill was withdrawn in 1846. In 1845, the Railway Clauses Consolidated Act, which required railway engines to consume their own smoke, was enacted. In 1846, a public health bill passed with a clause discussing the reduction of smoke emis- sions from furnaces and boilers. The clause, was removed following pressure by industry. Additional bills failed in 1849 and 1850.

In 1851, an emission clause passed in a sewer bill for the city of London, and this clause was enforced through citations. In 1853, a Smoke Nuisance Abatement (Metropolis) Act also passed through Parliament. This law was enforced only after several years of delay. In 1863, Parliament passed the Alkali Act, which reduced emis- sions of hydrochloric acid gas formed during soap production. Air pollution legislation also appeared in the Sanitary Acts of 1858 and 1866, the Public Health Acts of 1875 and 1891, and the Smoke Abatement Act of 1926.

4.1.5. Early Regulation in the United States, 1869–1940

Most pollution in the United States in the nineteenth century resulted from the burning of coal for manufacturing, home heating, and transportation. Early pollution control was not carried out by state or national agencies, but delegated to municipalities. The first clean air law in the United States may have been an 1869 ordinance by the city of Pittsburgh outlawing the burning of soft coal in locomotives within the city limits. This law, was not enforced. In 1881, Cincinnati passed a law requiring smoke reductions and the appointment of a smoke inspector. Again, the ordinance was not enforced, although the three primary causes of death in Cincinnati in 1886 were lung-related: tuberculosis, pneumonia, and bronchitis (Stradling, 1999). In 1881, a smoke-reduction law passed in Chicago. Although this law had the support of the judiciary, it had little effect.

St. Louis may have been the first city to pass effective legislation. In 1893, the city council passed a law forbidding the emission of "dense black or thick gray smoke" and a second law creating a commission to appoint an inspector and examine smoke-related issues. The ordinance, was overturned by the Missouri State Supreme Court in 1897. The Court stated that the ordinance exceeded the "power of the city under its charter" and was "wholly unreasonable" (Stradling, 1999). Nevertheless, by 1920, air pollution ordinances existed in 175 municipalities; by 1940, this number increased to 200 (Heinsohn and Kabel, 1999).

In 1910, Massachusetts became the first state to regulate air pollution by enacting smoke-control laws for the city of Boston. The first federal involvement in air pollution was probably the creation of an Office of Air Pollution by the Department of Interior's Bureau of Mines in the early 1900s. The purpose of this office was to control emission of coal smoke. The office was relatively inactive and was eliminated shortly thereafter.

Although early regulations in the United Kingdom and the United States did not reduce pollution, they led to pollution-control technologies, such as technologies for recycling chlorine from soda-ash factory emissions and the electrostatic precipitator, used for reducing particle emissions from stacks. Inventions unrelated to air pollution relocated some pollution problems. The advent of the electric motor in the twentieth century, for example, centralized sources of combustion at electric utilities, reducing air pollution at individual factories that had relied on energy from the steam engine.

4.1.6. London-type Smog

In 1905, the term smog was introduced by Harold Antoine Des Voeux, a member of the Coal Smoke Abatement Society in London, to describe the combination of smoke and fog that was visible in several cities throughout Great Britain. The term spread after Des Voeux presented a report at the Manchester Conference of the Smoke Abatement League of Great Britain in 1911 describing smog events in the autumn of 1909 in Glasgow and Edinburgh, Scotland, that killed more than 1,000 people.

The smoke in smog at the time was due to emissions from the burning of coal and other raw materials. Coal was combusted to generate energy, and raw materials were burned to produce chemicals, particularly soda ash [$Na_2CO_3(s)$], used in consumable products, such as soap, detergents, cleansers, paper, glass, and dyes. To produce soda ash, many materials, including charcoal, elemental sulfur, potassium nitrate, sodium chloride, and calcium carbonate were burned, emitting soot, sulfuric acid, nitric acid, hydrochloric acid, calcium sulfide, and hydrogen sulfide, among other compounds.

Emissions from soda-ash factories were added to the landscape of the United Kingdom and France starting in the early 1800s.

Today, pollution resulting from coal-and chemical-combustion smoke in the presence of fog or a low-lying temperature inversion is referred to as **London-type smog**. Some of the chemistry associated with London-type smog is described in Chapter 10. Several deadly London-type smog events have occurred in the nineteenth and twentieth centuries. These episodes provided motivation for modern-day air pollution regulation. The episodes are discussed briefly next.

4.1.6.1. London, United Kingdom

Several London-type smog events were recorded in London in the nineteenth and twentieth centuries. These include events in December 1873 (270–700 deaths more than the average rate), January 1880 (700–1,100 excess deaths), December 1892 (1,000 excess deaths), November 1948 (300 excess deaths), December 1952 (4,000 excess deaths), January 1956 (480 excess deaths), December 1957 (300–800 excess deaths), and December 1962 (340–700 excess deaths) (Brimblecombe, 1987). The worst of these episodes was in December 1952, when 4,000 excess deaths occurred. Excess deaths occurred in every age group, but the number was greater for people older than 45. People with a history of heart or respiratory problems made up 80 percent of those who died. During the episodes, temperature inversions coupled with fog and heavy emissions of pollutants, particularly from combustion of coal and other raw materials, were blamed for the disasters. During the 1952 episode, peak concentrations of $SO_2(g)$ and particulate smoke were estimated to be 1.4 ppmv and 4,460 $\mu g\ m^{-3}$, respectively. (This compares with 24-hour federal standards in the United States today of 0.14 ppmv for $SO_2(g)$ and 150 $\mu g\ m^{-3}$ for particulate matter less than 10 μm in diameter.) The particle and fog cover was so heavy during that event that the streets of London were dark at noontime, and it was necessary for buses to be guided by lantern light.

4.1.6.2. Meuse Valley, Belgium

In December 1930, a 5-day fog event in the presence of a strong temperature inversion and heavy emissions of sulfur dioxide (SO_2) from coal burning resulted in the death of 63 people and the illness of 6,000 others, mostly during the last two days of the pollution episode. The majority of those who died were elderly and previously had heart or lung disease. Symptoms of the victims included chest pain, cough, shortness of breath, and eye irritation.

4.1.6.3. Donora, Pennsylvania, United States

Pittsburgh, Pennsylvania, is located near large coal deposits and major river arteries. In 1758, coal was first burned in Pittsburgh to produce energy for iron and glass manufacturing. By 1865, half of all glass and 40 percent of all iron in the United States were produced in Allegheny County, where Pittsburgh is located (McNeill, 2000). In the late 1800s, the county emerged as a major industrial consumer of coal, particularly for steel. In 1875, Andrew Carnegie opened the Edgar Thomson Works at Braddock, Pennsylvania, introducing a source of high-volume steel to Allegheny County. Steel manufacturing spread to many other cities in the region, including Reading, Pennsylvania; Youngstown, Ohio; and Gary, Indiana. These cities burned coal not only for steel manufacturing, but also for iron manufacturing and railway

(a)

(b)

(c)

Figure 4.2. Panoramic views of (a) Reading, Pennsylvania, c. 1909; (b) Youngstown, Ohio, c. 1910; and (c) Indiana Steel Co.'s big mills, Gary Indiana, c. 1912. Photos (a) and (b) by O. Conneaut, (c) by Crose Photo Company, all available from the Library of Congress Prints and Photographs Division, Washington, DC.

transportation. Figure 4.2 shows that the burning of coal in Reading, Youngstown, and Gary darkened the skies of these cities.

Uncontrolled burning of coal in Allegheny County continued through 1941, when the first strong smoke abatement laws were passed in Pittsburgh. However, the 1941 laws were suspended until the end of World War II, and even then resulted in only minor improvements due to a lack of enforcement.

Donora, Pennsylvania, is a town south of Pittsburgh along the Monongahela River. Between October 26 and 31, 1948, heavy emission of soot and sulfur dioxide from steel mills and of metal fumes from the Donora Zinc Works a zinc smelter, under a strong temperature inversion resulted in the death of 20 people and the respiratory illness of 7,000 out of the town's 14,000 residents. Of the 20 people who died,14 had a known heart or lung problem. Young and old people became ill. Most of the illnesses arose by the third day. Symptoms included cough, sore throat, chest constriction, shortness of breath, eye irritation, nausea, and vomiting. The smog event darkened the city during peak daylight hours, as shown in Fig. 4.3.

4.1.7. Photochemical Smog

Although short, deadly air pollution episodes have attracted public attention, persistent pollution problems in sunny regions have also gained notoriety in the twentieth

Figure 4.3. Noontime photograph of Donora, Pennsylvania, on October 29, 1948, during a deadly smog event. Courtesy of the Pittsburgh Post-Gazette.

Figure 4.4. Panoramic view of Los Angeles, California, taken from Third and Olive Streets, December 3, 1909. Photo by Chas. Z. Bailey, available from Library of Congress Prints and Photographs Division, Washington, DC.

century. Most prominent was a layer of pollution that formed almost daily in Los Angeles, California.

In the early twentieth century, this layer was caused by a combination of directly emitted smoke (London-type smog) and chemically formed pollution, called **photochemical smog**. Sources of the smoke were factories, and sources of the chemically formed pollution were factories and automobiles. In 1903, the factory smoke was so thick one day that residents of Los Angeles thought they were observing an eclipse of the sun (SCAQMD, 2000). Figure 4.4 shows a panoramic view of Los Angeles taken in 1909. The figure shows clouds of dark smoke billowing out of stacks and traveling across the city. Between 1905 and 1912, regulations controlling smoke emissions were adopted by the Los Angeles City Council.

As automobile use increased, the relative fraction of photochemical versus London-type smog in Los Angeles increased. Between 1939 and 1943, visibility in Los Angeles declined precipitously. On July 26, 1943, a plume of pollution engulfed downtown Los Angeles, reducing visibility to three blocks. Even after a local Southern California Gas Company plant suspected of releasing butadiene was shut down, the pollution event continued, suggesting that the pollution was not from that source.

In 1945, the Los Angeles County Board of Supervisors banned the emission of dense smoke and designed an office called the Director of Air Pollution Control. The city of Los Angeles mandated emission controls in the same year. In 1945, Los Angeles County Health Officer **H. O. Swartout** suggested that pollution in Los Angeles originated not only from smokestacks, but also from other sources, namely locomotives, diesel trucks, backyard incinerators, lumber mills, city dumps, and auto-mobiles. In 1946, an air pollution expert from St. Louis, **Raymond R. Tucker**, was hired by the Los Angeles Times to suggest methods of ameliorating air pollution problems in Los Angeles. Tucker proposed 23 methods of reducing air pollution and suggested that a countywide air pollution agency be set up to enforce air pollution regulations (SCAQMD, 2000).

In the face of opposition from oil companies and the Los Angeles Chamber of Commerce, the Los Angeles County Board of Supervisors drafted legislation to be submitted to the State of California that would allow counties throughout the state to set up unified air pollution control districts. The legislation was supported by the League of California Cities, who felt that air pollution could be regulated more effec-tively at the county rather than at the city level. The bill passed 73 to 1 in the California State Legislature and 20 to 0 in the State Assembly and was signed by Governor Earl Warren on June 10, 1947. On October 14, 1947, the Board of Supervisors created the first regional air pollution control agency in the United States, the Los Angeles Air Pollution Control District. On December 30, 1947, the district issued its first mandate, requiring major industrial emitters to obtain emission permits. In the late 1940s and erly 1950s, the district further regulated open burning in garage dumps, emission of sulfur dioxide from refineries, and emission from industrial gasoline storage tanks (1953). In 1954, it banned the 300,000 backyard incinerators used in Los Angeles (Fig. 4.5). Nevertheless, smog problems in Los Angeles persisted (Fig. 4.6).

In 1950, 1957, and 1957, Orange, Riverside, and San Bernardino counties, respec-tively, set up their own air pollution control districts. These districts merged with the Los Angeles district in 1977 to form the South Coast Air Quality Management District (SCAQMD), which currently controls air pollution in the four-county Los Angeles region.

The chemistry of photochemical smog was first elucidated by **Arie Haagen-Smit** (1900–1977; Fig. 4.7), a Dutch professor of biochemistry at the California Institute of Technology. In 1948, Haagen-Smit began studying plants damaged by smog. In 1950, he found that when exposed to ozone sealed in a chamber, plants exhibited the same type of damage as did plants exposed to outdoor smog, suggesting that ozone was a constituent of photochemical smog. Haagen-Smit also found that ozone caused eye irritation, damage to materials, and respiratory problems. Other researchers at the California Institute of Technology found that rubber, exposed to high ozone levels, cracked within minutes. In 1952, Haagen-Smit discovered the

Figure 4.5. Backyard incinerator ban. W. G. Nye and Loy E. Moore, owners of the Peerless Incinerator Company, show their supply of backyard incinerators as they hear reports of banning all incinerators. On October 20, 1954, Moore said, "We're convinced we're being made the goats for some other industry." Courtesy Los Angeles Public Library, Herald-Examiner Photo Collection.

Figure 4.6. Smog bothers pedestrians. Three women on a sidewalk in downtown Los Angeles in the mid-1950s are affected by eye irritation due to smog. The building barely visible in the background is City Hall. Courtesy Los Angeles Public Library, Hollywood Citizens News Collection.

Figure 4.7. Arie Haagen-Smit (1900–1977). Courtesy of the Archives, California Institute of Technology.

mechanism of ozone formation in smog. In the laboratory, he produced ozone from oxides of nitrogen and reactive organic gases in the presence of sunlight. He suggested that ozone and its precursors were the main constituents of Los Angeles photochemical smog.

On the discovery of the source of ozone in smog, oil companies and business leaders argued that the ozone originated from the stratosphere. Subsequent measurements showed that ozone levels were low at nearby Catalina Island, proving that ozone in Los Angeles was local in origin.

Photochemical smog has since been observed in most cities of the world. Notable sites of photochemical smog include Mexico City, Santiago, Tokyo, Beijing, Johannesburg, and Athens. Unlike London-type smog, photochemical smog does not require smoke or a fog for its production. London-type and photochemical smog are exacerbated by a strong temperature inversion (Section 3.3.1.1). Figure 4.8 shows photochemical smog in Los Angeles during the summer of 2000.

In the following sections, gas chemistry of background tropospheric air and of photochemical smog are discussed. Regulatory efforts to control smog since the 1940s are discussed in Chapter 8.

Figure 4.8. Photochemical smog in Los Angeles, California, on July 23, 2000. The smog hides the high-rise buildings in downtown Los Angeles and the mountains in the background.

4.2. CHEMISTRY OF THE BACKGROUND TROPOSPHERE

Today, no region of the global atmosphere is unaffected by anthropogenic pollution. Nevertheless, the background troposphere is cleaner than are urban areas, and to understand photochemical smog, it is useful to examine the chemistry of the background troposphere. The background troposphere is affected by inorganic, light organic, and a few heavy organic gases. The heavy organics include isoprene, a hemiterpene, and other terpenes emitted from biogenic sources. Urban regions are affected by inorganic, light organic, and heavy organic gases. Important heavy organics in urban air, such as toluene and xylene, break down chemically over hours to a few days; thus, most of the free troposphere is not affected directly by these gases, although it is affected by their breakdown products. In the following subsections, inorganic and light organic chemical pathways important in the free troposphere are described.

4.2.1. Photostationary-State Ozone Concentration

In the background troposphere, the ozone $[O_3(g)]$ mixing ratio is determined primarily by a set of three reactions involving itself, nitric oxide $[NO(g)]$, and nitrogen dioxide $[NO_2(g)]$. These reactions are

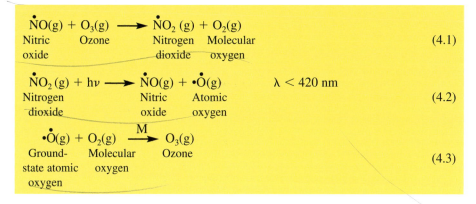

$$\overset{\bullet}{N}O(g) + O_3(g) \longrightarrow \overset{\bullet}{N}O_2(g) + O_2(g) \tag{4.1}$$
Nitric oxide   Ozone   Nitrogen dioxide   Molecular oxygen

$$\overset{\bullet}{N}O_2(g) + h\nu \longrightarrow \overset{\bullet}{N}O(g) + \bullet\overset{\bullet}{O}(g) \qquad \lambda < 420 \text{ nm} \tag{4.2}$$
Nitrogen dioxide   Nitric oxide   Atomic oxygen

$$\bullet\overset{\bullet}{O}(g) + O_2(g) \overset{M}{\longrightarrow} O_3(g) \tag{4.3}$$
Ground-state atomic oxygen   Molecular oxygen   Ozone

Background tropospheric, mixing ratios of $O_3(g)$ (20 to 60 ppbv) are much higher than are those of $NO(g)$ (1 to 60 pptv) or $NO_2(g)$ (5 to 70 pptv). Because the mixing ratio of $NO(g)$ is much lower than is that of $O_3(g)$, Reaction 4.1 does not deplete ozone during the day or night in background tropospheric air. In urban air Reaction 4.1 can deplete local ozone at night because $NO(g)$ mixing ratios at night may exceed those of $O_3(g)$.

If k_1 (cm^3 molecule^{-1} s^{-1}) is the rate coefficient of Reaction 4.1 and J (s^{-1}) is the photolysis rate coefficient of Reaction 4.2, the volume mixing ratio of ozone can be calculated from these two reactions as

$$\chi_{O_3(g)} = \frac{J}{N_d k_1} \frac{\chi_{NO_2(g)}}{\chi_{NO(g)}} \tag{4.4}$$

where χ is volume mixing ratio (molecule of gas per molecule of dry air) and N_d is the concentration of dry air (molecules of dry air per cubic centimeter). This equation is called the **photostationary-state relationship**. The equation does not state that ozone

EXAMPLE 4.1.

Find the photostationary-state mixing ratio of $O_3(g)$ at midday when $p_d = 1013$ mb, $T = 298$ K, $J \approx 0.01$ s^{-1}, $k_1 = 1.8 \times 10^{-14}$ cm^3 molecule^{-1} s^{-1}, $\chi_{NO(g)} = 5$ pptv, and $\chi_{NO_2(g)} = 10$ pptv (typical free-tropospheric mixing ratios).

Solution

From Equation 3.12, $N_d = 2.46 \times 10^{19}$ molecules cm^{-3}. Substituting this into Equation 4.4 gives $\chi_{O_3(g)}$ = 44.7 ppbv, which is a typical free-tropospheric ozone mixing ratio.

is affected by only NO(g) and NO$_2$(g). Indeed, other reactions affect ozone, including ozone photolysis. Instead, Equation 4.4 predicts a relationship among NO(g), NO$_2$(g), and O$_3$(g). If two of the three concentrations are known, the third can be found from the equation.

4.2.2. Daytime Removal of Nitrogen Oxides

During the day, NO$_2$(g) is removed slowly from the photostationary-state cycle by the reaction

$$\overset{\bullet}{N}O_2(g) + \overset{\bullet}{O}H(g) \xrightarrow{M} HNO_3(g) \qquad (4.5)$$

Nitrogen   Hydroxyl   Nitric
dioxide   radical   acid

Although HNO$_3$(g) photolyzes back to NO$_2$(g) + OH(g), its e-folding lifetime against photolysis is 15 to 80 days, depending on the day of the year and the latitude. Because this lifetime is fairly long, HNO$_3$(g) serves as a sink for **nitrogen oxides** [NO$_x$(g) = NO(g) + NO$_2$(g)] in the short term. In addition, because HNO$_3$(g) is soluble, much of it dissolves in cloud drops or aerosol particles before it photolyzes back to NO$_2$(g).

Reaction 4.5 requires the presence of the **hydroxyl radical** [OH(g)], an oxidizing agent that decomposes (scavenges) many gases. Given enough time, OH(g) breaks down every organic gas and most inorganic gases in the air.

The daytime average OH(g) concentration in the clean free troposphere usually ranges from 2 × 10^5 to 3 × 10^6 molecules cm^{-3}. In urban air, OH(g) concentrations typically range from 1 × 10^6 to 1 × 10^7 molecules cm^{-3}. The primary free-tropospheric source of OH(g) is the pathway

$$O_3(g) + h\nu \longrightarrow O_2(g) + \bullet\overset{\bullet}{O}(^1D)(g) \qquad \lambda < 310 \text{ nm} \qquad (4.6)$$

Ozone   Molecular Excited
                 oxygen   atomic
                           oxygen

$$\bullet\overset{\bullet}{O}(^1D)(g) + H_2O(g) \longrightarrow 2\overset{\bullet}{O}H(g) \qquad (4.7)$$

Excited   Water   Hydroxyl
atomic   vapor   radical
oxygen

Thus, OH(g) concentrations in the free troposphere depends on ozone and water vapor contents.

4.2.3. Nighttime Nitrogen Chemistry

During the night, Reaction 4.2 shuts off, eliminating the major chemical sources of O(g) and NO(g). Because O(g) is necessary for the formation of ozone, ozone production also shuts down at night. Thus, at night, neither O(g), NO(g), nor O_3(g) is produced chemically. If NO(g) is emitted at night, it destroys ozone by Reaction 4.1. Because NO_2(g) photolysis shuts off at night, NO_2(g) becomes available to produce NO_3(g), N_2O_5(g), and HNO_3(aq) at night by the sequence

$$\overset{\bullet}{N}O_2\,(g) + O_3\,(g) \longrightarrow \overset{\bullet}{N}O_3\,(g) + O_2(g)$$

| Nitrogen oxide | Ozone | Nitrate radical | Molecular oxygen | (4.8) |

$$\overset{\bullet}{N}O_2\,(g) + \overset{\bullet}{N}O_3\,(g) \overset{M}{\rightleftharpoons} N_2O_5\,(g)$$

| Nitrogen dioxide | Nitrate radical | Dinitrogen pentoxide | (4.9) |

$$N_2O_5\,(g) + H_2O(aq) \longrightarrow 2HNO_3\,(aq)$$

| Dinitrogen pentoxide | Liquid water | Dissolved nitric acid | (4.10) |

Reaction 4.10 occurs on aerosol or hydrometeor particle surfaces. In the morning, sunlight breaks down NO_3(g) within seconds, so NO_3(g) is not important during the day. Because N_2O_5(g) forms from NO_3(g) and decomposes thermally within seconds at high temperatures by the reverse of Reaction 4.9, N_2O_5(g) is also unimportant during the day.

4.2.4. Ozone Production from Carbon Monoxide

Daytime ozone production in the free troposphere is enhanced by carbon monoxide [CO(g)], methane [CH_4(g)], and certain nonmethane organic gases. CO(g) produces ozone by

$$CO(g) + \overset{\bullet}{O}H(g) \longrightarrow CO_2(g) + \overset{\bullet}{H}(g)$$

| Carbon monoxide | Hydroxyl radical | Carbon dioxide | Atomic hydrogen | (4.11) |

$$\overset{\bullet}{H}(g) + O_2(g) \overset{M}{\longrightarrow} H\overset{\bullet}{O}_2(g)$$

| Atomic hydrogen | Molecular oxygen | Hydroperoxy radical | (4.12) |

$$\overset{\bullet}{N}O(g) + H\overset{\bullet}{O}_2(g) \longrightarrow \overset{\bullet}{N}O_2(g) + \overset{\bullet}{O}H\,(g)$$

| Nitric oxide | Hydroperoxy radical | Nitrogen dioxide | Hydroxyl radical | (4.13) |

$$\overset{\bullet}{N}O_2\,(g) + h\nu \longrightarrow \overset{\bullet}{N}O(g) + \bullet\overset{\bullet}{O}(g) \qquad \lambda < 420 \text{ nm}$$

| Nitrogen dioxide | Nitric oxide | Atomic oxygen | (4.14) |

$$\overset{\bullet}{O}(g) + O_2(g) \xrightarrow{M} O_3(g)$$

Ground-   Molecular   Ozone
state atomic   oxygen

(4.15)

oxygen

Because the *e*-folding lifetime of CO(g) against breakdown by OH(g) in the free tropo-sphere is 28 to 110 days, the sequence does not interfere with the photostationary-state relationship among $O_3(g)$, NO(g), and $NO_2(g)$. The second reaction is almost instanta-neous. Reaction 4.11 not only leads to ozone formation, but also produces carbon dioxide.

4.2.5. Ozone Production from Methane

Methane [$CH_4(g)$], with a mixing ratio of 1.8 ppmv, is the most abundant organic gas in the Earth's atmosphere. Its free-tropospheric *e*-folding lifetime against chemical destruction is 8 to 12 years. This long lifetime has enabled it to mix uniformly up to the tropopause. Above the tropopause, its mixing ratio decreases. Methane's most important tropospheric loss mechanism is its reaction with the hydroxyl radical. The methane loss pathway produces ozone, but the incremental quantity of ozone produced is small compared with the photostationary quantity of ozone. The methane oxidation sequence producing ozone is

$$CH_4(g) + \overset{\bullet}{O}H(g) \longrightarrow \overset{\bullet}{C}H_3(g) + H_2O(g)$$

Methane   Hydroxyl   Methyl   Water
          radical   radical   vapor

(4.16)

$$\overset{\bullet}{C}H_3(g) + O_2(g) \xrightarrow{M} CH_3\overset{\bullet}{O}_2(g)$$

Methyl   Molecular   Methylperoxy
radical   oxygen   radical

(4.17)

$$\overset{\bullet}{N}O(g) + CH_3\overset{\bullet}{O}_2(g) \longrightarrow \overset{\bullet}{N}O_2(g) + CH_3\overset{\bullet}{O}(g)$$

Nitric   Methylperoxy   Nitrogen   Methoxy
oxide   radical   dioxide   radical

(4.18)

$$\overset{\bullet}{N}O_2(g) + h\nu \longrightarrow \overset{\bullet}{N}O(g) + \overset{\bullet}{O}(g) \qquad \lambda < 420 \text{ nm}$$

Nitrogen   Nitric   Atomic
dioxide   oxide   oxygen

(4.19)

$$\overset{\bullet}{O}(g) + O_2(g) \xrightarrow{M} O_3(g)$$

Ground-   Molecular   Ozone
state atomic   oxygen

(4.20)

oxygen

In the first reaction, OH(g) **abstracts** (removes) a hydrogen atom from methane, producing the **methyl radical** [$CH_3(g)$] and water. In the stratosphere, this reaction is an important source of water vapor. As with Reaction 4.12, Reaction 4.17 is fast. The remainder of the sequence is similar to the remainder of the carbon monoxide sequence, except that here, the **methylperoxy radical** [$CH_3O_2(g)$] converts NO(g) to

$NO_2(g)$, whereas in the carbon monoxide sequence, the **hydroperoxy radical** [$HO_2(g)$] performs the conversion.

4.2.6. Ozone Production from Formaldehyde

An important byproduct of the methane oxidation pathway is **formaldehyde** [$HCHO(g)$]. Formaldehyde is a colorless gas with a strong odor at higher than 0.05 to 1.0 ppmv. It is the most abundant aldehyde in the air and moderately soluble in water. Aside from gas-phase chemical reaction, the most important source of formaldehyde is emission from plywood, resins, adhesives, carpeting, particleboard, fiberboard, and other building materials (Hines et al., 1993). Mixing ratios of formaldehyde in urban air are generally less than 0.1 ppmv (Maroni et al., 1995). Indoor mixing ratios range from 0.07 to 1.9 ppmv, and typically exceed outdoor mixing ratios (Anderson et al., 1975; Jones, 1999).

Because formaldehyde is moderately soluble in water, it dissolves readily in the upper respiratory tract. Below mixing ratios of 0.05 ppmv, formaldehyde causes no known health problems; at 0.05 to 1.5 ppmv, it has neurophysiologic effects; at 0.01 to 2.0 ppmv, it causes eye irritation; at 0.1 to 25 ppmv, it causes irritation of the upper airway; at 5 to 30 ppmv, it causes irritation of the lower airway and pulmonary problems; at 50 to 100 ppmv, it causes pulmonary edema, inflammation, and pneumonia; and at greater than 100 ppmv, it can result in a coma or be fatal (Hines et al., 1993; Jones, 1999). Formaldehyde is but one of many eye irritants in photochemical smog. Compounds that cause eyes to swell, redden, and tear are **lachrymators**. The methoxy radical from Reaction 4.18 produces formaldehyde by

$$CH_3\dot{O}(g) + O_2(g) \longrightarrow HCHO(g) + H\dot{O}_2(g) \qquad (4.21)$$

Methoxy radical — Molecular oxygen — Formaldehyde — Hydroperoxy radical

The e-folding lifetime of $CH_3O(g)$ against destruction by $O_2(g)$ is 10^{-4} seconds. Once formaldehyde forms, it produces ozone precursors by

$$HCHO(g) + h\nu \longrightarrow \begin{cases} H\dot{C}O(g) + \dot{H}(g) & \lambda < 334 \text{ nm} \\ \\ CO(g) + H_2(g) & \lambda < 370 \text{ nm} \end{cases} \qquad (4.22)$$

Formaldehyde → Formyl radical + Atomic hydrogen; Carbon monoxide + Molecular hydrogen

$$HCHO(g) + \dot{O}H(g) \longrightarrow H\dot{C}O(g) + H_2O(g) \qquad (4.23)$$

Formaldehyde — Hydroxyl radical — Formyl radical — Water vapor

$$H\dot{C}O(g) + O_2(g) \longrightarrow CO(g) + H\dot{O}_2(g) \qquad (4.24)$$

Formyl radical — Molecular oxygen — Carbon monoxide — Hydroperoxy radical

$$\dot{H}(g) + O_2(g) \xrightarrow{M} H\dot{O}_2(g) \qquad (4.25)$$

Atomic hydrogen — Molecular oxygen — Hydroperoxy radical

$CO(g)$ forms ozone through Reactions 4.11 to 4.15, and $HO_2(g)$ forms ozone through Reactions 4.13 to 4.15.

4.2.7. Ozone Production from Ethane

The most concentrated nonmethane hydrocarbons in the free troposphere are **ethane** [$C_2H_6(g)$] and **propane** [$C_3H_8(g)$]. Background tropospheric mixing ratios of ethane are 0 to 2.5 ppbv and of propane are 0 to 1.0 ppbv. These hydrocarbons originate substantially from anthropogenic pollution sources and have relatively long lifetimes against photochemical destruction. The e-folding lifetime of ethane against chemical destruction is about 23 to 93 days, and that of propane is 5 to 21 days. The primary oxidant of ethane and propane is $OH(g)$. In both reactions, $OH(g)$ initiates the breakdown. The sequence of reactions, with respect to ethane, is

$$C_2H_6(g) + \overset{\bullet}{O}H(g) \longrightarrow \overset{\bullet}{C}_2H_5(g) + H_2O(g)$$

Ethane	Hydroxyl radical		Ethyl radical	Water vapor

(4.26)

$$\overset{\bullet}{C}_2H_5(g) + O_2(g) \xrightarrow{M} C_2H_5\overset{\bullet}{O}_2(g)$$

Ethyl radical   Molecular oxygen   Ethylperoxy radical

(4.27)

$$\overset{\bullet}{N}O(g) + C_2H_5\overset{\bullet}{O}_2(g) \longrightarrow \overset{\bullet}{N}O_2(g) + C_2H_5\overset{\bullet}{O}(g)$$

Nitric oxide   Ethylperoxy radical   Nitrogen dioxide   Ethoxy radical

(4.28)

$$\overset{\bullet}{N}O_2(g) + h\nu \longrightarrow \overset{\bullet}{N}O(g) + \bullet\overset{\bullet}{O}(g) \qquad \lambda < 420\ nm$$

Nitrogen dioxide   Nitric oxide   Atomic oxygen

(4.29)

$$\bullet\overset{\bullet}{O}(g) + O_2(g) \xrightarrow{M} O_3(g)$$

Ground-state atomic oxygen   Molecular oxygen   Ozone

(4.30)

The oxidation sequence for propane is similar.

4.2.8. Ozone and PAN Production from Acetaldehyde

An important byproduct of the ethane oxidation pathway is **acetaldehyde** [$CH_3CH(\!=\!O)(g)$], produced from the ethoxy radical formed in Reaction 4.28. The reaction producing acetaldehyde is

$$C_2H_5\overset{\bullet}{O}(g) + O_2(g) \longrightarrow CH_3CH(\!=\!O) + H\overset{\bullet}{O}_2(g)$$

Ethoxy radical   Molecular oxygen   Acetaldehyde   Hydroperoxy radical

(4.31)

This reaction is relatively instantaneous. Acetaldehyde is a precursor to **peroxyacetyl nitrate** (PAN), a daytime component of the background troposphere and, like formaldehyde, an eye irritant and lachrymator. Mixing ratios of PAN in clean air are typically 2 to 100 pptv. Those in rural air downwind of urban sites are up to 1 ppbv.

Polluted air mixing ratios increase to 35 ppbv, with typical values of 10 to 20 ppbv. PAN mixing ratios peak during the afternoon, the same time that ozone mixing ratios peak. PAN is not an important constituent of air at night or in regions of heavy cloudiness. Even in polluted air, PAN does not cause severe health effects, but it damages plants by discoloring their leaves. PAN was discovered during laboratory experiments of photochemical smog formation (Stephens et al., 1956). Its only source is chemical reaction in the presence of sunlight. The reaction pathway producing PAN is

$$CH_3CH(=O)(g) + \overset{\bullet}{O}H(g) \longrightarrow CH_3\overset{\bullet}{C}(=O)(g) + H_2O(g) \tag{4.32}$$

Acetaldehyde   Hydroxyl radical   Acetyl radical   Water vapor

$$CH_3\overset{\bullet}{C}(=O)(g) + O_2(g) \xrightarrow{M} CH_3C(=O)\overset{\bullet}{O}_2(g) \tag{4.33}$$

Acetyl radical   Molecular oxygen   Peroxyacetyl radical

$$CH_3C(=O)\overset{\bullet}{O}_2(g) + \overset{\bullet}{N}O_2(g) \underset{M}{\overset{M}{\rightleftharpoons}} CH_3C(=O)O_2NO_2(g) \tag{4.34}$$

Peroxyacetyl radical   Nitrogen dioxide   Peroxyacetyl nitrate (PAN)

The last reaction in this sequence is reversible and strongly temperature dependent. At 300 K and at surface pressure, PAN's e-folding lifetime against thermal decomposition by Reaction 4.34 is about 25 minutes. At 280 K, its lifetime increases to 13 hours.

Acetaldehyde also produces ozone. The peroxyacetyl radical in Reaction 4.34, for example, converts $NO(g)$ to $NO_2(g)$ by

$$CH_3C(=O)\overset{\bullet}{O}_2(g) + \overset{\bullet}{N}O(g) \longrightarrow CH_3C(=O)\overset{\bullet}{O}(g) + \overset{\bullet}{N}O_2(g) \tag{4.35}$$

Peroxyacetyl radical   Nitric oxide   Acetyloxy radical   Nitrogen dioxide

$NO_2(g)$ forms $O_3(g)$ through Reactions 4.2 and 4.3. A second mechanism of ozone formation from acetaldehyde is through photolysis,

$$CH_3CHO(g) + h\nu \longrightarrow \overset{\bullet}{C}H_3(g) + H\overset{\bullet}{C}O(g) \tag{4.36}$$

Acetaldehyde   Methyl radical   Formyl radical

The methyl radical from this reaction forms ozone through Reactions 4.17 to 4.20. The formyl radical forms ozone through Reaction 4.24 followed by Reactions 4.13 to 4.15.

4.3. CHEMISTRY OF PHOTOCHEMICAL SMOG

Photochemical smog is a soup of gases and aerosol particles. Some of the substances in smog are emitted, whereas others form chemically or physically in the air. In this section, the gas-phase components of smog are discussed. Aerosol particles in smog are discussed in Chapter 5.

Photochemical smog differs from background air in two ways. First, smog contains more high molecular weight organics, particularly aromatic compounds, than does background air. Because most high molecular weight and complex compounds break down quickly in urban air, they are unable to survive transport to the background troposphere. Second, the mixing ratios of nitrogen oxides and organic gases are higher in polluted air than in background air, causing mixing ratios of ozone to be higher in urban air than in background air.

Photochemical smog involves reactions among **nitrogen oxides** [$NO_x(g) = NO(g) + NO_2(g)$] and **reactive organic gases** (ROGs, total organic gases minus methane) in the presence of sunlight. The most recognized gas-phase by-product of smog reactions is ozone because ozone has harmful health effects (Section 3.6.5) and is an indicator of the presence of other pollutants.

On a typical day, ozone forms following emission of NO(g) and ROGs. Emitted pollutants are called **primary pollutants**. ROGs are broken down chemically into **peroxy radicals**, denoted by $RO_2(g)$. Peroxy radicals and NO(g) form ozone by the following sequence:

$$\dot{N}O(g) + R\dot{O}_2(g) \longrightarrow \dot{N}O_2(g) + R\dot{O}(g) \tag{4.37}$$
Nitric   Organic   Nitrogen   Organic
oxide   peroxy   dioxide   oxy
         radical   radical

$$\dot{N}O(g) + O_3(g) \longrightarrow \dot{N}O_2(g) + O_2(g) \tag{4.38}$$
Nitric   Ozone   Nitrogen   Molecular
oxide   dioxide   oxygen

$$\dot{N}O_2(g) + h\nu \longrightarrow \dot{N}O(g) + \bullet\dot{O}(g) \qquad \lambda < 420 \text{ nm} \tag{4.39}$$
Nitrogen   Nitric   Atomic
dioxide   oxide   oxygen

$$\bullet\dot{O}(g) + O_2(g) \xrightarrow{\text{M}} O_3(g) \tag{4.40}$$
Ground-   Molecular   Ozone
state atomic   oxygen
oxygen

Pollutants, such as ozone, that form chemically or physically in the air are called **secondary pollutants**.

Figure 4.9 shows a plot of ozone mixing ratios resulting from different initial mixtures of $NO_x(g)$ and ROGs. This plot is called an **ozone isopleth**. The figure shows that, for low mixing ratios of $NO_x(g)$, ozone mixing ratios are relatively insensitive to the quantity of ROGs. For high $NO_x(g)$, an increase in ROGs increases ozone. The plot also shows that, for low ROGs, increases in $NO_x(g)$ above 0.05 ppmv decrease ozone. For high ROGs, increases in $NO_x(g)$ always increase ozone.

The plot is useful for regulatory control of ozone. If ROG mixing ratios are high (e.g., 2 ppmC) and $NO_x(g)$ mixing ratios are moderate (e.g., 0.06 ppmv), the plot indicates that the most effective way to reduce ozone is to reduce $NO_x(g)$. Reducing ROGs under these conditions has little effect on ozone. If ROG mixing ratios are low (e.g., 0.7 ppmC), and $NO_x(g)$ mixing ratios are high (e.g., 0.2 ppmv), the most effective way to reduce ozone is to reduce ROGs. Reducing $NO_x(g)$ under these conditions actually

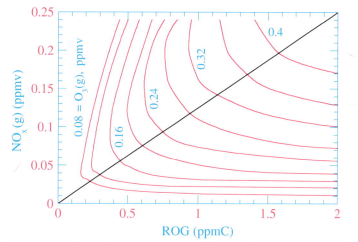

Figure 4.9. Peak ozone mixing ratios resulting from different initial mixing ratios of $NO_x(g)$ and ROGs. The ROG:$NO_x(g)$ ratio along the line through zero is 8:1. Adapted from Finlayson-Pitts and Pitts (1999).

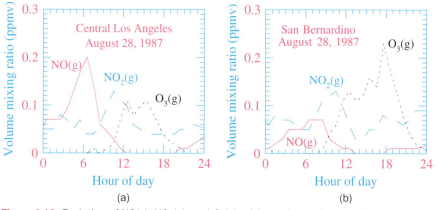

(a)   (b)

Figure 4.10. Evolution of $NO(g)$, $NO_2(g)$, and $O_3(g)$ mixing ratios at (a) central Los Angeles and (b) San Bernardino on August 28, 1987. Central Los Angeles is closer to the coast than is San Bernardino. A sea breeze sends primary pollutants, such as $NO(g)$, from the west side of the Los Angeles Basin (i.e., central Los Angeles) toward the east side (i.e., San Bernardino). As the pollutants travel, organic peroxy radicals convert $NO(g)$ to $NO_2(g)$. Photolysis of $NO_2(g)$ produces atomic oxygen, which forms ozone, a secondary pollutant.

increases ozone before further reaction decreases it. In many polluted urban areas, the ROG:$NO_x(g)$ ratio is lower than 8:1, indicating that limiting ROG emissions should be the most effective method of controlling ozone. Because ozone mixing ratios depend not only on chemistry but also on meteorology, deposition, and gas-to-particle conversion, such a conclusion is not always clearcut.

Figure 4.10 shows the evolution of $NO(g)$, $NO_2(g)$, and $O_3(g)$ during one day at two locations in the Los Angeles Basin – central Los Angeles and San Bernardino. In the basin, a daily sea breeze transfers primary pollutants [$NO(g)$ and ROGs], emitted on the west side of the basin (i.e., central Los Angeles), to the east side of the basin (i.e., San Bernardino), where they arrive as secondary pollutants [$O_3(g)$ and PAN]. Whereas $NO(g)$ mixing ratios peak on the west side of Los Angeles, as shown in Fig. 4.10(a) $O_3(g)$ mixing ratios peak on the east side, as shown in Fig. 4.10(b). Thus, the west side of the basin is a **source region** and the east side is a **receptor region** of photochemical smog.

A difference between ozone production in urban and clean air is that peroxy radicals convert $NO(g)$ to $NO_2(g)$ in urban air but less so in clean air. Because the photostationary state relationship is based on the assumption that only ozone converts $NO(g)$ to $NO_2(g)$, the relationship is usually not valid in urban air. In the afternoon in urban air, the relationship is more accurate than in the morning because ROG mixing ratios in urban air are lower in the afternoon than in the morning.

4.3.1. Emissions of Photochemical Smog Precursors

Gases emitted in urban air include nitrogen oxides, reactive organic gases, carbon monoxide $[CO(g)]$, and sulfur oxides $[SO_x(g) = SO_2(g) + SO_3(g)]$. Of these, $NO_x(g)$ and ROGs are the main precursors of photochemical smog. Table 4.1 shows the rates of primary pollutant gas emissions in the Los Angeles Basin for several species and groups on a summer day in 1987. $CO(g)$ was the most abundantly emitted gas. Of the $NO_x(g)$ emitted, about 85 percent was emitted as $NO(g)$. Almost all $SO_x(g)$ was emitted as $SO_2(g)$.

Of the ROGs toluene, pentane, butane, ethane, ethene, octane, and xylene were emitted in the greatest abundance. The abundance of a gas does not necessarily translate into proportional smog production. A combination of abundance and reactivity is essential for an ROG to be an important smog producer.

Table 4.1. Gas-Phase Emissions for August 27, 1987, in a 400 km × 150 km Region of the Los Angeles Basin

Substance	Emissions (tons day^{-1})	Percentage of Total
Carbon monoxide [$CO(g)$]	9,796	69.3
Nitric oxide [$NO(g)$]	754	
Nitrogen dioxide [$NO_2(g)$]	129	
Nitrous acid [$HONO(g)$]	6.5	
Total $NO_x(g)$ + $HONO(g)$	889.5	6.3
Sulfur dioxide [$SO_2(g)$]	109	
Sulfur trioxide [$SO_3(g)$]	4.5	
Total $SO_x(g)$	113.5	0.8
Alkanes	1399	
Alkenes	313	
Aldehydes	108	
Ketones	29	
Alcohols	33	
Aromatics	500	
Hemiterpenes	47	
Total ROGs	2,429	17.2
Methane [$CH_4(g)$]	904	6.4
Total Emissions	14,132	100

Source: Allen and Wagner (1992).

Table 4.2. Percentage Emission of Several Gases by Source Category

Source Category	$CO(g)$	$NO_x(g)$	$SO_x(g)$	ROG
Stationary	2	24	38	50
Mobile	98	76	62	50
Total	100	100	100	100

Source: Chang et al. (1991).

Table 4.2 shows the percentage emission of several gases by source category. Emissions originate from point, area, or mobile sources. A **point source** is an individual pollutant source, such as a smokestack, fixed in space. A **mobile source** is a moving individual pollutant source, such as the exhaust of a motor vehicle or an airplane. An **area source** is an area, such as a city block, an agricultural field, or an industrial facility, over which many fixed pollutant sources aside from point-source smokestacks exist. Together, point and area sources are **stationary sources**. Table 4.2 shows that $CO(g)$, the most abundantly emitted gas in the basin, originated almost entirely (98 percent) from mobile sources. Oxides of nitrogen were emitted mostly (76 percent) by mobile sources. The thermal combustion reaction in automobiles that produces nitric oxide at a high temperature is

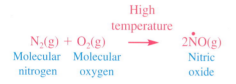

$$N_2(g) + O_2(g) \xrightarrow{\text{High temperature}} 2\overset{\bullet}{N}O(g) \qquad (4.41)$$

Molecular nitrogen   Molecular oxygen   Nitric oxide

Table 4.2 shows that stationary and mobile sources each accounted for 50 percent of ROGs emitted in the basin. Mobile sources accounted for 62 percent of $SO_x(g)$ emissions. The mass of $SO_x(g)$ emissions was one-eighth that of $NO_x(g)$ emissions. Sulfur emissions in Los Angeles are low relative to those in many other cities worldwide.

4.3.2. ROG Breakdown Processes

Once reactive organic gases are emitted, they are broken down chemically into free radicals. Six major processes break down ROGs – photolysis and reaction with $OH(g)$, $HO_2(g)$, $O(g)$, $NO_3(g)$, and $O_3(g)$. $OH(g)$ and $O(g)$ are present only during the day because they are short-lived and require photolysis for their production. $NO_3(g)$ is present only at night because it photolyzes quickly during the day. $O_3(g)$ and $HO_2(g)$ may be present during both day and night.

OH(g) is produced in urban air by some of the same reactions that produce it in the free troposphere. An early morning source of $OH(g)$ in urban air is photolysis of **nitrous acid** [$HONO(g)$]. $HONO(g)$ may be emitted by automobiles; thus, it is more abundant in urban air than in the free troposphere. Midmorning sources of $OH(g)$ in urban air are aldehyde photolysis and oxidation. The major afternoon source of $OH(g)$ in urban air is ozone photolysis. In sum, the three major reaction mechanisms that produce the hydroxyl radical in urban air are

Early Morning Source

$$HONO(g) + h\nu \longrightarrow \overset{\bullet}{O}H(g) + \overset{\bullet}{N}O(g) \qquad \lambda < 400 \text{ nm}$$

Nitrous   Hydroxyl   Nitric
acid   radical   oxide

(4.42)

Midmorning Source

$$HCHO(g) + h\nu \longrightarrow H\overset{\bullet}{C}O(g) + \overset{\bullet}{H}(g) \qquad \lambda < 334 \text{ nm}$$

Formal-   Formyl   Atomic
dehyde   radical   hydrogen

(4.43)

$$\overset{\bullet}{H}(g) + O_2(g) \overset{M}{\longrightarrow} H\overset{\bullet}{O}_2(g)$$

Atomic   Molecular   Hydroperoxy
hydrogen   oxygen   radical

(4.44)

$$H\overset{\bullet}{C}O(g) + O_2(g) \longrightarrow CO(g) + H\overset{\bullet}{O}_2(g)$$

Formyl   Molecular   Carbon   Hydroperoxy
radical   oxygen   monoxide   radical

(4.45)

$$\overset{\bullet}{N}O(g) + H\overset{\bullet}{O}_2(g) \longrightarrow \overset{\bullet}{N}O_2(g) + \overset{\bullet}{O}H(g)$$

Nitric   Hydroperoxy   Nitrogen   Hydroxyl
oxide   radical   dioxide   radical

(4.46)

Afternoon Source

$$O_3(g) + h\nu \longrightarrow O_2(g) + \bullet\overset{\bullet}{O}(^1D)(g) \qquad \lambda < 310 \text{ nm}$$

Ozone   Molecular   Excited
                oxygen   atomic
                          oxygen

(4.47)

$$\bullet\overset{\bullet}{O}(^1D)(g) + H_2O(g) \longrightarrow 2\overset{\bullet}{O}H(g)$$

Excited   Water   Hydroxyl
atomic   vapor   radical
oxygen

(4.48)

ROGs emitted in urban air include alkanes, alkenes, alkynes, aldehydes, ketones, alcohols, aromatics, and hemiterpenes. Table 4.3 shows lifetimes of these ROGs against breakdown by six processes. The table shows that photolysis breaks down aldehydes and ketones, $OH(g)$ breaks down all eight groups during the day, $HO_2(g)$ breaks down aldehydes during the day and night, $O(g)$ breaks down alkenes and terpenes during the day, $NO_3(g)$ breaks down alkanes, alkenes, aldehydes, aromatics, and terpenes during the night, and $O_3(g)$ breaks down alkenes and terpenes during the day and night.

The breakdown of ROGs produces radicals that lead to ozone formation. Table 4.4 shows the most important ROGs in Los Angeles during the summer of 1987 in terms of a combination of abundance and reactive ability to form ozone. The table shows that m- and p-xylene, both aromatic hydrocarbons, were the most important gases in terms of generating ozone. Although alkanes are emitted in greater abundance than are other organics, they are less reactive in producing ozone than are aromatics, alkenes, or aldehydes.

Table 4.3. Estimated *e*-Folding Lifetimes of Reactive Organic Gases Representing Alkanes, Alkenes, Alkynes, Aldehydes, Ketones, Alcohols, Aromatics, and Hemiterpenes against Photolysis and Oxidation by Gases at Specified Concentrations in Urban Air

ROG Species	Lifetime in Polluted Urban Air at Sea Level					
	Photolysis	OH(g) 5×10^6 Molecules cm^{-3}	HO$_2$(g) 2×10^9 Molecules cm^{-3}	O(g) 8×10^4 Molecules cm^{-3}	NO$_3$(g) 1×10^{10} Molecules cm^{-3}	O$_3$(g) 5×10^{12} Molecules cm^{-3}
n-Butane	—	22 h	1,000 y	18 y	29 d	650 y
trans-2-Butene	—	52 m	4 y	6.3 d	4 m	17 m
Acetylene	—	3.0 d	—	2.5 y	—	200 d
Formaldehyde	7 h	6.0 h	1.8 h	2.5 y	2.0 d	3,200 y
Acetone	23 d	9.6 d	—	—	—	—
Ethanol	—	19 h	—	—	—	—
Toluene	—	9.0 h	—	6 y	33 d	200 d
Isoprene	—	34 m	—	4 d	5 m	4.6 h

Lifetimes were obtained from rate- and photolysis-coefficient data. Gas concentrations are typical, but not necessarily average values for each region. Units: m = minutes, h = hours, d = days, y = years, — = insignificant loss.

In the following subsections, photochemical smog processes involving the chemical breakdown of organic gases to produce ozone are discussed.

4.3.3. Ozone Production from Alkanes

Table 4.4 shows that *i*-pentane and butane are the most effective alkanes in terms of concentration and reactivity in producing ozone in Los Angeles air. As in the free troposphere, the main pathway of alkane decomposition in urban air is OH(g) attack. Photolysis and reaction with O$_3$(g), HO$_2$(g), and NO$_3$(g) have little effect on alkane concentrations. Of all alkanes, methane is the least reactive and the least important with respect to urban air pollution. Methane is more important with respect to free tropospheric and stratospheric chemistry. The oxidation pathways of methane were given in Reactions 4.16 to 4.20, and those of ethane were shown in Reactions 4.26 to 4.30.

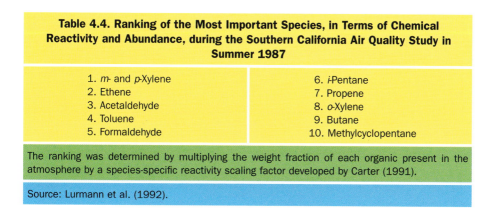

Table 4.4. Ranking of the Most Important Species, in Terms of Chemical Reactivity and Abundance, during the Southern California Air Quality Study in Summer 1987

1. *m*- and *p*-Xylene	6. *i*-Pentane
2. Ethene	7. Propene
3. Acetaldehyde	8. *o*-Xylene
4. Toluene	9. Butane
5. Formaldehyde	10. Methylcyclopentane

The ranking was determined by multiplying the weight fraction of each organic present in the atmosphere by a species-specific reactivity scaling factor developed by Carter (1991).

Source: Lurmann et al. (1992).

4.3.4. Ozone Production from Alkenes

Table 4.4 shows that alkenes, such as ethene and propene, are important ozone precursors in photochemical smog. Mixing ratios of ethene and propene in polluted air reach 1 to 30 ppbv. Table 4.3 indicates that alkenes react most rapidly with OH(g), O_3(g), and NO_3(g). In the following subsections, the first two of these reaction pathways are discussed.

4.3.4.1. Alkene Reaction with the Hydroxyl Radical

When ethene reacts with the hydroxyl radical, the radical substitutes into ethene's double bond to produce an **ethanyl radical** in an OH(g) **addition** process. The ethanyl radical then reacts to produce NO_2(g). The sequence is

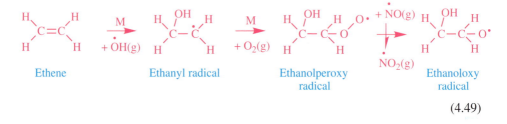

$$(4.49)$$

NO_2(g) produces ozone by Reactions 4.2 and 4.3. The **ethanoloxy radical**, a by-product of ethene oxidation, produces formaldehyde and **glycol aldehyde** [$HOCH_2CHO$(g)], both of which contribute to further ozone formation. The formaldehyde–ozone process was described in Section 4.2.6.

4.3.4.2. Alkene Reaction with Ozone

When ethene reacts with ozone, the ozone substitutes into ethene's double bond to form an unstable **ethene molozonide**. The molozonide decomposes to products that are also unstable. The reaction sequence of ethene with ozone is

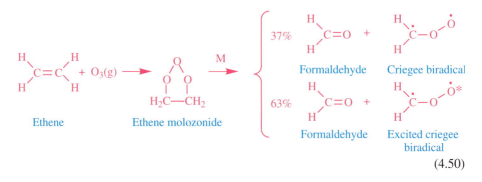

$$(4.50)$$

Formaldehyde produces ozone as described in Section 4.2.6. The **criegee biradical** forms NO_2(g) by

$$\text{H}_2\overset{\bullet}{\text{C}}\text{O}\overset{\bullet}{\text{O}}(g) + \overset{\bullet}{\text{N}}\text{O}(g) \longrightarrow \text{HCHO}(g) + \overset{\bullet}{\text{N}}\text{O}_2(g)$$

Criegee   Nitric   Formal-   Nitrogen
biradical   oxide   dehyde   dioxide

$$(4.51)$$

The **excited criegee biradical** isomerizes, and its product, excited formic acid, thermally decomposes by

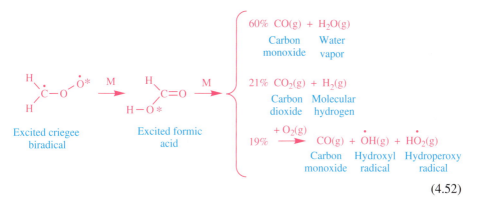

(4.52)

In sum, ozone attack on ethene produces HCHO(g), HO_2(g), CO(g), and NO_2(g). These gases not only reform the original ozone lost, but also produce new ozone.

4.3.5. Ozone Production from Aromatics

Toluene [$C_6H_5CH_3$(g)] originates from gasoline combustion, biomass burning, petroleum refining, detergent production, paint, and building materials. After methane, it is the second most abundantly emitted organic gas in Los Angeles air and the fourth most important gas in terms of abundance and chemical reactivity (Table 4.4). Mixing ratios of toluene in polluted air range from 1 to 30 ppbv (Table 3.3). Table 4.3 shows that toluene is decomposed almost exclusively by OH(g). OH(g) breaks down toluene by abstraction and addition. The respective pathways are

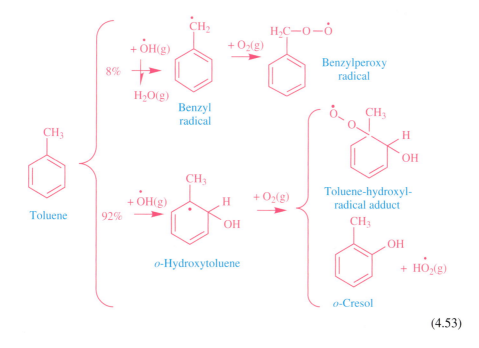

(4.53)

The benzylperoxy radical, formed from the abstraction pathway, converts NO(g) to NO$_2$(g). It also results in the formation of **benzaldehyde** [C$_6$H$_5$CHO(g)], which, like formaldehyde and acetaldehyde, decomposes to form ozone. The toluene-hydroxyl radical adduct, which is also a peroxy radical, converts NO(g) to NO$_2$(g). Cresol reacts with OH(g) to form the methylphenylperoxy radical [C$_6$H$_5$CH$_3$O$_2$(g)], which converts NO(g) to NO$_2$(g), resulting in O$_3$(g) formation.

The most important organic gas producing ozone in urban air is **xylene** [C$_6$H$_5$CH$_3$(g)] (Table 4.4). Xylene is present in gasoline, lacquers, and glues. Its mixing ratios in polluted air range from 1 to 30 ppbv (Table 3.3). As with toluene oxidation, xylene oxidation is primarily through reaction with OH(g). Oxidation of xylene by OH(g) produces peroxy radicals, which convert NO(g) to NO$_2$(g), resulting in ozone formation.

4.3.6. Ozone Production from Terpenes

The free troposphere and urban areas are affected by biogenic emissions of isoprene and other terpenes. **Biogenic emissions** are emissions produced from biological sources, such as plants, trees, algae, bacteria, and animals. Strictly speaking, **terpenes** are hydrocarbons that have the formula C$_{10}$H$_{16}$. Loosely speaking, they are a class of compounds that include hemiterpenes [C$_5$H$_8$(g)], such as **isoprene**; monoterpenes [C$_{10}$H$_{16}$(g)], such as **α-pinene**, **β-pinene**, and **d-limonene**; sesquiterpenes [C$_{15}$H$_{24}$(g)]; and diterpenes [C$_{20}$H$_{32}$(g)]. Isoprene is emitted by sycamore, oak, aspen spruce, willow, balsam, and poplar trees; α-pinene is emitted by pines, firs, cypress, spruce, and hemlock trees; β-pinene is emitted by loblolly pine, spruce, redwood, and California black sage trees; and d-limonene is emitted by loblolly pine, eucalyptus, and California black sage trees, and by lemon fruit.

Table 4.3 shows that OH(g), O$_3$(g), and NO$_3$(g) decompose isoprene. The reaction pathways of isoprene with OH(g) produce at least six peroxy radicals. The pathways are

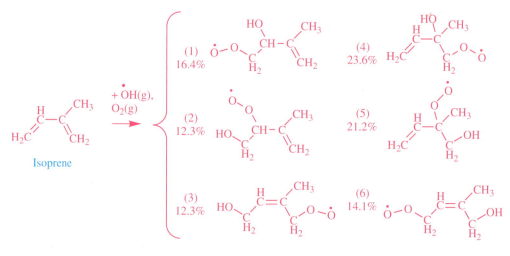

Isoprene peroxy radicals

(4.54)

(Paulson and Seinfeld, 1992). The *e*-folding lifetime of isoprene against reaction with OH(g) is about 30 minutes when [OH] = 5.0×10^6 molecules cm^{-3}. All six peroxy radicals convert NO(g) to $NO_2(g)$. The second and fifth radicals also create **methacrolein** and **methylvinylketone** by

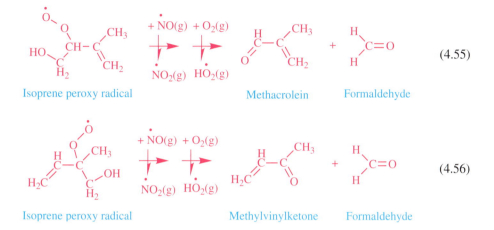

$$(4.55)$$

Isoprene peroxy radical   Methacrolein   Formaldehyde

$$(4.56)$$

Isoprene peroxy radical   Methylvinylketone   Formaldehyde

respectively. The $NO_2(g)$ from these reactions produces ozone. Methacrolein and methylvinylketone react with OH(g) and $O_3(g)$ to form additional products that convert NO(g) to $NO_2(g)$, resulting in more ozone.

The isoprene–ozone reaction is slower than is the isoprene–hydroxyl–radical reaction. Products of the isoprene–ozone reaction include methacrolein, methylvinylketone, the criegee biradical, and formaldehyde, all of which reproduce ozone.

In cities near forests, such as Atlanta, Georgia, terpenes can account for up to 40 percent of ozone above background levels. In less vegetated areas and in areas where anthropogenic emissions are large, such as in Los Angeles, they may account for only 3 to 8 percent of ozone above background levels.

4.3.7. Ozone Production from Alcohols

Alcohols, which can be distilled from corn, grapes, potatoes, sugarcane, molasses, and artichokes, among other farm products, have been used as an engine fuel since April 1, 1826, when Orford, New Hampshire, native **Samuel Morey** (1762–1843) patented the first internal combustion engine. His engine ran on **ethanol** [$C_2H_5OH(g)$] and turpentine. In September 1829, his engine was used to power a boat 5.8 m long up the Connecticut River at seven to eight miles per hour. Although alcohols were used in later prototype engines, they became relatively expensive in the United States due to a federal tax placed on alcohol following the Civil War of 1860 to 1865.

On August 27, 1859, **Edwin Laurentine Drake** (1819–1880) discovered oil after using a steam engine to power a drill through 21 m of rock in Titusville, Pennsylvania. This discovery is considered the beginning of the oil industry. Drake later died in poverty because of his poor business sense. Oil was soon refined to produce gasoline. The lower cost of gasoline in comparison with that of ethanol resulted in the comparatively greater use of gasoline than ethanol in early United States and European automobiles. The first practical gasoline-powered engine was constructed by **Étienne Lenoir** (1822–1900) of France in 1860 and ran on illuminating gas. In 1862, he built an automobile powered by this engine. In 1864, Austrian **Siegfried Marcus** (1831–1898) built the first of four four-wheeled vehicles he developed that were powered by internal combustion engines. In 1876, German **Nikolaus Otto** (1832–1891) developed the first four-stroke internal-combustion engine. In 1885, **Karl Benz** (1844–1929) of Germany designed and built the first practical automobile powered by an internal-combustion engine. The same year, **Gottlieb Daimler** (1834–1900) of Germany patented the first successful high-speed internal-combustion engine and developed a carburetor that allowed the use of gasoline as a fuel. In 1893, **J. Frank Duryea** (1869–1967) and **Charles E. Duryea** (1861–1938) produced the first successful gasoline-powered vehicle in the United States. In 1896, **Henry Ford** (1863–1947) completed his first successful automobile in Detroit, Michigan.

Whereas the United States had large oil reserves to draw on, France and Germany had few oil reserves and used ethanol as a fuel in automobiles to a greater extent. In 1906, 10 percent of engines in the Otto Gas Engine Works company in Germany ran on ethanol (Kovarik, 1998). The same year, the United States repealed the federal tax on ethanol, making it more competitive with gasoline. Soon after, however, oil fields in Texas were discovered, leading to a reduction in gasoline prices and the near-death of the alcohol-fuel industry.

Yet the alcohol-fuel industry continued to survive. Since the 1920s, every industrialized country except the United States has marketed blends of ethyl alcohol with gasoline in greater than nontrivial quantities. In the 1920s, I. G. Farben, a German firm, discovered a process to make synthetic **methanol** [$CH_3OH(g)$] from coal. Production of alcohol as a fuel in Germany increased to about 52 million gallons per year in 1937 as Hitler prepared for war (Egloff, 1940). Nevertheless, alcohol may never have represented more than 5 percent of the total fuel use in Europe in the 1930s (Egloff, 1940).

More recently, Brazil began a national effort in the 1970s to ensure that all gasoline sold contained ethanol (Section 8.2.3). In the United States, gasoline prices have always been much lower than alcohol-fuel prices, inhibiting the popularity of alcohol as an alternative to gasoline.

The chemical products of methanol oxidation are formaldehyde and ozone, and those of ethanol are acetaldehyde (a precursor to PAN) and ozone. Table 4.3 indicates that the only important loss process of alcohol is reaction with OH(g).

The reaction of methanol with OH(g) is

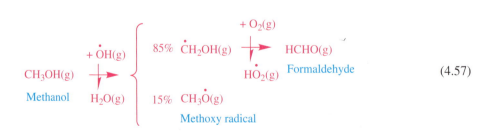

$$(4.57)$$

Methanol lost from these reactions has an *e*-folding lifetime of 71 days when [OH] = 5.0×10^6 molecules cm^{-3}; thus, the reaction is not rapid. The organic product of the first reaction is formaldehyde, and that of the second reaction is the methoxy radical, which produces formaldehyde by Reaction 4.21. Formaldehyde is an ozone precursor.

Ethanol oxidation by OH(g) produces three sets of possible products

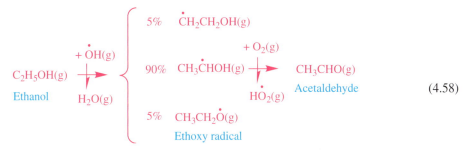

$$(4.58)$$

Ethanol lost from the most-probable (middle) reaction has an e-folding lifetime of about 19 hours when $[OH] = 5.0 \times 10^6$ molecules cm^{-3}. Acetaldehyde, formed from the middle reaction, produces PAN and ozone. Cities in Brazil have experienced high PAN mixing ratios since the introduction of their alcohol-fuel program.

4.4. POLLUTANT REMOVAL

Severe air pollution episodes generally last from a few days to more than a week, depending on the meteorology. During a pollution episode, air is usually confined so that pollution concentrations build up over successive days. The major loss processes of pollutants during an episode are chemical reaction and deposition to the ground. When meteorological conditions change, pollutants may diffuse upward, be swept away by winds to the background troposphere, or be rained out to the ground.

Organic gases emitted in polluted air are ultimately broken down to carbon dioxide and water. In the case of aromatics and other heavy organics, the initial breakdown steps are relatively fast. In the case of many simpler, lighter organic gases, the initial breakdown steps are often slower. Organic gases are also deposited to the ground and oceans, converted to aerosol particle constituents, and transported to the background troposphere. Oxides of nitrogen often evolve chemically to nitric acid, which converts to particulate matter or deposits to the soil or ocean water. Oxides of nitrogen also react with organic gases to form organic nitrate gases. Such gases decompose, convert to particulate matter, or deposit to the ground.

4.5. SUMMARY

Anthropogenic urban air pollution has been a problem for centuries. Before the twentieth century, most air pollution problems arose from the burning of wood, coal, and other raw materials without emission controls. Such burning resulted not only in smoky cities, but also in health problems. In the early-and mid-twentieth century, severe London-type smog events, during which emissions coupled with fog or a strong temperature inversion, were responsible for fatalities. Increased use of the automobile in the 1900s increased emissions of nitrogen oxides and reactive organic gases. In the presence of sunlight, these chemicals produce ozone, PAN, and a host of other products, giving rise to photochemical smog. Smog initiates when reactive organic gases photolyze or are oxidized by OH(g), HO_2(g), NO_3(g), O_3(g), or O(g) to produce organic radicals. The radicals convert $NO(g)$ to $NO_2(g)$, which photolyzes to $O(g)$, which reacts with $O_2(g)$ to form $O_3(g)$. The most important reactive organic gases in urban air are aromatics, alkenes, and aldehydes. Although alkanes are emitted in greater abundance than are the other organics, alkanes are less reactive and longer lived than are the others. Most organic gases are destroyed in urban air, but long-lived organics, particularly methane, ethane, and propane, are transported to the background troposphere, where they decay and produce ozone. Carbon monoxide is another gas emitted in abundance in urban air that escapes to the background troposphere because of its long lifetime. $CO(g)$ not only produces ozone in the background troposphere, but it is also a chemical source of $CO_2(g)$, a greenhouse gas. In the background troposphere, the concentrations of ozone, nitric oxide, and nitrogen dioxide are strongly coupled through the photostationary-state relationship. This relationship usually does not hold in urban air.

4.6. PROBLEMS

4.1.   Calculate the photostationary-state mixing ratio of ozone when $p_d = 1013$ mb, $T = 298$ K, $J \approx 0.01$ s^{-1}, $k_1 \approx 1.8 \times 10^{-14}$ cm^3 molecule^{-1} s^{-1}, $\chi_{NO}(g) = 180$ ppbv, and $\chi_{NO_2}(g) = 76$ ppbv. Perform the same calculation under the same conditions, except, $\chi_{NO}(g) = 9$ ppbv and $\chi_{NO_2}(g) = 37$ ppbv.

   (a) Ignoring the constant temperature and photolysis coefficient, which of the two cases do you think represents afternoon conditions? Why?
   (b) If the $NO(g)$ and $NO_2(g)$ mixing ratios were measured in urban air, do you think the morning or afternoon ozone mixing ratio calculated by the photostationary-state relationship would be closer to the actual mixing ratio of ozone? Why?

4.2.   Explain why the photostationary-state relationship is a useful relationship for background-tropospheric air but less useful for urban air.

4.3.   Why does ozone not form at night?

4.4.   Why does the hydroxyl radical not form at night?

4.5.   Why are nighttime ozone mixing ratios always nonzero in the background troposphere but sometimes zero in urban areas?

4.6.   If nighttime ozone mixing ratios in one location are zero and in another nearby location are nonzero, what do you think is the reason for the difference?

4.7.   If ozone mixing ratios are 0.16 ppmv and ROG mixing ratios are 1.5 ppmC, what is the best regulatory method of reducing ozone, if only the effects of $NO_x(g)$ and ROGs on ozone are considered?

4.8.   If $NO_x(g)$ mixing ratios are 0.2 ppmv and ROG mixing ratios are 0.3 ppmC, what is the best regulatory method of reducing ozone, if only the effects of NO_x and ROGs on ozone are considered?

4.9. If $NO_x(g)$ mixing ratios are 0.05 ppmv and ROG mixing ratios are 1 ppmC, what would be the resulting ozone mixing ratio if ROGs were increased by 1 ppmC and $NO_x(g)$ were increased by 0.1 ppmv?

4.10. In Fig. 4.10, why are ozone mixing ratios high and nitric oxide mixing ratios low in San Bernardino, whereas the reverse is true in central Los Angeles?

4.11. Why don't aromatic gases, emitted in urban air, reach the stratosphere to produce ozone?

4.12. What are the two fundamental differences between ozone production in the background troposphere and in urban air?

4.13. If the hydroxyl radical did not break down ROGs, would aromatics still be important smog producers? What about aldehydes? Explain.

4.14. Why is $CO(g)$, the most abundantly emitted gas in urban air, not an important smog producer?

4.15. Why should PAN mixing ratios peak at about the same time as ozone mixing ratios during the day?

4.16. In terms of chemical lifetimes and by-products, what are some of the costs and benefits of methanol and ethanol as alternative fuels?

4.17. Write out a chemical mechanism for the production of ozone from propane $[C_3H_8(g)]$ oxidation by $OH(g)$.

4.18. Write out a chemical mechanism showing how benzaldehyde $[C_6H_5CHO(g)]$ could form ozone.

AEROSOL PARTICLES IN SMOG AND THE GLOBAL ENVIRONMENT

A lthough most regulations of air pollution focus on gases, aerosol particles cause more visibility degradation and possibly more health problems than do gases. Particles smaller than 2.5 μm in diameter cause the most severe health problems. Particles enter the atmosphere by emissions and nucleation. In the air, their number concentrations and sizes change by coagulation, condensation, chemistry, water uptake, rainout, sedimentation, dry deposition, and transport. Particle concentration, size, and morphology affect the radiative energy balance in urban air and in the global atmosphere. In this chapter, compositions, concentrations, sources, transformation processes, sinks, and health effects of aerosol particles are discussed. The effects of aerosol particles on visibility are described in Chapter 7. Regulations relating to particles are given in Chapter 8.

5.1. SIZE DISTRIBUTIONS

Aerosol and hydrometeor particles are characterized by their size distribution and composition. A size distribution is the variation of concentration (i.e., number, surface area, volume, or mass of particles per unit volume of air) with size. Table 5.1 compares typical diameters, number concentrations, and mass concentrations of gases, aerosol particles, and hydrometeor particles under lower tropospheric conditions. The table indicates that the number and mass concentrations of gas molecules are much greater than are those of particles. The number concentration of aerosol particles decreases with increasing particle size. The number concentrations of hydrometeor particles are typically less than are those of aerosol particles, but the mass concentrations of hydrometeor particles are always greater than are those of aerosol particles.

Aerosol particle size distributions can be divided into modes, which are region of the size spectrum (in diameter space) in which distinct peaks in concentration occur. Usually, each mode can be described analytically with a lognormal function, which is a bell-curve distribution on a log–log scale, as shown in Fig. 5.1(a). Figure 5.1(b) shows the same distribution on a log-linear scale.

Aerosol particle distributions with 1, 2, 3, or 4 modes are called *unimodal*, *bimodal*, *trimodal*, or *quadrimodal*, respectively. Such modes may include a nucleation mode,

Table 5.1. Characteristics of Gases, Aerosol Particles, and Hydrometeor Particles

	Typical Diameter (μm)	Number Concentration (Molecules or Particles cm^{-3})	Mass Concentration ($\mu g \ m^{-3}$)
Gas molecules	0.0005	2.45×10^{19}	1.2×10^9
Aerosol particles			
Small	<0.2	10^3–10^6	<1
Medium	0.2–2.0	1–10^4	<250
Large	>2.0	<1–10	<250
Hydrometeor particles			
Fog drops	10–20	1–500	10^4–10^6
Cloud drops	10–200	1–1000	10^4–10^7
Drizzle	200–1,000	0.01–1	10^5–10^7
Raindrops	1,000–8,000	0.001–0.01	10^5–10^7

Data are for typical lower tropospheric conditions.

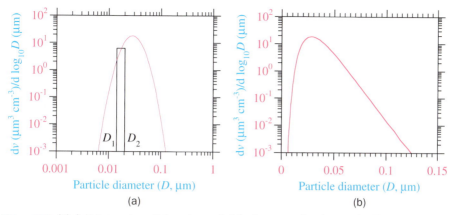

Figure 5.1. (a) A lognormal particle volume distribution on a log–log scale. The incremental volume concentration (dv, μm^3 cm^{-3}-air) of material between any two diameters (D_1 and D_2, μm) is estimated by multiplying the average value from the curve between the two diameters by $d \log_{10}D = \log_{10}D_2 - \log_{10}D_1$. Thus, for example, the volume concentration between diameters D_1 and D_2 is approximately 6 μm^3 cm^{-3} μm^{-1} × 0.15 μm = 0.9 μm^3 cm^{-3}. (b) The lognormal curve shown in (a), drawn on a log-linear scale.

two subaccumulation modes, and a coarse mode. The **nucleation mode** (mean diameters less than 0.1 μm) contains small emitted particles or newly nucleated particles (particles formed directly from the gas phase). Small nucleated or emitted particles increase in size by coagulation (collision and coalescence of particles) and growth (condensation of gases onto particles). Only a few gases, such as sulfuric acid, water, and some heavy organic gases, among others, condense onto particles. Molecular oxygen and nitrogen, which make up the bulk of the gas in the air, do not.

Growth and coagulation move nucleation mode particles into the **accumulation mode**, where diameters are 0.1 to 2 μm. Some of these particles are removed by rain, but they are too light to fall out of the air by **sedimentation** (dropping by their own weight against the force of drag). The accumulation mode sometimes consists of two submodes with mean diameters near 0.2 μm and 0.5 to 0.7 μm (Hering and Friedlander, 1982; John et al., 1989), possibly corresponding to newer and aged particles, respectively. The accumulation mode is important for two reasons. First, accumulation mode particles are likely to affect health by penetrating deep into the lungs. Second, accumulation mode particles are close in size to the peak wavelengths of visible light and, as a result, affect visibility (Chapter 7). Particles in the nucleation and accumulation modes together are **fine particles**.

The **coarse mode** consists of particles larger than 2 μm in diameter. These particles originate from windblown dust, sea spray, volcanos, plants, and other sources. Coarse mode particles are generally heavy enough to sediment out rapidly within hours to days. The emission sources and deposition sinks of fine particles differ from those of coarse mode particles. Fine particles usually do not grow by condensation to much larger than 1 μm, indicating that coarse mode particles originate primarily from emissions.

In general, the nucleation mode has the highest number concentration, the accumulation mode has the highest surface area concentration, and the coarse mode has the highest volume (or mass) concentration of aerosol particles. Figure 5.2 shows a quadrimodal distribution, fitted from data at Claremont, California, for the morning of

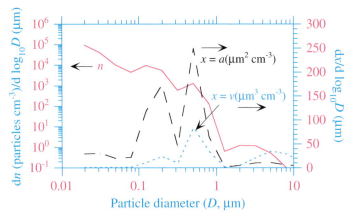

Figure 5.2. Number (n, particles cm^{-3}), area (a, μm^2 cm^{-3}), and volume (v, μm^3 cm^{-3}) concentration size distribution of particles at Claremont, California, on the morning of August 27, 1987. Sixteen model size bins and four lognormal modes were used to simulate the distribution (Jacobson 1997a).

August 27, 1987. All four modes (one nucleation, two subaccumulation, and one coarse particle mode) are most noticeable in the number concentration distribution. The nucleation mode is marginally noticeable in the area concentration distribution and invisible in the volume concentration distribution.

5.2. SOURCES AND COMPOSITIONS OF NEW PARTICLES

New aerosol particles originate from two sources: emissions and homogeneous nucleation. Emitted particles are called **primary particles**. Particles produced by homogenous nucleation, a gas-to-particle conversion process, are called **secondary particles**. Primary particles may originate from point, mobile, or area sources.

5.2.1. Emissions

Aerosol particle emission sources may be natural or anthropogenic. Natural emission processes include volcanic eruptions, soil-dust uplift, sea-spray uplift, natural biomass burning fires, and biological material release. Major anthropogenic sources include fugitive dust emissions (dust from road paving, passenger and agricultural vehicles, and building construction/demolition), fossil-fuel combustion, anthropogenic biomass burning, and industrial emissions.

In the United States in 1997, about 37 million short tons of particulates smaller than 10 μm were emitted (Table 3.9). Of these emissions, about 67 to 71 percent were anthropogenic in origin. On a global scale, more than half of all particle emissions are anthropogenic in origin.

Table 5.2 summarizes the natural and anthropogenic sources of the major components present in aerosol particles. These sources are discussed in more detail shortly.

5.2.1.1. Sea-Spray Emissions

The most abundant natural aerosol particles in the atmosphere in terms of mass are sea-spray drops. **Sea spray** forms when winds and waves force air bubbles to burst at the sea surface (Woodcock, 1953). Sea spray is emitted primarily in the

Table 5.2. Nontrivial Sources of Major Components of Aerosol-Particles

	Sea-Spray Emissions	Soil-Dust Emissions	Volcanic Emissions	Biomass Burning	Fossil-Fuel Combustion for Transportation and Energy	Fossil-Fuel and Metal Combustion for Industrial Processes
Black carbon (C)			X	X	X	X
Organic matter (C,H,O,N)	X	X	X	X	X	X
Ammonium (NH_4^+)				X	X	X
Sodium (Na^+)	X	X	X	X		X
Calcium (Ca^{2+})	X	X	X	X		X
Magnesium (Mg^{2+})	X	X	X	X		X
Potassium (K^+)	X	X	X	X		X
Sulfate (SO_4^{2-})	X	X	X	X	X	X
Nitrate (NO_3^-)				X		X
Chloride (Cl^-)	X	X	X	X		X
Silicon (Si)		X	X		X	X
Aluminum (Al)		X	X		X	X
Iron (Fe)		X	X	X	X	X

coarse mode of the particle size distribution. Winds also tear off wave crests to form larger drops, called **spume drops**, but these drops fall back to the ocean quickly. Sea spray initially contains all the components of sea water. About 96.8 percent of sea-spray weight is water and 3.2 percent is sea salt, most of which is sodium chloride. Table 5.3 shows the relative composition of the major constituents in sea water. The chlorine-to-sodium mass ratio in sea water, which can be obtained from the table, is about 1.8:1.

When sea water is emitted as sea-spray or spume drops, the chlorine-to-sodium mass ratio, originally 1.8:1, sometimes decreases because the chlorine is removed by sea-spray acidification (e.g., Ericksson, 1960; Duce, 1969; Hitchcock et al., 1980). **Sea-spray acidification** occurs when sulfuric or nitric acid enters a sea-spray drop and forces chloride to evaporate as hydrochloric acid [HCl(g)]. Some sea-spray drops lose all of their chloride in the presence of nitric or sulfuric acid.

The size of a sea-spray drop is also affected by dehydration. **Dehydration** (loss of water) occurs when water from a drop evaporates due to a decrease in the relative humidity between the air just above the ocean surface and that a few meters higher. Dehydration increases the concentration of solute in a drop.

Table 5.3. Mass Percentage of Major Constituents in Sea Water[a]

Constituent	Mass Percentage in Sea Water	Constituent	Mass Percentage in Sea Water
Water	96.78	Sulfur	0.0876
Sodium	1.05	Calcium	0.0398
Chlorine	1.88	Potassium	0.0386
Magnesium	0.125	Carbon	0.0027

[a] The concentration of these constituents, including water, in sea water is 1,033,234 mg/L.

Source: Lide (1998).

5.2.1.2. Soil-Dust and Fugitive-Dust Emissions

Soil is the natural, unconsolidated mineral and organic matter lying above bedrock on the surface of the Earth. A **mineral** is a natural, homogeneous, inorganic, solid substance with a characteristic chemical composition, crystalline structure, color, and hardness. Minerals in soil originate from the breakdown of rocks. Organic matter originates from the decay of dead plants and animals. **Rocks** are consolidated or unconsolidated aggregates of minerals or biological debris.

Three generic rock families exist: sedimentary, igneous, and metamorphic. **Sedimentary rocks** cover about 75 percent of the Earth's surface and form primarily on land and the floors of lakes and seas by the slow, layer by layer deposition and cementation of carbonates, sulfates, chlorides, shell fragments, and fragments of existing rocks. An example of a sedimentary rock is **chalk**, made of skeletons of microorganisms. **Igneous rocks** ("rocks from fire" in Greek) form by the cooling of magma, which is a molten, silica-rich mixture of liquid and crystals. Igneous rocks can form either on the Earth's surface following a volcanic eruption or beneath the surface when magma is present. **Granite** is an igneous rock. **Metamorphic rocks** (rocks that "change in form") result from the structural transformation of existing rocks due to high temperatures and pressures in the Earth's interior. During metamorphosis, no rock melting or change in chemical composition occurs; elements merely rearrange themselves to form new mineral structures that are stable in the new environment. **Marble** is a metamorphic rock.

Breakdown of Rocks to Soil Material

Breakdown of rocks to soil material occurs primarily by physical weathering. **Physical weathering** is the disintegration of rocks and minerals by processes that do not involve chemical reactions. Disintegration may occur when a stress is applied to a rock, causing it to break into blocks or sheets of different sizes and ultimately into fine soil minerals. Stresses arise when rocks are subjected to high pressure by soil or other rocks lying above. Stresses also arise when rocks freeze then thaw or when saline solutions enter cracks and cause rocks to disintegrate or fracture. Salts have higher thermal expansion coefficients than do rocks, so when temperatures increase, salts within rock fractures expand, forcing the rock to open and break apart. One source of salt for rock disintegration is sea spray transported from the oceans. Another source is desert salt (from deposits) transported by winds. Physical weathering of rocks on the Earth's surface can occur by their constant exposure to winds or running water.

Another process that causes some rocks to break down and others to reform is **chemical weathering**. Equation 3.20, which showed calcium carbonate breaking down on reaction with dissolved carbonic acid, was a chemical-weathering reaction. Another chemical weathering reaction is

$$CaSO_4 - 2H_2O(s) \rightleftharpoons Ca^{2+} + SO_4^{2-} + 2H_2O(aq) \tag{5.1}$$

Calcium sulfate   Calcium   Sulfate   Liquid
dihydrate (gypsum)   ion   ion   water

which occurs when gypsum dissolves in water. Alternatively, when high sulfate and calcium ion concentrations are present in surface or groundwater, Reaction 5.1 can proceed to the left, producing gypsum. Gypsum can form when sulfate-containing particles from the air dissolve in stream water containing calcium ions. Crystalline gypsum can also form when sea-spray drops, which contain calcium, collide with volcanic aerosol particles, which contain sulfate.

Types of Minerals in Soil Dust

Table 2.2 showed the most abundant elements in the Earth's continental crust. Soil-dust particles, emitted from the top layer of the crust, contain minerals made of these elements and organic matter. Minerals in soil include quartz, feldspars, hematite, calcite, dolomite, gypsum, epsomite, kaolinite, illite, smectite, vermiculite, and chlorite.

Pure **quartz** [$SiO_2(s)$] is a clear, colorless mineral that is resistant to chemical weathering. The Greeks thought it was a frozen water, and its name originates from the Saxon word *querkluftertz,* which means "cross-veined ore."

Feldspars, which make up at least 50 percent of the rocks on the Earth's surface, are by far the most abundant minerals on Earth. The name *feldspar* originates from the Swedish words *feld* ("field") and *spar* (the name of a mineral commonly found overlying granite). Two common types of feldspars are **potassium feldspar** [$KAlSi_3O_8(s)$] and **plagioclase feldspar** [$NaAlSi_3O_3$-$CaAl_2Si_2O_8(s)$].

Hematite [$Fe_2O_3(s)$] (Greek for "bloodlike stone") is an oxide mineral because it includes a metallic element bonded with oxygen. The iron within it causes it to appear reddish-brown.

Calcite [$CaCO_3(s)$] and **dolomite** [$CaMg(CO_3)_2(s)$] are carbonate minerals. The name *calcite* is derived from the word **calcspar**, which was derived from the Greek word for limestone, *khálix.* Dolomite is similar in form to calcite. It was named after French geologist and mineralogist Silvain de Dolomieu (1750–1801).

Gypsum [$CaSO_4$-$2H_2O(s)$] and **epsomite** [$MgSO_4$-$7H_2O(s)$] are two of only a handful of sulfate-containing minerals. Gypsum, which is colorless to white, was named after the Greek word *gypsos* ("plaster"). Epsomite was named after the location it was first found, Epsom, England. As seen in Table 2.2, sulfur is not an abundant component of the Earth's crust.

Kaolinite, illite, smectite, vermiculite, and chlorite are all **clays**, which are odorous minerals resulting from the weathering of rocks. Clays are usually soft, compact, and composed of aggregates of small crystals. The major components of clays are oxygen, silicon, aluminum, iron, and magnesium. **Kaolinite** [$Al_4Si_4O_{10}(OH)_8(s)$ in pure form] was named after the Chinese word *kauling* ("high ridge"), which is the name of a hill near Jauchu Fa, where clay for the manufacture of porcelain was dug. Early porcelain clay was called kaolinite. **Illite** is a mineral group name, originating from the state name for Illinois, where many illite minerals were first studied. **Smectite** is a mineral group name, originating from *smectis*, a Greek name for "fuller's earth." **Vermiculite** is a mineral group name that originates from the Latin word *vermiculari* ("to breed worms"), which refers to the mineral's ability to exude wormlike structures when rapidly heated. **Chlorite** is a mineral group name, originating from the Greek word for green, which is the common color of this group of clays.

In addition to containing minerals, soils contain organic matter, such as plant litter or animal tissue broken down by bacteria.

Soil dust, which consists of the minerals and organic material making up soil, is lifted into the air by winds. The extent of lifting depends on the wind speed and particle mass. Most mass of soil dust lifted into the air is in particles larger than 1 μm in diameter; thus, soil-dust particles are predominantly coarse-mode particles. Those larger than 10 μm fall out quite rapidly, but those between 1 and 10 μm can stay in the air for days to weeks or longer, depending on the height to which they are originally lifted. Table 5.4 shows the time required for particles of different diameters to fall 1 km in the air by sedimentation. The table indicates that particles 1 μm in diameter fall 1 km in 328 days, whereas those 10 μm in diameter fall 1 km in

3.6 days. Although most soil-dust particles are not initially lofted more than 1 km in the air, many are, and these particles can travel long distances before falling to the ground.

Source regions of soil dust on a global scale include deserts (e.g., the Sahara in North Africa, Gobi in Mongolia, Mojave in southeastern California) and regions where foliage has been cleared by biomass burning and plowing. Figure 5.3 shows a satellite image of soil-dust plumes originating from the Sahara Desert. Soil dust also enters the air when off-road and agricultural vehicles drive over loose soil or when on-road vehicles resuspend soil dust. Dust is also resuspended during construction or demolition. Fifty percent of all soil dust emitted may be due to anthropogenic activities (Tegen et al., 1996).

Table 5.4. Time for Particles to Fall 1 km in the Atmosphere by Sedimentation Under Near-Surface Conditions

Particle Diameter (μm)	Time to Fall 1 km
0.02	228 years
0.1	36 years
1.0	328 days
10.0	3.6 days
100.0	1.1 hours
1,000.0	4 minutes
5,000.0	1.8 minutes

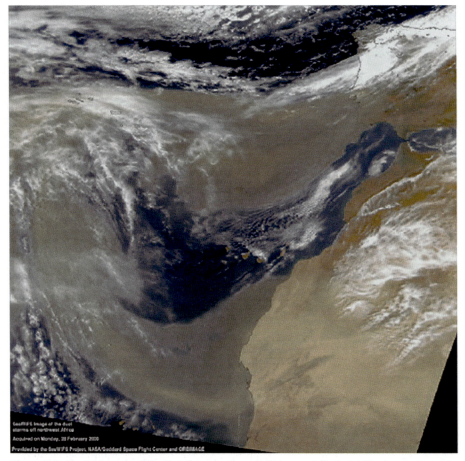

Figure 5.3. Dust storms originating from northwest Africa and Portugal/Spain, captured by SeaWiFS satellite, February 28, 2000. Courtesy of SeaWiFS Project, NASA/Goddard Space Flight Center and ORBIMAGE.

5.2.1.3. Volcanic Eruptions

The word volcano originates from the ancient Roman god of fire *Vulcan*, after whom the Romans named an active volcano, *Vulcano*. Today, more than 500 volcanos are active on the Earth's surface. Figure 5.4 shows emissions from the Mount St. Helens eruption in 1982. Volcanos result from the sudden release of gases dissolved in magma, which contains 1 to 4 percent gas by mass. Water vapor is the most abundant gas in magma, making up 50 to 80 percent of its mass. Volcanic magma also contains carbon dioxide [$CO_2(g)$], sulfur dioxide [$SO_2(g)$], carbonyl sulfide [$OCS(g)$], and molecular nitrogen [$N_2(g)$]. Lesser gases include carbon monoxide [$CO(g)$], molecular hydrogen [$H_2(g)$], molecular sulfur [$S_2(g)$], hydrochloric acid [$HCl(g)$], molecular chlorine [$Cl_2(g)$], and molecular fluorine [$F_2(g)$].

Volcanos emit particles that contain the elements of the Earth's mantle (e.g., Table 2.2). The most abundant particle components in volcanic eruptions are silicate minerals (minerals containing Si). Emitted volcanic particles range in diameter from smaller than 0.1 to larger than 100 μm. As seen in Table 5.4, particles 100 μm in diameter take 1.1 hour to fall 1 km by sedimentation. The only volcanic particles that survive more than a few months before falling to the ground are those smaller than 4 μm. Such particles require no less than 23 days to fall 1 km. Larger volcanic particles fall to the ground quickly. Volcanic particles of all sizes are also removed by rain. Volcanic particles contain the components listed in Table 5.2.

Figure 5.4. Dome-shattering eruption from Mount St. Helens in the fall of 1982. Photo by Peter Frenzen, available from Mount St. Helens National Monument photo gallery.

Volcanic gases, such as carbonyl sulfide [$OCS(g)$] and sulfur dioxide [$SO_2(g)$], are sources of new particles. When $OCS(g)$ is injected volcanically into the stratosphere, some of it photolyzes, and its products react to form $SO_2(g)$, which oxidizes to gas-phase sulfuric acid [$H_2SO_4(g)$], which nucleates to form new sulfuric acid–water aerosol particles. A layer of such particles, called the Junge layer, has formed in the stratosphere by this mechanism (Junge, 1961). The average diameter of these particles is 0.14 μm. Sulfuric acid is the dominant particle constituent, aside from liquid water, in the stratosphere and upper troposphere. More than 97 percent of particles in the lower stratosphere and 91 to 94 percent of particles in the upper troposphere contain oxygen and sulfur in detectable quantities (Sheridan et al., 1994).

5.2.1.4. Biomass Burning

Biomass burning is the burning of evergreen forests, deciduous forests, woodlands, grasslands, and agricultural lands, either to clear land for other use, stimulate grass growth, manage forest growth, or satisfy a ritual. Biomass fires are responsible for a large portion of particle emissions. Such fires may be natural or anthropogenic in origin. Figure 5.5 shows smoke from a fire set intentionally to burn dry vegetation in South Africa.

Figure 5.5. Aircraft photograph of smoldering fire set to burn dry vegetation near Kruger National Park, South Africa, September 7, 2000. Photo by NASA/GSFC/JPL, MISR, and Air MISR teams.

Biomass burning produces gases, such as $CO_2(g)$, $CO(g)$, $CH_4(g)$, $NO_x(g)$, and ROGs, and particles, such as ash, plant fibers, and soil dust (Fig. 5.6) as well as organic matter and soot. **Ash** is the primarily inorganic solid or liquid residue left after biomass burning, but may also contain organic compounds oxidized to different degrees. **Organic matter** (OM) consists of carbon- and hydrogen-based compounds and often contains oxygen (O), nitrogen (N), etc., as well. It is usually grey (e.g., Fig. 5.5), yellow, or brown. **Soot** contains **black carbon** (BC) (carbon atoms bonded together) coated by OM in the form of aliphatic hydrocarbons, polycyclic aromatic hydrocarbons (PAHs), and small amounts of O and N (Chang *et al.*, 1982; Reid and Hobbs, 1998; Fang *et al.*, 1999). The ratio of BC to OM produced in smoke depends on temperature. High-temperature flames produce more BC than OM; low-temperature flames (such as in smoldering biomass) produce less. Because BC is black, the more BC produced, the blacker the smoke (Fig. 5.7).

Vegetation contains low concentrations of metals, including titanium (Ti), manganese (Mn), zinc (Zn), lead (Pb), cadmium (Cd), copper (Cu), cobalt (Co),

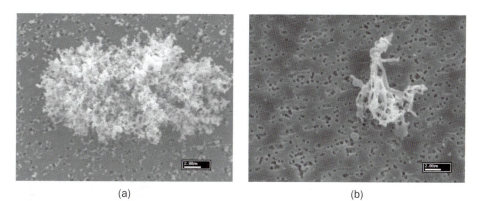

(a)   (b)

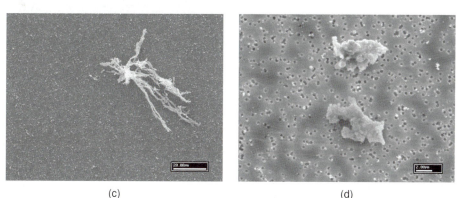

(c)   (d)

Figure 5.6. Scanning electron microscopy (SEM) images of (a) an ash aggregate, (b) a combusted plant fiber, (c) an elongated ash particle, and (d) soil-dust particles collected from forest-fire emissions in Brazil by Reid and Hobbs (1998).

antimony (Sb), arsenic (As), nickel (Ni), and chromium (Cr). These substances vaporize during burning, then quickly recondense onto soot or ash particles. Young smoke also contains K^+, Ca^{2+}, Mg^{2+}, Na^+, NH_4^+, Cl^-, NO_3^-, and SO_4^{2-} (Reid et al., 1998; Ferek et al., 1998; Andreae et al., 1998). Young smoke particles are found in the upper-nucleation mode and lower-accumulation mode.

5.2.1.5. Fossil-Fuel Combustion

Fossil-fuel combustion is an anthropogenic emission source. Fossil fuels that produce aerosol particles include coal, oil, natural gas, gasoline, kerosene, and diesel.

Coal is a combustible brown-to-black carbonaceous sedimentary rock formed by compaction of partially decomposed plant material. Metamorphisis of plant material to black coal goes through several stages, from unconsolidated brown-black peat to consolidated, brown-black peat coal, to brown-black lignite coal, to dark-brown-to-black bituminous (soft) coal, to black anthracite (hard) coal. Countries with the largest bituminous and anthracite coal reserves are the United States (29.8 percent of worldwide reserves in 2000), Russia (20.5 percent), China (13.4 percent), India (10.2 percent), South Africa (7.7 percent), and Australia (6.9 percent). The United States doubled coal production between 1923 and 1998. Even at the present rate of recovery, U.S. reserves will last at least 250 years. Today, more than 25 percent of the world's electricity is generated from coal (NMA, 2000).

Figure 5.7. Soot emissions from a prescribed burn at Horse Creek Mesa, Big Horn National Forest, Wyoming, October 9, 1981. Photo by U.S. Forest Service staff, available from the National Renewable Energy Laboratory.

Oil (or petroleum) is a natural greasy, viscous, combustible liquid (at room temperature) that is insoluble in water. It forms from the geological-scale decomposition of plants and animals and is made of hydrocarbon complexes. Oil is found in the Earth's continental and ocean crusts. Natural gas is a colorless, flammable gas, made primarily of methane but also of other hydrocarbon gases, that is often found near petroleum deposits. As such, natural gas is often mined along with petroleum. Gasoline is a volatile mixture of liquid hydrocarbons derived by refining petroleum. Kerosene is a combustible, oily, water-white liquid with a strong odor that is distilled from petroleum and used as a fuel and solvent. Diesel fuel is a combustible liquid distilled from petroleum after kerosene. Diesel fuel is named after Rudolf Diesel (1858–1913), the German automotive engineer who designed and built the first engine to run on diesel fuel.

Particulate components emitted during the combustion of fossil fuels include soot (BC and OM), OM alone, sulfate (SO_4^{2-}), metals, and fly ash. Coal combustion results in the emission of submicron soot, OM, and sulfate and sub/supermicron fly ash. The fly ash consists of oxygen, silicon, aluminum, iron, calcium, and magnesium in the form of quartz, hematite, gypsum, and clays. Combustion in gasoline engines usually results in the emission of submicron OM, sulfate, and elemental silicon, iron, zinc, and sulfur. Combustion in diesel engines usually results in the emission of these components plus soot and ammonium. Diesel-powered vehicles emit 10 to 100 times more particulate mass than do gasoline-powered vehicles; thus, it is harder for diesel-powered vehicles than for gasoline-powered vehicles to meet particulate emission standards. In the United States, 99 percent of heavy-duty trucks and buses but only 0.1 percent of light-duty vehicles run on diesel. In Europe, 99 percent of heavy-duty trucks and buses and 25 percent of light-duty vehicles run on diesel (Cohen and Nikula, 1999). In the United Kingdom, 80–95 percent of soot originates from diesel engines (Pooley and Mille, 1999).

Table 5.5. Metals Present in Fly Ash of Different Industrial Origin

Source	Metals Present in Fly Ash
Smelters	Fe, Cd, Zn
Oil-fired power plants	V, Ni, Fe
Coal-fired power plants	Fe, Zn, Pb, V, Mn, Cr, Cu, Ni, As, Co, Cd, Sb, Hg
Municipal waste incineration	Zn, Fe, Hg, Pb, Sn, As, Cd, Co, Cu, Mn, Ni, Sb
Open-hearth furnaces at steel mills	Fe, Zn, Cr, Cu, Mn, Ni, Pb

Sources: Pooley and Mille, 1999; Henry and Knapp, 1980; Schroeder et al., 1987; Ghio and Samet, 1999.

Most soot from fossil-fuels originates from coal, diesel-fuel, and jet-fuel combustion. Soot emitted during fossil-fuel combustion contains BC covered with a layer of polycyclic aromatic hydrocarbons (PAHs) and coated by a shell of organic and inorganic compounds (Steiner et al., 1992). In the case of diesel exhaust, the PAHs include nitro-PAHs (PAHs with nitrogen-containing functional groups), which may be harmful. Most fossil-fuel soot from vehicles is emitted in particles less than 0.2 μm in diameter (Venkataraman et al., 1994; Maricq et al., 1999; ACEA, 1999). Thus, fossil-fuel combustion usually results in the emission of upper nucleation mode and lower accumulation mode particles.

5.2.1.6. Industrial Sources

Many industrial processes involve burning of fossil fuels together with metals. As such, industrial processes emit soot, sulfate, fly ash, and metals. Fly ash from industrial processes usually contains $Fe_2O_3(s)$, $Fe_3O_4(s)$, $Al_2O_3(s)$, $SiO_2(s)$, and various carbonaceous compounds that have been oxidized to different degrees (Greenberg et al., 1978). Fly ash is emitted primarily in the coarse mode (>2.0 μm diameter). Metals are emitted during high-temperature industrial processes, such as waste incineration, smelting, cement kilning, and power-plant combustion. In such cases, heavy metals vaporize at high temperatures, then recondense onto soot and fly-ash particles that are emitted simultaneously. Table 5.5 lists some metals present in fly ash of different industrial origin.

Of all the metals emitted into the air industrially, iron is by far the most abundant. Lead, a criteria air pollutant, is emitted industrially from lead-ore smelting, lead-acid battery manufacturing, lead-ore crushing, and solid waste disposal. Because lead is no longer used as an additive in gasoline in the United States, its ambient concentrations have declined since the early 1980s. Lead is still used in gasoline in several countries.

5.2.1.7. Miscellaneous Sources

Additonal particle types in the air include tire-rubber particles, pollens, spores, plant debris, viruses, and meteoric debris. **Tire-rubber particles** are emitted due to the constant erosion of a tire at the tire-road interface. Such particles are generally larger than 2 μm in diameter. **Pollens**, **spores**, **plant debris**, and **viruses** are biological particles lifted by the wind. They often serve as sites on which cloud drops and ice crystals form. A stratosphere source of new particles is **meteoric debris**. Most meteorites disintegrate before they drop to an altitude of 80 km. Those that reach the stratosphere contain iron (Fe), titanium (Ti), and aluminum (Al), among other elements. The net contribution of meteorites to particles in the stratosphere is small (Sheridan et al., 1994).

5.2.2. Homogeneous Nucleation

Aside from emissions, homogeneous nucleation is the only source of new particles in the air. *Nucleation* is a process by which gas molecules aggregate to form clusters. If the radius of the cluster reaches a critical size, the cluster becomes stable and can grow further.

Nucleation is either homogeneous or heterogeneous. **Homogeneous nucleation** occurs when gases nucleate without the aid of an existing surface. Thus, homogeneous nucleation is a source of new particles. **Heterogeneous nucleation** occurs when gases nucleate on a preexisting surface. Thus, it does not result in new particles. Homogeneous or heterogeneous nucleation must occur before a particle can grow by condensation, a process discussed shortly.

Homogeneous and heterogeneous nucleation are either homomolecular, binary, or ternary. **Homomolecular nucleation** occurs when molecules of only one gas nucleate; **binary nucleation** occurs when molecules of two gases, such as sulfuric acid and water, nucleate; and **ternary nucleation** occurs when molecules of three gases, such as sulfuric acid, water, and ammonia, nucleate.

The most important homogeneous nucleation process in the air is binary nucleation of sulfuric acid with water. Homogeneously nucleated sulfuric acid–water particles are typically 3 to 20 nm in diameter. In the remote atmosphere (e.g., over the ocean), homogenous nucleation events can produce more than 10^4 particles cm^{-3} in this size range over a short period. Homogenous nucleation of water vapor does not occur under typical atmospheric conditions. Water vapor nucleation is always heterogeneous. Indeed, all cloud drops in the atmosphere consist of water that condensed onto aerosol particles following the heterogeneous nucleation of these particles. Aerosol particles that become cloud drops following heterogeneous nucleation by and condensation of water vapor are called **cloud condensation nuclei** (CCN).

5.3. PROCESSES AFFECTING PARTICLE SIZE

Once in the air, particles increase in size by coagulation and growth. Growth can occur by condensation, vapor deposition, dissolution, or chemical reaction. These processes are discussed in the following subsections.

5.3.1. Coagulation

Coagulation occurs when two particles collide and stick (coalesce) together (Fig. 5.8), reducing the number concentration but conserving the volume concentration of particles in the air. Coagulation can occur between two small particles, between a small and a large particle, or between two large particles. Five important mechanisms that drive particles to collide are Brownian motion, enhancement to Brownian motion due to convection, gravitational collection, turbulent inertial motion, and turbulent shear.

Brownian motion is the random movement of particles suspended in a fluid. Coagulation due to Brownian motion is the process by which particles diffuse, collide, and coalesce due to random motion. When two particles collide due to Brownian motion, they may or may not stick together, depending on the efficiency of coalescence, which, in turn, depends on particle shape, composition, and surface characteristics. Because the kinetic energy of a small particle is small relative to that of a large particle at a given temperature, the likelihood that bounce-off occurs when small particles collide is low, and the coalescence efficiency of small particle coagulation is often assumed to be unity (Pruppacher and Klett, 1997).

When particles fall through the air, eddies created in their wake enhance diffusion of other particles to their surfaces. The coagulation mechanism due to this process is called **convective Brownian diffusion enhancement**.

A third mechanism causing coagulation is **gravitational collection**. When two particles of different size fall, the larger one may catch up and collide with the smaller one. The kinetic energy of the larger particle is higher, increasing the chance that collision will result in a bounce-off rather than a coalescence; thus, the coalescence efficiency of this process is not unity. Gravitational collection is an important mechanism for producing raindrops.

Two additional mechanisms that drive particles to collide are turbulent inertial motion and turbulent shear (Saffman and Turner, 1956). Coagulation due to **turbulent inertial motion** occurs when turbulence enhances the rate by which particles of

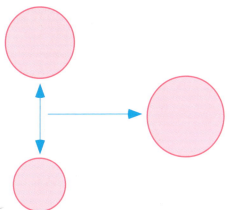

Figure 5.8. Schematic showing coagulation. When two particles collide, they may coalesce to form one large particle, thereby reducing the number concentration but conserving the volume concentration of particles.

different size falling through the air coagulate. Coagulation due to **turbulent shear** occurs when wind shear allows particles at different heights to move at different velocities, causing faster particles to catch up and coagulate with slower particles.

Brownian motion dominates all five coagulation processes when at least one of the two colliding particles is small. When both particles are large (but not exactly the same size), gravitational collection (settling) is the dominant coagulation process.

Outside of clouds, small aerosol particles are affected more by coagulation than are large aerosol particles because the air contains many more small than large aerosol particles. In urban regions, coagulation affects the number concentration of aerosol particles primarily smaller than 0.2 μm in size over the course of a day (Jacobson, 1997a), as can be seen in Fig. 5.9. This figure shows a model calculation of the change

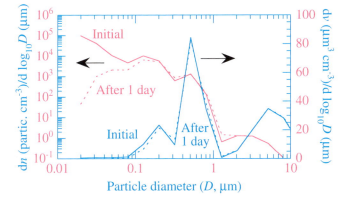

Figure 5.9. Estimated change in aerosol number and volume concentrations at Claremont over a 24-hour period when coagulation alone was considered. Number concentration is shown in red; volume concentration, in blue.

in the number and volume concentration of particles in polluted urban air over a 24-hour period. Whereas the number concentration of small particles was affected, changes in the volume concentration of such particles were affected less.

Over the ocean, coagulation is an important mechanism by which sea-spray drops become internally mixed with other aerosol constituents, such as soil-dust particles. Andreae et al. (1986), for example, found that 80 to 90 percent of silicate particles over the equatorial Pacific Ocean between Ecuador and Hawaii contained sea-spray constituents. Murphy et al. (1998) found that almost all aerosol particles larger than 0.13 μm in the boundary layer in a remote South Pacific Ocean site contained sea-spray components. Pósfai et al. (1999) found that almost all soot particles in the North Atlantic contained sulfate. The internal mixing of aerosols by coagulation is supported by model simulations that show that on a global scale, about half the increase in size of soot particles following their emissions may be due to coagulation with nonsoot particles, such as sulfate, organic matter, sea spray, and soil, whereas the rest may be due to growth processes, such as those discussed next (Jacobson, 2001a).

5.3.2. Growth Processes

Coagulation is a process that involves two particles, whereas condensation, vapor deposition, dissolution, and surface reaction are gas-to-particle conversion processes. These processes are discussed next.

5.3.2.1. Condensation/Evaporation

Condensation and evaporation occur only after homogeneous or heterogeneous nucleation. On a nucleated liquid surface, gas molecules continuously **condense** (change state from gas to liquid) and liquid molecules continuously evaporate (change state from liquid to gas). In equilibrium, transfer rates in both directions are equal, and the resulting partial pressure of the gas immediately over the particle's surface is the gas's **saturation vapor pressure** (SVP). If the partial pressure of the gas away from the surface increases above the SVP over the surface, excess molecules diffuse to the

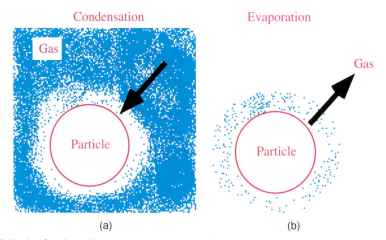

Figure 5.10. (a) Condensation occurs when the partial pressure of a gas away from a particle surface (represented by the thick cloud of gas away from the surface) exceeds the SVP of the gas over the surface (represented by the thin cloud of gas near the surface). (b) Evaporation occurs when the SVP exceeds the partial pressure of the gas. The schematics are not to scale.

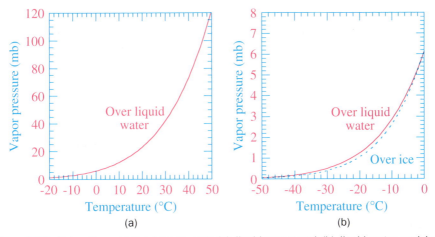

Figure 5.11. Saturation vapor pressure over (a) liquid water and (b) liquid water and ice, versus temperature.

surface [Fig. 5.10(a)] and condense. If the gas's partial pressure away from the surface decreases below the SVP, gas molecules over the surface diffuse away from the surface [Fig. 5.10(b)], and liquid molecules on the surface evaporate to maintain saturation over the surface. In sum, if the ambient partial pressure of a gas exceeds the gas's SVP, condensation occurs. If the ambient partial pressure of the gas falls below the gas's SVP, evaporation occurs.

The most abundant condensing gas is water vapor. Figure. 5.11(a) shows the SVP of water over a liquid surface versus temperature. Figure. 5.11(b) shows the same over a liquid and an ice surface at temperatures below 0 °C. Because water vapor's partial pressure cannot exceed its SVP without the excess vapor condensing, the SVP is the maximum possible partial pressure of water vapor in the air at a given temperature. Near the poles, where temperatures are below 0 °C, the SVP can be as low as 0.0003 percent of sea-level air pressure. Near the equator, where temperatures are close to 30 °C, the SVP can increase to 4 percent or more of sea-level air pressure.

EXAMPLE 5.1

Determine the maximum partial pressure and percentage water vapor in the atmosphere at 0 and 30°C.

Solution

From Fig. 5.11(a), the SVP and, therefore, the maximum partial pressure of water vapor at 0 and 30°C are 6.1 and 42.5 mb, respectively. Because sea-level dry-air pressure is 1013 mb, water vapor comprises no more than 0.6 and 4.2 percent of total air by volume, respectively, at these two temperatures.

The **relative humidity** (RH) is the partial pressure of water vapor divided by the SVP of water over a liquid surface, all multiplied by 100 percent. When the relative humidity exceeds 100 percent, the partial pressure of water vapor exceeds the SVP of water over a liquid surface, and the excess water condenses onto heterogeneously nucleated cloud condensation nuclei to form **cloud drops**. When the RH drops below 100 percent, liquid water on a surface evaporates, leaving the residual aerosol particle.

EXAMPLE 5.2.

If the partial pressure of water vapor is 20 mb and the temperature is 30 °C, what is the relative humidity?

Solution

From Fig. 5.11(a), the SVP is 42.5 mb. Therefore, the relative humidity is 100 percent $\times$ 20 mb/42.5 mb = 47 percent.

Sulfuric acid gas, which has a low SVP, also condenses onto particles. Once condensed, sulfuric acid rarely evaporates because its SVP is so low. When sulfuric acid condenses, it condenses primarily onto accumulation mode particles because the accumulation mode has a larger surface area concentration (surface area per volume of air) than do other modes. Some other condensable gases (with low SVP) include high-molecular-weight organic gases, such as by-products of toluene, xylene, alkylbenzene, alkane, alkene, and biogenic hydrocarbon oxidation (Pandis et al., 1992).

5.3.2.2. Water Vapor Deposition/Sublimation

Water vapor deposition is the process by which water vapor diffuses to an aerosol particle surface and deposits (changes state from a gas to a solid) on the surface as ice. It occurs in clouds only at subfreezing temperatures (below 0 °C) and when the partial pressure of water exceeds the SVP of water over ice. Figure. 5.11(b) shows the SVP of water vapor over ice at subfreezing temperatures. The reverse of water vapor deposition is **sublimation**, the conversion of ice to water vapor.

5.3.2.3. Dissolution, Dissociation, and Hydration

Dissolution is the process by which a gas, suspended over an aerosol particle surface, diffuses to and dissolves in a liquid on the surface. The liquid in which the gas dissolves is a **solvent**. In aerosol and hydrometeor particles, liquid water is most often the solvent. Any gas, liquid, or solid that dissolves in a solvent is a **solute**. One or more solutes plus the solvent make up a **solution**. The ability of a gas to dissolve in water depends on the **solubility** of the gas, which is the maximum amount of a gas that can dissolve in a given amount of solvent at a given temperature.

In a solution, dissolved molecules may **dissociate** (break into simpler components, namely ions). Positive ions, such as H^+, Na^+, K^+, Ca^{2+}, and Mg^{2+}, are **cations**. Negative ions, such as OH^-, Cl^-, NO_3^-, HSO_4^-, SO_4^{2-}, HCO_3^-, and CO_3^{2-}, are **anions**. The dissociation process is reversible, meaning that ions can reform a dissolved molecule. Substances that undergo partial or complete dissociation in solution are **electrolytes**. The degree of dissociation of an electrolyte depends on the acidity of the solution, the strength of the electrolyte, and the concentrations of ions in solution.

The **acidity** of a solution is a measure of the concentration of **hydrogen ions** (protons or H^+ ions) in solution. Acidity is measured in terms of **pH**, where

$$pH = -\log_{10}[H^+] \qquad (5.2)$$

$[H^+]$ is the **molarity** of H^+ (moles of H^+ per liter of solution). The more acidic a solution, the greater the relative number (higher the molarity) of protons, and the lower the pH. The pH scale (Fig. 10.3) ranges from <0 (highly acidic) to >14 (highly basic or alkaline). In pure water, the only source of H^+ is

$$H_2O(aq) \rightleftharpoons H^+ + OH^-$$

Liquid   Hydrogen Hydroxide

water   ion   ion (5.3)

where OH^- is the **hydroxide ion** and arrows in both directions indicate that the reaction is reversible. Because the product $[H^+][OH^-]$ must equal 10^{-14} moles^2 L^{-2}, and $[H^+]$ must equal $[OH^-]$ to balance charge, the pH of pure water is 7 ($[H^+] = 10^{-7}$ moles L^{-1}).

Acids are substances that, when added to a solution, dissociate, increasing the molarity of H^+. The more H^+ added, the stronger the acid and the lower the pH. Common acids include sulfuric $[H_2SO_4(aq)]$, hydrochloric $[HCl(aq)]$, nitric $[HNO_3(aq)]$, and carbonic $[H_2CO_3(aq)]$ acid. When the pH is low (<2), $HCl(aq)$, $HNO_3(aq)$, and $H_2SO_4(aq)$ dissociate readily, whereas $H_2CO_3(aq)$ does not. The former acids are **strong acids**, and the latter acid is a **weak acid**.

Bases (alkalis) are substances that, when added to a solution, remove H^+, increasing pH. Some bases include ammonia $[NH_3(aq)]$ and slaked lime $[Ca(OH)_2(aq)]$.

When anions, cations, or certain undissociated molecules are dissolved in water, the water can bond to the ion in a process called **hydration**. Several water molecules can hydrate to each ion. Hydration increases the liquid water content of particles. The higher the relative humidity and the greater the quantity of solute in solution, the greater the liquid water content of aerosol particles due to hydration. At relative humidities above 100 percent, however, the volume of water added to a particle by hydration is small compared with that added by water vapor condensation.

Next, dissolution and reaction of some strong acids and a base in aerosol particles are discussed.

Hydrochloric Acid

Gas-phase **hydrochloric acid** $[HCl(g)]$ is abundant over the ocean, where it originates from sea-spray and sea-water evaporation. Over land, it is emitted anthropogenically during coal combustion. If $HCl(g)$ becomes supersaturated in the gas phase (if its partial pressure exceeds its saturation vapor pressure), $HCl(g)$ dissolves into water-containing particles and dissociates by the reversible process,

$$HCl(g) \rightleftharpoons HCl(aq) \rightleftharpoons H^+ + Cl^-$$

Hydrochloric Dissolved   Hydrogen Chloride

acid gas   hydrochloric ion   ion (5.4)

              acid

Dissociation of $HCl(aq)$ is complete so long as the pH exceeds -6, which occurs nearly always. The pH of fresh sea-spray drops, which are primarily in the coarse mode, ranges from $+7$ to $+9$. The pH of such drops decreases during dehydration and sea-spray acidification. **Sea-spray acidification**, briefly discussed in Section 5.2.1.1, occurs when nitric acid or sulfuric acid enters a particle and dissociates, adding H^+ to the solution and forcing Cl^- to reassociate with H^+ and evaporate as $HCl(g)$. A net sea-spray acidification process involving nitric acid is

$$HNO_3(g) + Cl^- \rightleftharpoons HCl(g) + NO_3^-$$

Nitric   Chloride   Hydrochloric   Nitrate   (5.5)
acid gas   ion   acid gas   ion

Sea-spray acidification is most severe along coastal regions near pollution sources and can result in a depletion of chloride ions from sea-spray drops. Figure 5.12 shows the effect of sea-spray acidification. The figure shows the measured composition of aerosol particles 3.3 to 6.3 μm in diameter at Riverside, California, about 60 km from the Pacific Ocean. Sodium in the particles originated from the ocean. In clean air over the ocean, the mass ratio of chloride to sodium is typically 1.8:1. The figure shows that, over Riverside, the ratio was about 0.18:1, one-tenth the clean air ratio. The fact that Riverside particles contained lots of nitrate and sulfate suggests that acidification by these ions was responsible for the near depletion of chloride in the particles.

Nitric Acid

Gas-phase **nitric acid** [$HNO_3(g)$] forms from chemical oxidation of nitrogen dioxide (Reaction 4.5). Because emitted aerosol particles generally do not contain nitric acid, nitric acid enters aerosol particles almost exclusively from the gas phase. The process is

$$HNO_3(g) \rightleftharpoons HNO_3(aq) \rightleftharpoons H^+ + NO_3^-$$

Nitric   Dissolved   Hydrogen Nitrate   (5.6)
acid gas   nitric acid   ion   ion

Because it dissociates when the pH exceeds -1, nitric acid is a strong acid.

When nitric acid dissolves in sea-spray drops containing chloride, it displaces the chloride to the gas phase by sea-spray acidification, as shown in Reaction 5.5. Similarly, when nitric acid dissolves in soil-particle solutions containing calcium carbonate and water, it dissociates the calcium carbonate, causing the carbonate to reform carbon dioxide gas, which evaporates. The process, called **soil-particle acidification**, is described by

$$CaCO_3(s) + 2HNO_3(g) \rightleftharpoons Ca^{2+} + NO_3^- + CO_2(g) + H_2O(aq)$$

Calcium   Nitric   Calcium Nitrate   Carbon   Liquid   (5.7)
carbonate   acid gas   ion   ion   dioxide gas water

(e.g., Hayami and Carmichael, 1997; Dentener et al., 1997; Tabazadeh et al., 1998, Jacobson, 1999d). The net result of this process is that nitrate ions build up in soil-dust particles that contain calcite. A similar result occurs when nitric acid gas is exposed to soil-dust particles that contain magnesite [$MgCO_3(s)$]. Because nitric acid readily enters soil-dust and sea-spray particles during acidification and these particles are primarily in the coarse mode, nitrate is usually in the coarse mode. The high coarse mode nitrate concentration in Fig. 5.12 was most likely due to acidification of sea-spray and soil-dust particles.

Sulfuric Acid

Gas-phase **sulfuric acid** [$H_2SO_4(g)$] is condensable due to its low saturation vapor pressure. Once it condenses, it does not readily evaporate, so it is **involatile**. As sulfuric acid condenses, water vapor molecules simultaneously hydrate to it. Thus, condensation of sulfuric acid produces a solution of sulfuric acid and water, even if a solution did not preexist.

Once condensed irreversibly, sulfuric acid dissociates reversibly. Condensation and dissociation are represented by

$$H_2SO_4(g) \longrightarrow H_2SO_4(aq) \rightleftharpoons H^+ + HSO_4^- \rightleftharpoons 2H^+ + SO_4^{2-}$$

Sulfuric acid gas · Dissolved sulfuric acid · Hydrogen ion · Bisulfate ion · Hydrogen ion · Sulfate ion

(5.8)

The first dissociation [producing the **bisulfate ion** (HSO_4^-)] occurs when the pH exceeds -3, so sulfuric acid is a strong acid. The second dissociation [producing the **sulfate ion** (SO_4^{2-})] occurs when the pH exceeds $+2$, so the bisulfate ion is also a strong acid.

Condensation of sulfuric acid occurs most readily over the particle size mode with the most surface area, which is the accumulation mode. When sulfuric acid condenses on coarse mode sea-spray drops, it displaces chloride to the gas phase. When it condenses on soil-dust particles, it displaces carbonate. In a competition with nitrate, sulfuric acid also displaces nitrate to the gas phase.

Ammonia

Ammonia gas [$NH_3(g)$] is emitted during bacterial metabolism in domestic and wild animals, humans, fertilizers, natural soil, and the oceans. It is also emitted during biomass burning and fossil-fuel combustion. Figure 5.13 summarizes the relative contributions of different sources of ammonia in Los Angeles in 1984.

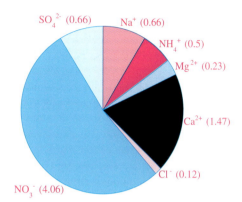

Figure 5.12. Example of sea-spray and soil-particle acidification. The pie chart shows measured mass concentration ($\mu g \ m^{-3}$, in parentheses) of inorganic ions summed over particles with diameters between about 3.3 and 6.3 μm on August 29, 1987, from 05:00 to 08:30 PST at Riverside, California. Chloride associated with the sodium in sea spray and carbonate associated with the calcium in soil-dust particles were most likely displaced by the addition of nitrate and sulfate to the particles. Data from the seventh stage of eight-stage impactor measurements by John et al. (1990).

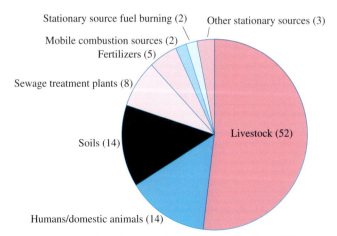

Figure 5.13. Percentage of total ammonia gas emissions from different sources in the Los Angeles Basin in 1984.

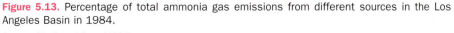

Source: Gharib and Cass (1984).

When ammonia dissolves in water, it combines with the hydrogen ion to form the **ammonium ion** [NH_4^+] by

$$NH_3(g) \rightleftharpoons NH_3(aq)$$

Ammonia   Dissolved
  gas   ammonia

(5.9)

$$NH_3(aq) + H^+ \rightleftharpoons NH_4^+$$

Dissolved   Hydrogen   Ammonium
ammonia   ion   ion

(5.10)

To maintain charge balance, NH_4^+ enters primarily aerosol particles that contain anions. Thus, ammonium is more likely to enter particles containing sulfate, nitrate, or chloride than those containing sodium, potassium, calcium, or magnesium. Because sea-spray and soil-particle solutions contain high concentrations of cations, ammonia rarely enters these particles. An exception is when high concentrations of sulfate or nitrate are also present (Fig. 5.12, for example).

Ammonia is frequently found in particles containing sulfate, and such particles are usually found in the accumulation mode. When nitrate concentrations in the accumulation mode are high, the nitrate is also balanced by ammonium. Figure 5.14 illustrates this point. The figure shows measured aerosol particle compositions versus size at Long Beach and Riverside, California. Long Beach is a coastal site and Riverside is 60 km inland in the Los Angeles Basin. At Long Beach sulfate dominated, but nitrate was present in the accumulation mode. At Riverside nitrate dominated, but sulfate was present in the accumulation mode. Ammonium balanced charge with nitrate and sulfate in the accumulation mode at both Long Beach and Riverside.

5.3.2.4. Solid Precipitation

When their concentrations in aerosol particle solution are high, ions may precipitate to form **solid electrolytes**. Indeed, many soils of the world have formed from the deposition of minerals originating from aerosol particles. **Precipitation** is the formation of an insoluble solid compound due to the buildup in concentration of dissolved ions in a solution. Solids can be suspended throughout a solution but are not part of the solution. If the water content of a solution suddenly increases, solid electrolytes often dissociate back to ions. Solid electrolytes generally do not form in cloud drops because these drops are too dilute. For a similar reason, solid formation is often inhibited in aerosol particles when the relative humidity is high.

The most abundant sulfate-containing electrolyte in aerosol particles on a global scale may be **gypsum** [$CaSO_4 \cdot 2H_2O(s)$] (Jacobson, 2001a), which can form at relative humidities below 98 percent. Gypsum forms when calcium and sulfate react in sea-spray drops or soil-dust particles, so it is present primarily in the coarse mode. **Ammonium sulfate** [$(NH_4)_2SO_4(s)$], which forms in accumulation mode particles, may be less abundant than is gypsum, but it is more important in terms of its effects on visibility because the accumulation mode affects radiative fields more than does the coarse mode.

In urban regions, where nitrate production and ammonia gas emissions are high, concentrations of solid **ammonium nitrate** [$NH_4NO_3(s)$] may build up. In Fig. 5.14, for example, some of the ammonium and nitrate in accumulation mode particles may have formed ammonium nitrate. Ammonium nitrate, in either liquid or solid form, is considered to be one of the major causes of visibility reduction in Los Angeles smog.

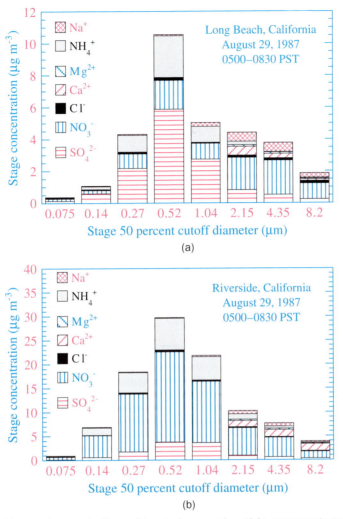

Figure 5.14. Measured concentrations of inorganic aerosol particle components versus particle diameter at (a) Long Beach and (b) Riverside, California, on the morning of August 29, 1987. Data were obtained by John et al. (1990) with an eight-stage Berner impactor.

5.3.3. Removal Processes

Aerosol particles are removed from the air by sedimentation, dry deposition, and rainout. Gases also sediment, but their weights are so small that their sedimentation velocities are negligible. A typical gas molecule has a diameter of 0.5 to 1 nm (nanometer). Such diameters result in sedimentation (fall) velocities of only 1 to 3 km per 10,000 years. Table 5.4 indicates that particles smaller than 0.5 μm in diameter stay in the air several years before falling even 1 km. For these and smaller particles, sedimentation is a long-term removal process. If small particles are near the ground, dry deposition can usually remove them more efficiently than can sedimentation. **Dry deposition** is a process by which gases and particles are carried by molecular diffusion, turbulent diffusion, or advection to trees, buildings, grass, ocean surfaces, or car windows, then rest on, bond to, or react with the surface. Dry deposition is more efficient for particles than for gases because particles are heavier than are gases. As such, particles fall and tend to stay on a surface more readily than do gases, unless wind speeds are high. Gases, especially if they are chemically unreactive, are more readily resuspended into the air.

Aerosol particle **rainout** is a process by which aerosol particles coagulate with raindrops, which subsequently fall to the ground. Rainout is an important removal process for aerosol particles. Because rain clouds occur only in the troposphere, rainout is not a process by which stratospheric particles are removed. Rainout is an effective removal process for volcanic particles.

5.4. SUMMARY OF THE COMPOSITION OF AEROSOL PARTICLES

The composition of aerosol particles varies with particle size and location. Some generalities about composition include the following:

- Newly nucleated particles usually contain sulfate and water, although they may also contain ammonium.
- Biomass burning and fossil-fuel combustion produce primarily small accumulation mode particles, but coagulation and gas-to-particle conversion move these particles to the middle and high accumulation modes. Coagulation also moves some particles to the coarse mode.
- Metals that evaporate during industrial emissions recondense, primarily onto accumulation mode soot particles and coarse mode fly-ash particles. The metal emitted in the greatest abundance is usually iron.
- Sea spray and soil particles are primarily in the coarse mode.

Table 5.6. Dominant Sources and Components of Nucleation, Accumulation, and Coarse Mode Particles

Nucleation Mode	Accumulation Mode	Coarse Mode
Nucleation $H_2O(aq)$, SO_4^{2-}, NH_4^+	Fossil-fuel emissions BC, OM, SO_4^{2-}, Fe, Zn	Sea-spray emissions H_2O, Na^+, Ca^{2+}, Mg^{2+}, K^+, Cl^-, SO_4^{2-}, Br^-, OM
Fossil-fuel emissions BC, OM, SO_4^{2-}, Fe, Zn	Biomass-burning emissions BC, OM, K^+, Na^+, Ca^{2+}, Mg^{2+}, SO_4^{2-}, NO_3^-, Cl^-, Fe, Mn, Zn, Pb, V, Cd, Cu, Co, Sb, As, Ni, Cr	Soil-dust emissions Si, Al, Fe, Ti, P, Mn, Co, Ni, Cr, Na^+, Ca^{2+}, Mg^{2+}, K^+, SO_4^{2-}, Cl^-, CO_3^{2-}, OM
Biomass-burning emissions BC, OM, K^+, Na^+, Ca^{2+}, Mg^{2+}, SO_4^{2-}, NO_3^-, Cl^-, Fe, Mn, Zn, Pb, V, Cd, Cu, Co, Sb, As, Ni, Cr	Industrial emission BC, OM, Fe, Al, S, P, Mn, Zn, Pb, Ba, Sr, V, Cd, Cu, Co, Hg, Sb, As, Sn, Ni, Cr, H_2O, NH_4^+, Na^+, Ca^{2+}, K^+, SO_4^{2-}, NO_3^-, Cl^-, CO_3^{2-}	Biomass burning ash, industrial fly-ash, tire-particle emissions
Condensation/dissolution $H_2O(aq)$, SO_4^{2-}, NH_4^+, OM	Condensation/dissolution $H_2O(aq)$, SO_4^{2-}, NH_4^+, OM Coagulation of all components from nucleation mode	Condensation/dissolution $H_2O(aq)$, NO_3^- Coagulation of all components from smaller modes

- When sulfuric acid condenses, it usually condenses onto accumulation-mode particles because these particles have more surface area, when averaged over all particles in the mode, than do nucleation-or coarse-mode particles.
- Once in accumulation-mode particles, sulfuric acid dissociates primarily to sulfate $[SO_4^{2-}]$. To maintain charge balance, ammonia gas $[NH_3(g)]$, dissolves and dissociates, producing the ammonium ion $[NH_4^+]$; the major-cation in particles. Thus, ammonium and sulfate often coexist in accumulation-mode particles.
- Because sulfuric acid has a lower SVP and a greater solubility than does nitric acid, nitric acid is inhibited from entering accumulation-mode particles that already contain sulfuric acid.
- Nitric acid tends to dissolve in coarse-mode particles and displace chloride in sea-spray drops and carbonate in soil-dust particles during acidification. Sulfuric acid also displaces chloride and carbonate during acidification.

Table 5.6 summarizes the predominant components and their sources in each the nucleation, accumulation, and coarse particle modes.

5.5. AEROSOL PARTICLE MORPHOLOGY AND SHAPE

The morphologies (structures) and shapes of aerosol particles vary with composition. The older an aerosol particle, the greater the number of layers and attachments the particle is likely to have. If the aerosol particle is hygroscopic, it absorbs liquid water at high relative humidities and becomes spherical. If ions are present and the relative humidity decreases, solid crystals may form within the particle. Some observed aerosol particles are flat, others are globular, others contain layers, and still others are fibrous.

Of particular interest is the morphology and shape of soot particles, which contain BC, OM, O, N, and H. Soot particles have important optical effects. The only source of soot is emissions. Globally, about 55 percent of soot originates from fossil-fuel combustion, and the rest originates from biomass burning (Cooke and Wilson, 1996; Liousse et al., 1996). An emitted soot particle is irregularly shaped and mostly solid, containing from 30 to 2000 graphitic spherules aggregated with random orientation by collision during combustion (Katrlnak et al., 1993). An example of a soot aggregate is shown in Fig. 5.15(b).

Once emitted, soot particles can coagulate or grow. Because soot particles are porous and have a large surface area, they serve as sites on which condensation occurs. Although BC in soot is hydrophobic, some organics in soot attract water, in

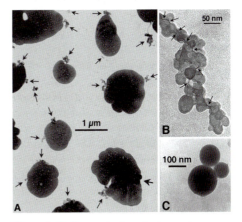

Figure 5.15. Transmission electron microscopy (TEM) images of (a) ammonium sulfate particles containing soot (arrows point to soot inclusions), (b) a chainlike soot aggregate, and (c) fly-ash spheres consisting of amorphous silica collected from a polluted marine boundary layer in the North Atlantic Ocean by Pósfai et al. (1999).

which inorganics gases dissolve (Andrews and Larson, 1993). Evidence of condensation and coagulation is abundant because traffic tunnel studies (Venkataraman et al., 1994) and test vehicle studies (Maricq et al., 1999; ACEA, 1999) indicate that most fossil-fuel BC is emitted in particles smaller than 0.2 μm in diameter, but ambient measurements in Los Angeles, the Grand Canyon, Glen Canyon, Chicago, Lake Michigan, Vienna, and the North Sea show that accumulation mode BC often exceeds emissions mode BC (McMurry and Zhang, 1989; Hitzenberger and Puxbaum, 1993; Venkataraman and Friedlander, 1994; Berner et al., 1996; Offenberg and Baker, 2000). The most likely way that ambient BC redistributes so dramatically is by coagulation and growth. Similarly, the measured mean number diameter of biomass-burning smoke less than 4 minutes old is 0.10 to 0.13 μm (Reid and Hobbs, 1998), yet the mass of such aerosol particles increases by 20 to 40 percent during aging, with one-third to one-half the growth occurring within hours after emissions (Reid et al., 1998).

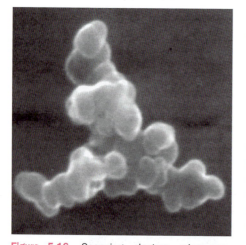

Transmission electron microscopy (TEM) images support the theory that soot particles can become coated once emitted. Katrlnak et al. (1992, 1993) show TEM images of soot from fossil-fuel sources coated with sulfate or nitrate. Martins et al. (1998) show a TEM image of a coated biomass-burning soot particle. Pósfai et al. (1999) show TEM images of North Atlantic soot particles containing ammonium sulfate. This image is reproduced in Fig. 5.15(a). They found that internally mixed soot and sulfate appear to comprise a large fraction of aerosol particles in the troposphere. Almost all soot particles found in the North Atlantic contained sulfate. Strawa et al. (1999) took scanning electron microscopy (SEM) images of black carbon particles in the Arctic stratosphere. One such image is reproduced in Fig. 5.16. The rounded edges of the particle seems to indicate that the particle is coated. Katrlnak et al. (1993) report that rounded grains on black carbon aggregates indicate a coating.

Figure 5.16. Scanning electron microscopy (SEM) image of a coated soot particle from the Arctic stratosphere by Strawa et al. (1999).

In sum, whereas emitted soot particles are relatively distinct, or **externally mixed** from other aerosol particles, soot particles typically coagulate or grow to become **internally mixed** with other particle components. Although soot becomes internally mixed, it does not become "well mixed" (diluted) in an internal mixture because soot consists of a solid aggregate of many graphite spherules. Thus, soot is a distinct component in a mixed particle. Katrlnak et al. (1992) found that coated soot aggregates have carbon structures similar to uncoated aggregates; thus, individual spherules in BC structures do not readily break off or compress during coating.

5.6. HEALTH EFFECTS OF AEROSOL PARTICLES

Aerosol particles contain a variety of hazardous inorganic and organic substances. Some hazardous organic substances include benzene, polychlorinated biphenyls, and polycyclic aromatic hydrocarbons (PAHs). Hazardous inorganic substances include metals and sulfur compounds. Metals cause lung injury, bronchioconstriction, and increased incidence of infection (Ghio and Samet, 1999). Particles smaller than 10 μm in diameter (PM_{10}) have been correlated with asthma and chronic obstructive pulmonary disease (MacNee and Donaldson, 1999).

A major worldwide health problem directly linked to aerosol particles is Coal Workers' Pneumoconiosis (CWP), more commonly known as **black-lung disease**.

Coal workers develop black-lung disease over many years of exposure to coal dust. The dust first builds up in air sacs in the lungs, then scars the sacs, making breathing difficult. During the last 30 years alone in the United States, black lung disease, first identified in 1831, has killed an average of 2000 coal workers per year (NIOSH, 2001).

A more deadly health hazard linked to aerosol particles is indoor-burning of biomass and coal. Such burning is carried out daily for cooking and home heating by large segments of the population in many developing countries. The World Health Organization estimates that, of 2.7 million people who die each year from air pollution, 1.8 million die in rural areas, where the largest source of mortality is indoor-burning of biomass and coal (WHO, 2000).

With respect to outdoor air, some studies have found that there may be no low threshold for PM_{10}-related health problems (Pope et al., 1995). Because most mass of PM_{10} is not hazardous, damage from PM_{10} may be due primarily to small particles, particularly ultrafine particles (smaller than 0.1 μm in diameter). Such particles may be toxic to the lungs, even when they contain components that are not toxic when present in larger particles (MacNee and Donaldson, 1999).

Studies in the 1970s found a link between cardiopulmonary disease and high concentrations of aerosol particles and sulfur oxides. Other studies in the 1970s and 1980s found a link between low concentrations of aerosol particles and health. A summary of several studies by Pope (2000) concluded that short-term (acute) increases of 10 μg m^{-3} PM_{10} were associated with a 0.5 to 1.5 percent increase in daily mortality, higher hospitalization and health-care visits for respiratory and cardiovascular disease, and enhanced outbreaks of asthma and coughing. Increased death rates usually occurred within 1 to 5 days following an air pollution episode. Long-term exposures to 5 μg m^{-3} of particles smaller than 2.5 μm in diameter ($PM_{2.5}$) above background levels resulted in a variety of cardiopulmonary problems, including increased mortality, increased disease, and decreased lung function in adults and children (Pope and Dockery, 1999).

Additional studies have shown that $PM_{2.5}$ results in more respiratory illness and premature death than do larger aerosol particles (Özkatnak and Thurston, 1987; U.S. EPA, 1996). One six-city, 16-year study concluded that people living in areas where aerosol particle concentrations were lower than even the federal PM_{10} standard had a lifespan two years shorter than people living in cleaner air (Dockery - et al., 1993). Air pollution was correlated with death from lung cancer and cardiopulmonary disease. Mortality was correlated with fine particulates, including sulfates.

Finally, several studies have examined the effect of both gas and particle pollution on health. A study by Hall et al. (1992) found that 98 percent of the 12 million people living in the Los Angeles Basin experienced ozone-related symptoms for 17 days each year, and air pollution in the basin caused 1600 premature deaths per year. Children and people working outdoors were most likely to experience respiratory problems. A study by Dr. Kay Kilburn of the University of Southern California found that the lung function of children raised in Los Angeles was 10 to 15 percent lower than that of children raised in cleaner environments. A study by Dr. David Abbey of Loma Linda University found that people living in areas of Los Angeles that violated federal particulate standards at least 42 days per year had a 33 percent greater risk of bronchitis and 74 percent greater risk of asthma than a control group. Women living in high-particulate areas had a 37 percent higher risk of developing cancer. A 1987 study by Drs. Russell

Sherwin and Valda Richters of the University of Southern California found that 75 percent of a group of young people who died accidentally had airspace inflammation, 27 percent of the group had severe damage to their lungs, 39 percent had severe illness to the bronchial glands, and 29 percent had severe illness in their bronchial linings (SCAQMD, 2000).

Although epidemiological studies have found an association between short-term exposure to outdoor particulate air pollution and health problems, people spend most of their time indoors, and concentrations of aerosol particles are often greater indoors than they are outdoors. Concentrations of gases and aerosol particles measured in the vicinity of an individual indoors are even greater than are concentrations measured from a stationary indoor monitor away from the individual. The relatively high concentration of pollution measured near an individual is called the **personal cloud** (Rodes et al., 1991; McBride et al., 1999). A personal cloud may arise when a person's movement stirs up gases and particles on clothes and nearby surfaces, increasing pollutant concentrations. People also release thermal-IR radiation, which rises, stirring and lifting pollutants. Personal cloud concentrations can range from 3 to 67 μg m^{-3} for PM$_{10}$ and from 6 to 27 μg m^{-3} for PM$_{2.5}$ (Wallace, 2000). Personal cloud concentrations must be separated from background outdoor or indoor air concentrations when determining the effects of particles on health.

5.7. SUMMARY

Aerosol particles appear in a variety of shapes and compositions and vary in size from a few gas molecules to the size of a raindrop. Natural sources of aerosol particles include sea-spray uplift, soil-dust uplift, volcanic eruptions, natural biomass burning, plant material emissions, and meteoric debris. Anthropogenic sources include fugitive dust emissions, biomass burning, fossil-fuel combustion, and industrial sources. Aerosol particle size distributions are usually trimodal or quadrimodal, consisting of a nucleation, one or two subaccumulation, and a coarse particle mode. Homogeneous nucleation and emissions from combustion and biomass burning dominate the nucleation mode. Emissions of sea-spray, natural soil dust, and fugitive soil dust dominate the coarse mode. Aerosol particles coagulate and grow by condensation or dissolution from the nucleation mode to the accumulation mode. Chemistry within aerosol particles and between gases and aerosol particles affects growth. Growth does not move accumulation mode particles to the coarse mode, except when water vapor grows onto aerosol particles to form cloud drops. The main removal processes of aerosol particles from the atmosphere are rainout, sedimentation, and dry deposition. Aerosol particles are responsible for a variety of health problems.

5.8. PROBLEMS

5.1.   Why do accumulation mode aerosol particles not grow readily into the coarse mode?

5.2.   Why do accumulation mode particles generally contain more sulfate than do coarse mode particles?

5.3. On a global scale, why is most chloride observed in the coarse mode?

5.4. Why is most ammonium found in accumulation mode particles?

5.5. Write an equilibrium reaction showing nitric acid gas reacting with magnesite, a solid. What particle size mode should this reaction most likely occur in?

5.6. Why is the carbonate ion not abundant in aerosol particles?

5.7. Why might a sea-spray drop over midocean lose all its chloride when it reaches the coast?

5.8. Why is more nitrate than sulfate generally observed in particles containing soil minerals?

5.9. Why is coagulation not an important process for moving particle mass from the lower to the upper accumulation mode?

5.10. Visibility is affected primarily by particles with diameter close to the wavelength of visible light, 0.5 μm. Which particle mode does this correspond to, and which three particle components in this mode do you think affect visibility in the background troposphere the most?

5.11. Particles smaller than 2.5 μm in diameter affect human health more than do larger particles. Identify five chemicals that you might expect to see in high concentrations in these particles in polluted air. Why did you pick these chemicals?

New book

5.10
5.11
5.14

EFFECTS OF METEOROLOGY ON AIR POLLUTION

In this chapter, the effects of meteorology on air pollution are discussed. The concentrations of gases and aerosol particles are affected by winds, temperatures, vertical temperature profiles, clouds, and the relative humidity. These meteorological parameters are influenced by large-and small-scale weather systems. Large-scale weather systems are controlled by large-scale regions of high and low pressure. Small-scale weather systems are controlled by ground temperatures and small-scale variations in pressure. The first section of the chapter examines the forces acting on air. The second section examines how forces combine to form winds. The third section discusses how radiation, coupled with forces and the rotation of the Earth, generates the global circulation of the atmosphere. Sections 6.4 and 6.5 discuss the two major types of large-scale pressure systems. Section 6.6 discusses the effects of such pressure systems on air pollution. The last section focuses on the effects of local meteorology on air pollution.

6.1. FORCES

Winds arise due to forces acting on the air. Next, the major forces are described.

6.1.1. Pressure Gradient Force

When high air pressure exists in one location and low pressure exists nearby, air moves from high to low pressure. The force causing this motion is the **pressure gradient force** (PGF). The force is proportional to the difference in pressure divided by the distance between the two locations and always acts from high to low pressure.

6.1.2. Apparent Coriolis Force

When air is in motion over a rotating Earth, it appears to an observer fixed in space to be deflected to the right in the Northern Hemisphere and to the left in the Southern Hemisphere by the **apparent Coriolis force** (ACoF). The ACoF is not a real force; it is an apparent or fictitious force that, to an observer fixed in space, is really an acceleration of moving air to the right in the Northern Hemisphere and left in the Southern Hemisphere and arises when the Earth rotates under a body (air in this case) in motion. The ACoF is zero at the equator, maximum at the poles, zero for bodies at rest, proportional to the speed of the air, and always acts 90° to the right (left) of the moving body in the Northern (Southern) Hemisphere.

Figure 6.1 illustrates the ACoF. If the Earth did not rotate, an object thrown from point A directly north would be received at point B, along the same longitude as point A. Because the Earth rotates, objects thrown to the north have a west-to-east velocity equal to that of the Earth's rotation rate at the latitude they originate from. The Earth's rotation rate near the equator (low latitude) is greater than that near the poles (high latitudes); thus, objects thrown from low latitudes have a greater west-to-east velocity than does the Earth below them when they reach a high latitude. For example, an object thrown from point A toward point B in the north will end up at point C, instead of at point B' by the time the person at point A reaches point A', and the object will appear as if it has been deflected to the right (from point B' to point C'). Similarly, an object thrown from point B toward point A in the south will end up at point D, instead of at point A' by the time the person at point B reaches point B'. The Coriolis effect, therefore, appears to deflect

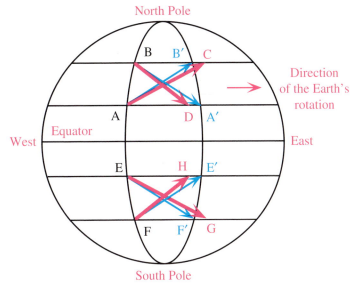

Figure 6.1. Example of the apparent Coriolis force (ACoF), described in the text. Thin arrows in the figure are intended paths, thick arrows are actual paths.

moving bodies to the right in the Northern Hemisphere. Moving bodies in the Southern Hemisphere appear to be deflected to the left.

6.1.3. Friction Force

A third force that acts on moving air is the **friction force** (FF). This force is important near the surface only. The FF slows the wind. Its magnitude is proportional to the wind speed, and it acts in exactly the opposite direction from the wind. The rougher the surface, the greater the FF. The FF over oceans and deserts is small, whereas the FF over forests and buildings is large.

6.1.4. Apparent Centrifugal Force

A fourth force, which also acts on moving air, is the **apparent centrifugal force** (ACfF). This force is also a fictitious force. The force arises when an object rotates around an axis. The apparent force is directed outward, away from the axis of rotation. When a passenger in a car rounds a curve, for example, a viewer travelling with the passenger sees the passenger being pulled outward, away from the axis of rotation, by this force. By contrast, a viewer fixed in space sees the passenger accelerating inward due to a **centripetal acceleration**, which is equal in magnitude to but opposite in direction from the apparent centrifugal force.

6.2. WINDS

The major forces acting on the air in the horizontal are the PGF, ACoF, FF, and ACfF. In the vertical, the major forces are the upward-directed vertical pressure-gradient force and the downward-directed force of gravity. These forces drive winds. Examples of horizontal winds arising from force balances are given next.

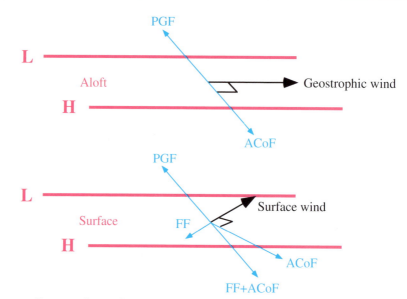

Figure 6.2. Forces acting to give winds aloft and at the surface in the Northern Hemisphere.

6.2.1. Geostrophic Wind

The type of wind involving the least number of forces is the **geostrophic** ("Earth-turning") **wind**. This wind involves only the PGF and ACoF. It arises above the boundary layer, where surface friction is negligible, and along straight isobars. An **isobar** is a line of constant pressure. Suppose a horizontal pressure gradient, represented by two parallel isobars, exists, such as in the top diagram of Fig. 6.2. The PGF causes still air to move from high to low pressure. As the air moves, the ACoF deflects the air to the right. The ACoF continues deflecting the air until the ACoF exactly balances the magnitude and direction of the PGF **(geostrophic balance)**. Figure 6.2 shows that the resulting geostrophic wind flows parallel to isobars. The closer the isobars are together, the faster the geostrophic wind. In reality, geostrophic balance occurs following a process called **geostrophic adjustment**, during which the wind overshoots then undershoots its ultimate path in an oscillatory fashion. In the Southern Hemisphere, the geostrophic wind flows in the opposite direction from that shown in the top of Fig. 6.2.

6.2.2. Surface Winds along Straight Isobars

When isobars are straight near the surface, the FF affects the equilibrium wind speed and direction. The bottom of Fig. 6.2 shows the wind direction that results from a balance among the PGF, ACoF, and FF at the surface in the Northern Hemisphere. Friction, which acts in the opposite direction from the wind, slows the wind. Because the magnitude of the ACoF is proportional to the wind speed, a reduction in wind speed reduces the magnitude of the ACoF. Because the sum of the FF and the ACoF must balance the PGF, the equilibrium wind direction shifts toward low pressure. On average, surface friction turns winds 15 to 45° toward low pressure, with lower values corresponding to smooth surfaces and higher values corresponding to rough surfaces.

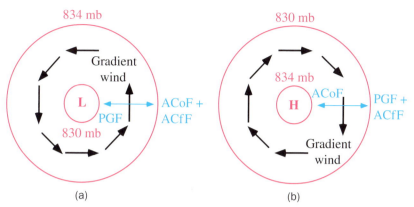

Figure 6.3. Gradient winds around a center of (a) low and (b) high pressure in the Northern Hemisphere, and the forces affecting them.

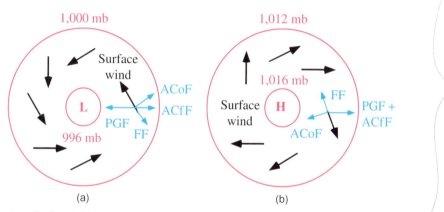

Figure 6.4. Surface winds around centers of (a) low and (b) high pressure in the Northern Hemisphere and the forces affecting them.

6.2.3. Gradient Wind

When centers of low and high pressure exist aloft, the wind is controlled by the PGF, ACoF, and ACfF. The resulting wind is the **gradient wind**. For a low-pressure center, the PGF acts toward and the ACfF acts away from the center of the low. Balance requires that the ACoF also oppose the PGF. Figure 6.3(a) shows the resulting counterclockwise **(cyclonic)** gradient wind in the Northern Hemisphere. Around a high-pressure center aloft in the Northern Hemisphere, the PGF and ACfF act away from and the ACoF acts toward the center of the high. Figure 6.3(b) shows the resulting clock-wise **(anticyclonic)** wind. In the Southern Hemisphere, gradient winds flow clockwise around low-pressure centers and counterclockwise around high-pressure centers.

6.2.4. Surface Winds along Curved Isobars

Near the surface, the FF affects the flow around centers of low and high pressure. A surface low-pressure center is a **cyclone**, and a surface high-pressure center is an **anticyclone**. Figure 6.4 shows the force balances and resulting winds in the presence of a surface (a) low-pressure system and (b) high-pressure system in the Northern Hemisphere. In the low-pressure case, surface winds rotate counterclockwise (cyclonically), but the FF causes them to converge toward the center of the low. In the high-pressure case, surface winds rotate clockwise (anticylonically), but the FF causes them to diverge away from the center of the high. In the Southern Hemisphere, winds converge clockwise around a surface low-pressure center and diverge counter-clockwise around a surface high-pressure center.

6.3. GLOBAL CIRCULATION OF THE ATMOSPHERE

Air pollution is affected by winds, winds are affected by large-scale pressure systems, and large-scale pressure systems are affected by the global circulation of the atmosphere. Figure 6.5 shows features of the global circulation, including the major circulation cells, the belts of low and high pressure, and the predominant wind directions.

Winds have a west-east **(zonal)**, south-north **(meridianal)**, and vertical component. The three circulation cells in each hemisphere shown in Fig. 6.5 represent the meridianal and vertical components of the Earth's winds, averaged zonally (over all longitudes) and over a long time period. The cells are symmetric about the equator and extend up to the tropopause (Section 3.3.1.2), which is near 18 km altitude over the equator and near 8 km altitude over the poles.

Two cells, called **Hadley cells**, extend from 0 to 30°N and S latitude, respectively. These cells were named after **George Hadley** (1685–1768), an English physicist and

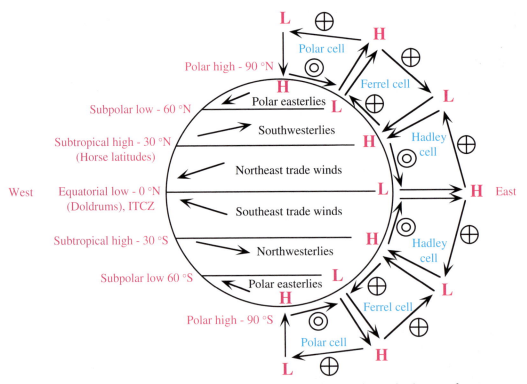

Figure 6.5. Diagram of the three major circulation cells in the atmosphere, the predominant surface pressure systems, and the predominant surface wind systems on the Earth; H, high pressure; L, low pressure. Elevated high and low pressures are relative to pressures at the same altitude. The circled X's denote winds going into the page (west to east). The circled dots denote winds coming out of the page (east to west).

meteorologist who, in 1735, first proposed the cells in a paper called, "Concerning the Cause of the General Trade Winds," presented to the Royal Society of London. Hadley's original cells, however, extended between the equator and poles.

In 1855, William Ferrel (1817–1891; Fig. 6.6) an American school teacher, meteorologist, and oceanographer, published an article in the *Nashville Journal of Medicine* pointing out that Hadley's one-cell model did not fit observations so well as did the three-cell model shown in Fig. 6.5. In 1860, Ferrel went on to publish a collection of papers showing the first application of mathema-tical theory to fluid motions on a rotating Earth. Today, the middle cell in the three-cell model is named the Ferrel cell. Ferrel cells extend from 30 to 60° in the Northern and Southern Hemispheres. Two Polar cells extend from 60 to 90°N and S latitude, respectively.

Figure 6.6. William Ferrel (1817–1891).

6.3.1. Equatorial Low-Pressure Belt

Circulation in the three cells is controlled by heating at the equator, cooling at the poles, and the rotation of the Earth. Air rises over the equator because the sun heats this region intensely. Much of the heating occurs over water, some of which evapo-rates. As air containing water vapor rises, the air expands and cools, and the water vapor recondenses to form clouds of great vertical extent. Condensation of water vapor releases latent heat, providing the air with more buoyancy. Over the equator, the air can rise up to about 18 km before it is decelerated by the stratospheric inver-sion. Once the air reaches the tropopause, it cannot rise much farther, so it diverges to the north and south. At the surface, air is drawn in horizontally to replace the rising air. So long as divergence aloft exceeds convergence at the surface, surface air pres-sure decreases and air pressure aloft increases (relative to pressures at the same altitude but other latitudes). The surface low-pressure belt at the equator is called the equatorial low-pressure belt. Because pressure gradients are weak, winds are light, and the weather is often rainy over equatorial waters, this region is also called the doldrums.

6.3.2. Winds Aloft in the Hadley Cells

As air diverges toward the north in the elevated part of the Northern Hemisphere Hadley cell, the ACoF deflects much of it to the right (to the east), giving rise to westerly winds aloft (winds are generally named after the direction that they originate from). Westerly winds aloft in the Northern Hemisphere Hadley cell increase in magnitude with increasing distance from the equator until they meet equatorward-moving air from the Ferrel cell at 30°N, the subtropical front. The front is a region of sharp temperature contrast. The winds at the front are strongest at the tropopause, where they are called the subtropical jet stream. Winds aloft in the Southern Hemisphere Hadley cell are also westerly and culminate in a tropopause subtropical jet stream at 30°S.

6.3.3. Subtropical High-Pressure Belts

As air converges at the subtropical fronts at 30°N and S, much of it descends. Air is drawn in horizontally aloft to replace the descending air. So long as inflow aloft exceeds outflow at the surface, surface air pressure builds up. The surface high-pressure belts at 30°N and S are called **subtropical high-pressure belts**. Because descending air compresses and warms, evaporating clouds, and because pressure gradients are relatively weak around high-pressure centers, surface high-pressure systems are characterized by sunny skies and light winds. Sunny skies and the lack of rainfall at 30°N and S are two reasons why most deserts of the world are located at these latitudes. The light winds forced many ships sailing at 30°N to lighten their cargo, the heaviest and most dispensable component of which was often horses. Thus the 30°N latitude band is also called the **horse latitudes**.

6.3.4. The Trade Winds

At the surface at 30°N and S, descending air diverges both equatorward and poleward. Most of the air moving equatorward is deflected by the ACoF to the right (toward the west) in the Northern Hemisphere and to the left (toward the west) in the Southern Hemisphere, except that friction reduces the extent of ACoF turning. The resulting winds in the Northern Hemisphere are called the **northeast trade winds** because they originate from the northeast. Those in the Southern Hemisphere are called the **southeast trade winds** because they originate from the southeast. Sailors from Europe have used the northeast trades to speed their voyages westward since the fifteenth century. The trade winds are consistent winds. As seen in Fig. 6.5, the northeast and southeast trade winds converge at the **Intertropical Convergence Zone** (ITCZ), which moves north of the equator in the Northern Hemisphere summer and south of the equator in the Southern Hemisphere summer, generally following the direction of the sun. At the ITCZ, air convergence and surface heating lead to the rising arm of the Hadley cells.

6.3.5. Subpolar Low-Pressure Belts

As surface air moves poleward in the Ferrel cells, the ACoF turns it toward the east in both hemispheres, but surface friction reduces the extent of turning, so that near-surface winds at **midlatitudes** (30 to 60°N and S) are generally westerly to southwesterly (from the west or southwest) in the Northern Hemisphere and westerly to northwesterly in the Southern Hemisphere. In both hemispheres, poleward-moving near-surface air in the Ferrel cell meets equatorward-moving air from the Polar cell at the **polar front**, which is a region of sharp temperature contrast between these two cells. Converging air at the surface front rises and diverges aloft, reducing surface air pressure and increasing air pressure aloft relative to pressures at other latitudes. The surface low-pressure regions at 60°N and S are called **subpolar low-pressure belts**. Regions of rising air and surface low pressure are associated with storms. Thus, the intersection of the Ferrell and Polar cells is associated with stormy weather. Unlike at the equator, surface pressure gradients and winds at the polar front are relatively strong. West–east wind speeds also increase with increasing height at the polar fronts. At the tropopause in each hemisphere, they culminate in the **polar front jet streams**. Whereas the subtropical jet streams do not meander to the north or south over great distances, the polar front jet streams do. Their predominant direction is still from west to east.

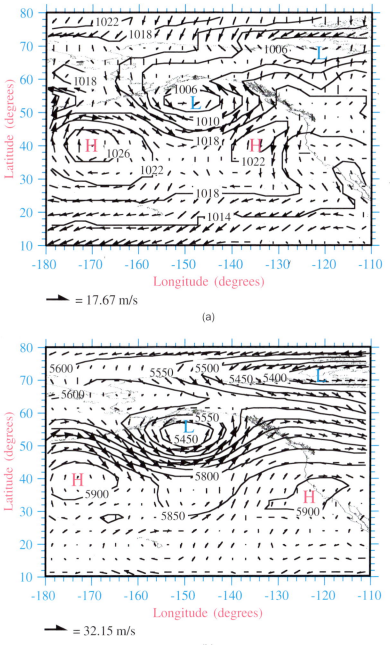

$\longrightarrow$ = 17.67 m/s

(a)

$\longrightarrow$ = 32.15 m/s

(b)

Figure 6.7. Map of (a) sea-level pressures (mb) and near-surface winds (m/s) and (b) 500-mb heights (m) and winds (m/s) obtained from NCEP (2000) for August 3, 1990, at 12 GMT for the northern Pacific Ocean. Height contours on a constant pressure chart are analogous to isobars on a constant height chart; thus, high (low) heights in map (b) correspond to high (low) pressures on a constant height chart. The surface low-pressure system at $-148°$W, 53°N in (a) is the Aleutian low. The surface high-pressure system at $-134°$W, 42°N in (a) is the Pacific high.

6.3.6. Westerly Winds Aloft at Midlatitudes

One would expect the ACoF to deflect air moving equatorward aloft in the Ferrel cells to the west, creating easterly winds (from east to west) aloft at midlatitudes in both hemispheres. In fact, the Coriolis force does act on the winds in this way, but the winds aloft in these regions are accelerated in the opposite direction (from west to east) by the movement of air around pressure centers.

Descending air at 30°N and S creates centers of surface high pressure, and rising air at 60°N and S creates bands or centers of surface low pressure. Figure 6.7(a) shows an example of surface high- and low-pressure centers formed over the Pacific Ocean in the Northern Hemisphere at these respective latitudes. Surface winds moving around a Northern Hemisphere surface high-pressure center travel clockwise (diverging away from the center of the high), and surface winds moving around a surface low-pressure center travel counterclockwise (converging into the center of the low). Indeed, these characteristics can be seen in Fig. 6.7(a), which shows winds traveling clockwise around the highs and counterclockwise around the lows. The reason winds move the way they do around pressure centers was described in Sections 6.2.3 and 6.2.4. The positions of the highs and lows shown in Fig. 6.7(a) create a near-surface west-to-east flow that meanders sinusoidally around the globe. The flow created by these highs and lows is consistent with the expectation that near-surface winds in the Ferrel cell are predominantly westerly.

Figure 6.7(b) shows an elevated map corresponding to the surface map in Fig. 6.7(a). The elevated map shows height contours on a surface of constant pressure (500 mb) and winds traveling around centers of low and high heights. Height contours on a constant pressure chart are analogous to isobars on a constant altitude chart; thus, high (low) heights in Fig. 6.7(b) correspond to high (low) pressures on a constant height chart. The lows aloft lie slightly to the west of the surface lows. The figure indicates that winds traveling around the highs and lows aloft connect, as they do at the surface, resulting in sinusoidal west-to-east flow around the globe. Thus, high- and low-pressure systems aloft are responsible for westerly winds aloft in the Ferrel cell.

6.3.7. Polar Easterlies

Air moving poleward aloft in the Polar cells is turned toward the east in both hemispheres by the ACoF, causing elevated winds in the Polar cells to be westerly. At the poles, air aloft descends, increasing surface air pressures. The surface high-pressure regions are called **Polar highs**. Air at the polar surface diverges equatorward. The ACoF turns this air toward the west. Friction is weak because polar surfaces are either snow or ice, and the resulting surface winds in the Polar cells are easterly and called **polar easterlies**.

6.4. SEMIPERMANENT PRESSURE SYSTEMS

The subtropical high-pressure belts in the Northern and Southern Hemispheres are dominated by surface high-pressure centers over the oceans. These high-pressure centers are called **semipermanent surface high-pressure centers** because they are usually visible on a sea level isobar map most of the year. These pressure systems tend to move northward in the Northern Hemisphere summer and southward in the winter. On average, they are centered near 30°N or S. In the Northern Hemisphere, the two semipermanent surface high-pressure systems are the Pacific high (in the Pacific Ocean) and the Bermuda–Azores high (in the Atlantic Ocean). In Fig. 6.7(a), the surface high-pressure center at $-134°W$, $42°N$ is the Pacific high. In the Southern Hemisphere, semipermanent high-pressure systems are located at $30°S$ in the south Pacific, south Atlantic, and Indian Oceans.

The subpolar low-pressure belt in the Northern Hemisphere is dominated by semipermanent surface low-pressure centers. The subpolar low-pressure belt in the Southern Hemisphere is dominated by a band of low pressure. In the Northern Hemisphere, the two semipermanent low-pressure centers are the Aleutian low (in the Pacific Ocean) and the Icelandic low (in the Atlantic Ocean). These pressure systems tend to move northward in the Northern Hemisphere summer and southward in the winter, but generally stay between 40 and $65°N$. In Fig. 6.7(a), the surface low-pressure center at $-148°W$, $53°N$ is the Aleutian low.

6.5. THERMAL PRESSURE SYSTEMS

Whereas semipermanent surface high- and low-pressure centers exist over the oceans all year, thermal surface high- and low-pressure systems form over land seasonally. Thermal pressure systems form in response to surface heating and cooling, which depend on properties of soil and water, such as specific heat. Specific heat $(J kg^{-1} K^{-1})$ is the energy required to increase the temperature of 1 g of a substance 1 K. Soil has a lower specific heat than does water, as shown in Table 6.1. During the day, the addition of the same amount of sunlight increases the temperature of soil more than it does that of water. During the night, the release of the same amount of thermal-IR energy decreases soil temperature more than it does that of water. As such, land heats during the day and cools during the night more than does water. Similarly, land heats during the summer and cools during the winter more than does water.

Specific heats vary not only between land and water, but also between different soil types, as shown in Table 6.1. Because sand has a lower specific heat than does clay, sandy soil heats to a greater extent than does clayey soil during the day and summer. The preferential heating of sand in comparison with that of clay and the preferential heating of land in comparison with that of water is an important factor that results in thermal low-pressure centers.

When a region of soil heats, air above the soil warms, rises, and diverges aloft, creating low pressure at the surface. Because the low-pressure system forms by heating, it is called a thermal low-pressure system. Thermal low-pressure centers form in the summer over deserts and other sunny areas (e.g., the Mojave Desert in southern California, the plateau of Iran, the north of India). These areas are all located near $30°N$, the same latitude as the semipermanent highs. Descending air at $30°N$ due to the highs helps to form the thermal lows by evaporating clouds, clearing the skies. The descending air also keeps the thermal lows shallow (air rising in the thermal lows

Table 6.1. Specific Heats of Four Media at 298.15 K

Substance	Specific Heat ($J kg^{-1} K^{-1}$)
Dry air at constant pressure	1004.67
Liquid water	4185.5
Clay	1360
Dry sand	827

diverges horizontally at a low altitude). In some cases, such as over the Mojave Desert, the thermal lows do not produce clouds due to the lack of water vapor and shallowness of the low. In other cases, such as over the Indian continent in the summer, rising air in a low sucks warm, moist surface air in horizontally from the ocean, ultimately producing heavy rainfall. The strong sea breeze due to this thermal low is the summer monsoon, where a **monsoon** is a seasonal wind caused by a strong seasonal variation in temperature between land and water.

During the winter, when temperatures cool over land, air densities increase, causing air to descend. Aloft, air is drawn in horizontally to replace the descending air, building up surface air pressure. The resulting surface high-pressure system is called a **thermal high-pressure system**. In the Northern Hemisphere, two thermal high-pressure systems are the **Siberian high** (over Siberia) and the **Canadian high** (over the Rocky Mountains between Canada and the United States). As with thermal low-pressure systems, thermal high-pressure systems are often shallow.

6.6. EFFECTS OF LARGE-SCALE PRESSURE SYSTEMS ON AIR POLLUTION

Semipermanent and thermal pressure systems affect air pollution. Table 6.2 compares characteristics of such pressure systems, including their effects on pollution. Semipermanent low-pressure systems are associated with cloudy skies, stormy weather, and fast surface winds. Thermal low-pressure systems, which are often shallow, may or may not produce clouds. Air rises in both types of low-pressure systems, dispersing near-surface pollution upward. When clouds form in low-pressure systems, they block sunlight that would otherwise drive photochemical reactions, reducing pollution further.

Surface high-pressure systems are characterized by relatively slow surface winds, sinking air, and cloud-free skies. In such pressure systems, air sinks, confining near-surface pollution. In addition, the slow near-surface winds associated with high-pressure systems prevent horizontal dispersion of pollutants, and the cloud-free skies caused by the pressure systems maximize the sunlight available to drive photochemical smog formation. In sum, the major effects of pressure systems on pollution are through vertical pollutant transfer, horizontal pollutant transfer, and cloud cover. Each of these effects is discussed in turn.

Table 6.2. Summary of Characteristics of Northern Hemisphere Surface Low- and High-Pressure Systems

Characteristic	Surface Low-Pressure Systems		Surface High-Pressure Systems	
	Semipermanent	Thermal	Semipermanent	Thermal
Latitude range	45–65°N	25–45°N	25–45°N	45–65°N
Surface pressure gradients	Strong	Varying	Weak	Varying
Surface wind speeds	Fast	Varying	Slow	Varying
Surface wind directions	Converging, counterclockwise	Converging, counterclockwise	Diverging, clockwise	Diverging, clockwise
Vertical air motions	Upward	Upward	Downward	Downward
Cloud cover	Cloudy	Cloud-free or cloudy	Cloud-free, sunny	Cloud free
Storm formation?	Yes	Sometimes	No	No
Effect on air pollution	Reduces	Reduces	Enhances	Enhances

6.6.1. Vertical Pollutant Transport

Pressure systems affect vertical air motions and, therefore, pollutant dispersion by forced and free convection (Section 3.2.2). In a semipermanent low-pressure system, for example, near-surface winds converge and rise, dispersing near-surface pollutants upward. In a semipermanent high-pressure system, winds aloft converge and sink, confining near-surface pollutants. Both cases illustrate forced convection. In thermal low-pressure systems, surface warming causes near-surface air to become buoyant and rise. In thermal high-pressure systems, surface cooling causes near-surface air to become negatively buoyant and sink or stagnate. Both cases illustrate free convection. To understand better how free convection in thermal pressure systems and forced convection in semipermanent pressure systems affect pollutant dispersion, it is necessary to discuss adiabatic processes and atmospheric stability.

6.6.1.1. Adiabatic and Environmental Lapse Rates

Whether air rises or sinks buoyantly in a thermal pressure system depends on atmospheric stability, which depends on adiabatic and environmental lapse rates. These terms are discussed next.

Imagine a balloon filled with air. The air pressure inside the balloon exactly equals that outside the balloon. Imagine also that no energy (such as solar or thermal-IR energy or latent heat energy created by condensation of water vapor) can enter or leave the balloon, but that the balloon's membrane is flexible enough for it to expand and contract due to changes in air pressure outside the balloon. Suppose now that the balloon rises. Because air pressure always decreases with increasing altitude, the balloon must rise into decreasing air pressure. For the air pressure inside the balloon to decrease to the air pressure outside the balloon, the balloon must expand in volume. This type of expansion, caused by a change in air pressure alone, is called an **adiabatic expansion**. Solar heating and latent heat release are **diabatic heating processes** and do not contribute to an adiabatic expansion.

During an adiabatic expansion, kinetic energy of air molecules is converted to work to expand the air. Because temperature is proportional to the kinetic energy of air molecules (Equation 3.1), an adiabatic expansion cools the air. In sum, rising air expands and expanding air cools; thus, rising air cools during an adiabatic expansion. The rate of cooling during an adiabatic expansion near the surface of the Earth is approximately 9.8 K or °C per kilometer increase in altitude. This rate is called the **dry or unsaturated adiabatic lapse rate (Γ_d)**. Lapse rates are opposite in sign to changes in temperature with height; thus, a positive lapse rate indicates that temperature decreases with increasing height.

If the balloon in our example rises in air saturated with water vapor, the resulting adiabatic expansion cools the air and decreases the saturation vapor pressure of water in the balloon, causing the relative humidity to increase to more than 100 percent and water vapor to condense to form cloud drops, releasing latent heat. The rate of temperature increase with increasing height due to this latent heat release is typically 4 K km^{-1}, but it increases to 8 K km^{-1} in the tropics. Subtracting this latent-heat release rate from the dry adiabatic lapse rate gives a net lapse rate during cloud formation of between 6 and 2 K km^{-1}. This rate is called the **wet-, saturated-, or pseudoadiabatic lapse rate (Γ_w)**. It is the negative rate of change of temperature with increasing altitude during an adiabatic expansion in which condensation also occurs. The wet adiabatic lapse rate is applicable only in clouds.

The opposite of an adiabatic expansion is an **adiabatic compression**, which occurs when a balloon sinks adiabatically from low to high pressure. The compression is due to the increased air pressure around the balloon. During an adiabatic compression, work is converted to kinetic energy (which is proportional to temperature), warming the air in the balloon.

Whereas dry and wet adiabatic lapse rates describe the extent of cooling of a balloon rising adiabatically, the **environmental lapse rate** describes the actual change in air temperature with altitude in the environment outside a balloon. It is defined as

$$\Gamma_e = -\frac{\Delta T}{\Delta z} \tag{6.1}$$

where $\Delta T/\Delta z$ is the actual change in air temperature with altitude. An increasing temperature with increasing altitude gives a negative environmental lapse rate.

EXAMPLE 6.1.

If the observed temperature cools 14 K between the ground and 2 km above the ground, what is the environmental lapse rate, and is it larger or smaller than the dry adiabatic lapse rate?

Solution

The environmental lapse rate in this example is $\Gamma_e = +7\text{K/km}$, which is less than the dry adiabatic lapse rate of $\Gamma_d = 9.8$ K/km.

6.6.1.2. Stability

One purpose of examining the dry, wet, and environmental lapse rates is to determine the stability of the air, where the **stability** is a measure of whether pollutants emitted will convectively rise and disperse or build up in concentration near the surface.

Figure 6.8 illustrates the concept of stability. When an unsaturated parcel of air (the balloon, in our example) is displaced vertically, it rises, expands, and cools dry adiabatically (along the dashed line in the figure). If the environmental temperature profile is stable (right thick line), the rising parcel is cooler and more dense than is the air in the environment around it at every altitude. As a result of its lack of buoyancy, the parcel sinks, compresses, and warms until its temperature (and density) equal that

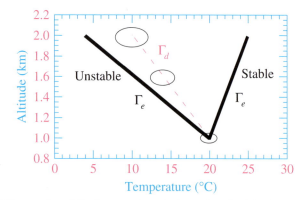

Figure 6.8. Stability and instability in unsaturated air, as described in the text.

of the air around it. In reality, the parcel overshoots its original altitude as it descends, but eventually obtains that altitude in an oscillatory manner. In sum, a parcel of pollution at the temperature of the environment that is accelerated vertically in **stable air** returns back to its original altitude and temperature. Stable air is associated with near-surface pollution buildup because pollutants perturbed vertically in stable air cannot rise and disperse.

In **unstable air**, an unsaturated parcel that is perturbed vertically continues to accelerate in the direction of the perturbation. Unstable air is associated with near-surface pollutant cleansing. If the environmental temperature profile is unstable (left thick line in Fig. 6.8), a parcel rising adiabatically (along the dashed line) is warmer and less dense than is the environment around it at every altitude, and the parcel continues to accelerate. The parcel stops accelerating only when it encounters air with the same temperature (and density) as the parcel. This occurs when the parcel reaches a layer with a new environmental lapse rate.

In neutral air (when the dry and environmental lapse rates are equal), an unsaturated parcel that is perturbed vertically neither accelerates nor decelerates, but continues along the direction of its initial perturbation at a constant velocity. Neutral air results in pollution dilution slower than in unstable air but faster than in stable air.

Whether unsaturated air is stable or unstable can be determined by comparing the dry adiabatic lapse rate with the environmental lapse rate. Symbolically, the stability criteria are

$$\Gamma_e \begin{cases} > \Gamma_d & \text{dry unstable} \\ = \Gamma_d & \text{dry neutral} \\ < \Gamma_d & \text{dry stable} \end{cases} \tag{6.2}$$

If the air is saturated, such as in a cloud, the wet adiabatic lapse rate is used to determine stability. In such a case, the stability criteria are

$$\Gamma_e \begin{cases} > \Gamma_w & \text{wet unstable} \\ = \Gamma_w & \text{wet neutral} \\ < \Gamma_w & \text{wet stable} \end{cases} \tag{6.3}$$

Although stability at any point in space and time depends on Γ_d or Γ_w, but not both, generalized stability criteria for all temperature profiles are often summarized as follows:

$$\begin{cases} \Gamma_e > \Gamma_d & \text{absolutely unstable} \\ \Gamma_e = \Gamma_d & \text{dry neutral} \\ \Gamma_d > \Gamma_e > \Gamma_w & \text{conditionally unstable} \\ \Gamma_e = \Gamma_w & \text{wet neutral} \\ \Gamma_e < \Gamma_w & \text{absolutely stable} \end{cases} \tag{6.4}$$

These conditions indicate that when $\Gamma_e > \Gamma_d$, the air is **absolutely unstable**, or unstable regardless of whether the air is saturated or unsaturated. Conversely, if $\Gamma_e < \Gamma_w$, the air is **absolutely stable**, or stable regardless of whether the air is saturated. If the air is **conditional unstable**, stability depends on whether the air is saturated. Figure 6.9 illustrates the stability criteria in Equation 6.4.

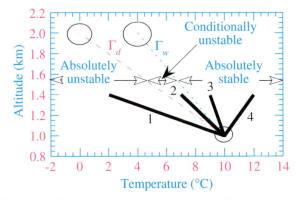

Figure 6.9. Stability criteria for unsaturated and saturated air. If air is saturated, the environmental lapse rate is compared with the wet adiabatic lapse rate to determine stability. Environmental lapse rates 3 and 4 are stable and 1 and 2 are unstable with respect to saturated air. Environmental lapse rates 2, 3, and 4 are stable and 1 is unstable with respect to unsaturated air. A rising or sinking air parcel follows the Γ_d line when the air is unsaturated and the Γ_w line when the air is saturated.

EXAMPLE 6.2.

Given the environmental lapse rate from Example 6.1, determine the stability class of the atmosphere.

Solution

The environmental lapse rate in the example was $\Gamma_e = +7\text{K/km}$. Because the wet adiabatic lapse rate ranges from $\Gamma_w = +2$ to $+6$ K/km, the atmosphere in this example is conditionally unstable (Equation 6.4).

In thermal low-pressure systems, sunlight warms the surface. The surface energy is conducted to the air, warming the lower boundary layer, decreasing the stability of the boundary layer. When the temperature profile near the surface becomes unstable, convective thermals buoyantly rise from the surface, carrying pollution with them.

In thermal high-pressure systems, radiative cooling of the surface stabilizes the temperature profile, preventing near-surface air and pollutants from rising. In many cases, air near the surface becomes so stable that a temperature inversion forms, further inhibiting pollution dispersion. Inversions are discussed next.

6.6.1.3. Temperature Inversions

The stable environmental profile in Fig. 6.8 and Profile 4 in Fig. 6.9 are **temperature inversions**, increases in air temperature with increasing height. Inversions are always stable, but a stable profile is not necessarily an inversion (e.g., Profile 3 in Fig. 6.9). Inversions are important because they trap pollution near the surface to a greater extent than does a stable temperature profile that is not an inversion.

An inversion is characterized by its strength, thickness, top/base height, and top/base temperatures. The **inversion strength** is the difference between the temperature at the inversion top and that at its base. The **inversion thickness** is the difference between the inversion's top and base heights. The **inversion base height** is the height from the ground to the bottom of the inversion. It is also called the **mixing depth** because it is the estimated height to which pollutants released from the surface mix. In reality, pollutants often mix into the inversion layer itself. Inversion layer characteristics are illustrated in Fig. 6.10.

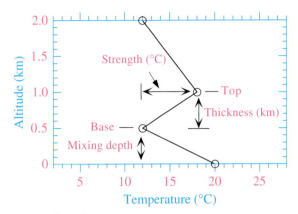

Figure 6.10. A temperature inversion.

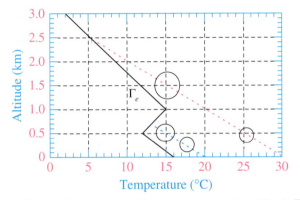

Figure 6.11. Schematic of pollutants trapped by an inversion and stable air. The parcel of air released at 20 °C rises at the unsaturated adiabatic lapse rate of 10 °C/km until its temperature equals that of the environment. This parcel is trapped by the inversion. The parcel of air released at 30° also rises at 10 °C/km. It escapes the inversion, but stops rising in stable free-tropospheric air, where the environmental lapse rate is 6.5 °C/km.

Stable air and inversions, in particular, trap pollutants, preventing them from dispersing into the free troposphere and causing pollutant concentrations to build up near the surface. Figure 6.11 illustrates how such trapping occurs. The figure shows two air parcels with different initial temperatures released at the surface under an inversion. Suppose the parcels represent exhaust plumes that are initially warmer than the environment. Due to their buoyancy, both parcels rise, expand, and cool at the dry adiabatic lapse rate of near 10 °C/km. The parcel released at 20 °C rises and cools until its temperature approaches that of the air around it. At that point the parcel decelerates, then comes to rest after oscillating around its final altitude. The path of this parcel illustrates how an inversion traps pollutants emitted from the surface. It also illustrates that pollutants often penetrate into the inversion layer. The parcel released at 30 °C also rises and cools, but passes easily through the inversion layer. Because the free troposphere above the inversion is stable, the parcel ultimately comes to rest above the inversion.

Inversions form whenever a layer of air becomes colder than the layer of air above it. Common inversion types include the radiation inversion, the large-scale subsidence inversion, the marine inversion, the frontal inversion, and the small-scale subsidence inversion. These are discussed next.

Radiation Inversion

Radiation (nocturnal) inversions occur nightly as land cools by emitting thermal-IR radiation. During the day, land also emits thermal-IR radiation, but this loss is exceeded by a gain in solar radiation. At night, thermal-IR emissions cool the ground, which in turn cools molecular layers of air above the ground, creating an inversion. The strength of a radiation inversion is maximized during long, calm, cloud-free nights when the air is dry. Long nights maximize the time during which thermal-IR cooling occurs, calm nights minimize downward turbulent mixing of energy, cloud-free nights minimize absorption of thermal-IR energy by cloud drops, and dry air minimizes absorption of thermal-IR energy by water vapor. The morning temperature profile in Fig. 6.12 shows a radiation inversion. Radiation inversions also form in the winter during the day in regions that are not exposed to much sunlight. They do not form regularly over the ocean because ocean water cools only slightly at night.

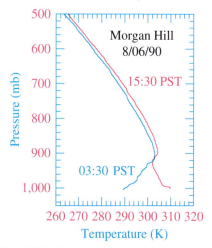

Figure 6.12. Observed temperature profiles in the early morning and late afternoon at Morgan Hill, California, on August 6, 1990. The morning sounding shows a radiation inversion coupled with a large-scale subsidence inversion. The afternoon sounding shows a large-scale subsidence inversion.

Large-Scale Subsidence Inversion

A **large-scale subsidence inversion** occurs within a surface high-pressure system. In such a system, air descends, compressing and warming adiabatically. When a layer of air descends adiabatically, the entire layer becomes more stable, often to the point that an inversion forms. The descension of air and creation of an inversion in a high-pressure system both contribute to near-surface pollution buildup.

Figure 6.13 illustrates the formation of a subsidence inversion. The figure shows a 1.37-km thick layer of air based at 3 km. At this altitude, the pressure thickness of the layer is 114 mb. The initial temperature profile of the layer is stable, but not an inversion. As the layer descends, both its top and bottom compress and warm adiabatically at the rate of about 10 °C/km. Whereas the pressure thickness of the layer remains constant during descension to conserve mass, the height thickness of the layer

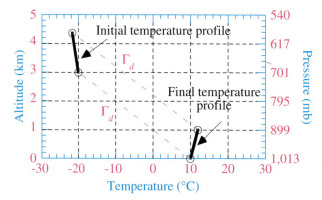

Figure 6.13. Formation of a subsidence inversion in sinking unsaturated air, as described in the text.

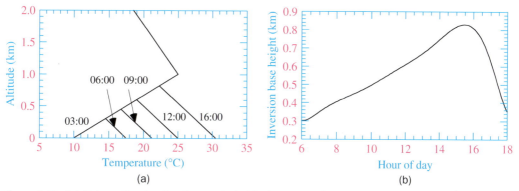

Figure 6.14. (a) Schematic showing the gradual chipping away of a morning radiation/large-scale subsidence inversion to produce an afternoon large-scale subsidence inversion. (b) Change in the inversion base height during the day corresponding to the chipping-away process in (a).

decreases with decreasing altitude (due to compression) so that at the surface, the pressure thickness of the layer is still 114 mb, but the height thickness is 1 km. Figure 6.13 shows that the final temperature profile in the layer is more stable than is the initial profile. In fact, the sinking of air created an inversion. Conversely, the lifting of an unsaturated layer of air decreases the layer's stability.

Over land at night, air in a large-scale pressure system can descend and warm on top of near-surface air that has been cooled radiatively, creating a combined radiation/large-scale subsidence inversion. The morning inversion in Fig. 6.12 shows such a case. The afternoon profile in the same figure shows the contribution of the large-scale subsidence inversion. The subsidence inversion in Fig. 6.12 was due to the Pacific high, well-known for producing large-scale subsidence inversions. Such inversions are present 85 to 95 percent of the days of the year in Los Angeles, which is one reason this city has historically had severe air pollution problems.

Figure 6.12 indicates that between morning and afternoon the radiation inversion eroded, so that by the afternoon all that remained was the large-scale subsidence inversion. This process is illustrated by Fig. 6.14. The schematic shows that, at 03:00, a strong radiation/large-scale subsidence inversion exists. At 06:00, after the sun rises, the ground and near-surface air heat sufficiently to chip away at the bottom of the inversion. The chipping continues until the late afternoon, when the inversion base height (mixing depth) reaches a maximum. As the sun goes down and surface temperature cools, the inversion base height decreases again. If the ground in Fig. 6.14(a) were heated to 35 °C instead of 30 °C in the late afternoon, the inversion in the figure would disappear. The elimination of an inversion due to surface heating is called **popping the inversion**.

A large-scale subsidence inversion serves as a lid to confine pollution beneath it, as shown in Fig. 6.15. The mixing depth in which pollution is trapped is generally thickest in the late afternoon and thinnest during the night and early morning. Mixing ratios of primary pollutants, such as $CO(g)$, $NO(g)$, primary ROGs, and primary aerosol particles usually peak during the morning, when mixing depths are thin and rush hour emission rates are high. Although emission rates are also high during afternoon rush hours, afternoon mixing depths are thick, diluting primary pollutants. Afternoon rush hours are also spread over more hours than are morning rush hours. Despite thick afternoon mixing depths, some secondary pollutants, such as $O_3(g)$, $PAN(g)$, and secondary (aged) aerosol particles, reach their peak mixing ratios in the afternoon, when their chemical formation rates peak.

Figure 6.15. Photograph of pollution trapped under a large-scale subsidence inversion in Los Angeles, California, on the afternoon of July 23, 2000.

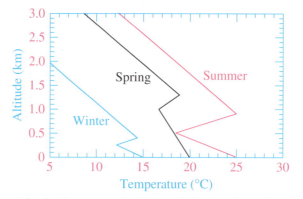

Figure 6.16. Schematic showing seasonal variation of afternoon inversions in Los Angeles.

Figure 6.16 illustrates the seasonal variation of afternoon inversion profiles in Los Angeles. During the winter, the Pacific high is further from Los Angeles than it is during any other season, and the large-scale subsidence inversion strength is weak. During the summer, the inversion is strong because the center of the Pacific high is closer to Los Angeles than it is during any other season.

Marine Inversion

A marine inversion occurs over coastal areas. During the day, land heats faster than does water. The rising air over land decreases near-surface air pressures, drawing ocean air inland. The resulting breeze between ocean and land is the **sea breeze**. As cool, marine air moves inland during a sea breeze, it forces warm, inland air to rise, creating warm air over cold air and a **marine inversion**, which contributes to the inversion strength dominated by the large-scale subsidence inversion in Los Angeles.

Small-Scale Subsidence Inversion

As air flows down a mountain slope, it compresses and warms adiabatically, just as in a large-scale subsidence inversion. When the air compresses and warms on top of cool air, a **small-scale subsidence inversion** forms. In Los Angeles, air sometimes flows from east to west, over the San Bernardino Mountains, into the Los Angeles Basin. Such air compresses and warms as it descends the mountain onto cool marine air below it, creating an inversion.

Frontal Inversion

A **cold front** is the leading edge of a cold air mass and the boundary between a cold air mass and a warm air mass. An **air mass** is a large body of air with similar temperature and moisture characteristics. Cold fronts and warm fronts form around low-pressure centers, particularly in the cyclones that predominate at 60 °N. Cold fronts rotate counterclockwise around a surface cyclone. At a cold front, cold, dense air acts as a wedge and forces air in the warm air mass to rise, creating warm air over cold air and a **frontal inversion.** Because frontal inversions occur in low-pressure systems, where air generally rises and clouds form, frontal inversions are not usually associated with pollution buildup.

6.6.2. Horizontal Pollutant Transport

Large-scale pressure systems affect the direction and speed of winds, which, in turn, affect pollutant transport. In this subsection, large-scale winds and their effects on pollution are examined. Small-scale winds are discussed in Section 6.7.4.

6.6.2.1. Effects of Wind Speeds on Pollutants

Winds around a surface low-pressure systems are generally faster than are those around a surface high-pressure systems, partly because pressure gradients in a low-pressure system are generally stronger than are those in a high-pressure system. Fast winds tend to clear out chemically produced pollution faster than do slow winds, but fast winds resuspend more soil dust and other aerosol particles from the ground than do slow winds. Most soil-dust resuspension occurs when the soil is bare, as illustrated in Fig. 6.17.

6.6.2.2. Effects of Wind Direction on Pollutants

Large-scale pressure systems redirect air pollution. When a surface low-pressure center in the Northern Hemisphere is located to the west of a location, it produces southerly to southwesterly winds at the location, as seen in Fig. 6.4(a). Similarly, surface high-pressure centers located to the west of a location produces northwesterly to northerly winds at the location, as seen in Fig. 6.4(b). In summers, the Pacific high is sometimes located to the northwest of the San Francisco Bay Area. Under such conditions, air traveling clockwise around the high passes through the Sacramento Valley, where temperatures are hot and pollutant concentrations are high, into the Bay Area, where temperatures are usually cooler and pollution also exists. The transport of hot, polluted air from the Sacramento Valley increases temperatures and pollution levels in the Bay Area.

6.6.2.3. Santa Ana Winds

In autumn and winter, the Canadian high-pressure system forms over the Great Basin due to cold surface temperatures. Winds around the high flow clockwise down

Figure 6.17. Dust storm approaching Spearman, Texas, on April 14, 1935. (Courtesy National Oceanic and Atmospheric Administration Central Library.)

the Rocky Mountains, compressing and warming adiabatically. The winds then travel through Utah, New Mexico, Nevada, and into southern California. As the winds travel over the Mojave Desert, they heat further and pick up desert soil dust. The winds then reach the San Gabriel and San Bernardino Mountain Ranges, which enclose the Los Angeles Basin. Two passes into the basin are Cajon Pass and Banning Pass. As winds are compressed through these passes, they speed up to conserve momentum (mass times velocity). The fast winds resuspend soil dust. The winds, called the Santa Ana winds, then enter the Los Angeles Basin, bringing dust with them. If the winds are strong, they overpower the sea breeze (which blows from ocean to land), clearing pollution out of the basin to the ocean. Warm, dry, strong Santa Ana Winds are also responsible for spreading brush fires in the basin. When the Santa Ana Winds are weak, they are countered by the sea breeze, resulting in stagnation. Some of the worst air pollution events in the Los Angeles Basin occur under weak Santa Ana conditions. In such cases, pollution builds up over a period of several days. Santa Ana Winds are generally strongest between October and December.

6.6.2.4. Long-Range Transport of Air Pollutants

Winds carry air pollution, sometimes over great distances. The chimney was developed centuries ago, not only to lift pollution above the ground, but also to take advantage of winds aloft that disperse pollution horizontally. Chimneys exacerbate pollution downwind of the point of emission. Starting in the seventeenth century, for example, the Besshi copper mine and smelter on Shikoku Island, Japan, was a pollution source that damaged crops downwind. In the eighteenth century, smoke from soda ash factory chimneys in France and Great Britain devastated nearby countrysides

(Chapter 10). In the 1930s to 1950s, a smelter in Trail, British Columbia, Canada, released pollution that traveled to Washington State, in the United States. This last case is an example of **transboundary air pollution**, which occurs when pollution crosses political boundaries.

Since the 1950s, it has been recognized that sulfur dioxide, emitted from tall smokestacks, is often carried long distances before it deposits to the ground as sulfuric acid. Because most anthropogenic $SO_2(g)$ is emitted in midlatitudes, where the prevailing near-surface winds are southwesterly and the prevailing elevated winds are westerly, $SO_2(g)$ is transported to the northeast or east. If it is emitted high enough, $SO_2(g)$ can travel hundreds to thousands of kilometers. The largest smokestack in the world is located in **Sudbury**, Ontario, Canada, north of Lake Huron. This nickel-smelting stack, which is 380 m tall, was designed to carry $SO_2(g)$ emissions far from the local region. Not only have emissions from the stack devastated vast areas of land immediately downwind of it, but the great height of the stack has enabled $SO_2(g)$ to be transported long distances, including to the United States.

Long-range transport affects not only pollutants emitted from stacks, but also photochemical smog closer to the ground. In 1987, Wisconsin felt that high mixing ratios of ozone there were exacerbated by ozone transport from Illinois and Indiana. Wisconsin filed a lawsuit to force Illinois and Indiana to control pollutant emissions better. The lawsuit led to a settlement mandating a study of ozone transport pathways (Gerritson, 1993).

Pollutants travel long distances along many other well-documented pathways. Pollutants from New York City, for example, travel to Mount Washington, Vermont. Pollutants travel along the **BoWash corridor** between Boston and Washington DC. Pollutants from the northeast United States travel to the clean north Atlantic Ocean (Liu et al., 1987; Dickerson et al., 1995; Moody et al., 996; Levy et al., 1997; Prados et al., 1999). Pollutants from Los Angeles travel northward to Santa Barbara, northeastward to the San Joaquin Valley, southward to San Diego, and eastward to the Mojave Desert. Such pollutants have also been traced to the Grand Canyon, Arizona (Poulos and Pielke, 1994). Pollutants from the San Francisco Bay Area spill into the San Joaquin Valley through Altamont Pass.

An example of transboundary pollution is the transport of forest-fire smoke from Indonesia to six other Asian countries in September 1997. Sulfur dioxide emissions from China are also suspected of causing a portion of acid deposition problems in Japan. Aerosol-particles and ozone precursors from Asia travel long distances over the Pacific Ocean to North America (Prospero and Savoie, 1989; Zhang et al., 1993; Song and Carmichael, 1999; Jacob et al., 1999). Hydrocarbons, ozone, and PAN travel long distances across Europe (Derwent and Jenkin, 1991) as do pollutants from Europe to Africa (Kallos et al., 1998). Pollutants also travel between the United States and Canada and between the United States and Mexico.

6.6.3. Cloud Cover

Clouds affect pollution in two major ways. First, they reduce the penetration of UV radiation, therefore decreasing rates of photolysis below them. Second, pollutants dissolve in cloud water and are either rained out or returned to the air upon cloud evaporation. Thus, rain-forming clouds help to cleanse the atmosphere. Because cloud cover is often greater and mixing depths, higher in surface low-pressure systems than they are in surface high-pressure systems, photochemical smog concentrations are usually lower in the boundary layer of a low-pressure system than in that of a high-pressure system.

6.7. EFFECTS OF LOCAL METEOROLOGY ON AIR POLLUTION

Whereas large-scale pressure systems control the prevailing meteorology of a region, local factors also affect meteorology and, thus, air pollution. Some of these factors are discussed briefly.

6.7.1. Ground Temperatures

Ground temperatures affect local meteorology and air pollution in at least three ways. Figure 6.14 gives insight into the first mechanism. The figure shows that warm ground surfaces produce high inversion base heights (thick mixing depths) and low pollution mixing ratios. Conversely, cold ground surfaces produce thin mixing depths and high pollution mixing ratios.

A second mechanism by which ground temperatures affect pollution is through their effects on wind speeds. Warm surfaces enhance convection, causing surface air to mix with air aloft and vice-versa. Because horizontal wind speeds at the ground are zero and those aloft are faster, the vertical mixing of horizontal winds speeds up winds near the surface and slows them down aloft. Faster near-surface winds result in greater dispersion of near-surface pollutants but may also increase the resuspension of loose soil dust and other aerosol particles from the ground. Cooler ground temperatures have the opposite effect, slowing down near-surface winds and enhancing near-surface pollution buildup.

Third, changes in ground temperatures change near-surface air temperatures, and air temperatures affect rates of several processes. For example, rates of biogenic gas emissions from trees, carbon monoxide emissions from vehicles, chemical reactions, and gas-to-particle conversion are all temperature-dependent.

6.7.2. Soil Liquid Water Content

An important parameter that affects ground temperatures, and therefore pollutant concentrations, is soil liquid water. Increases in soil liquid water cool the ground, reducing convection, decreasing mixing depths, and slowing near-surface winds. Thinner mixing depths and slower winds enhance pollutant buildup. Conversely, decreases in soil liquid water increase convection, increasing mixing depths and speeding near-surface winds, reducing pollution.

Soil water cools the ground in two major ways. First, evaporation of liquid water in soil cools the soil. Therefore, the more liquid water a soil has, the greater the evaporation and cooling of the soil during the day. Second, liquid water in soil increases the average specific heat of a soil–air–water mixture. The wetter the soil, the less the soil can heat up when solar radiation is added to it.

In a study of the effects of soil liquid water on temperatures, winds, and pollution, it was found that increases in soil water of only 4 percent decreased peak near-surface air temperatures by up to 6 °C, decreased wind speeds by up to 1.5 m/s, delayed the times of peak ozone mixing ratio by up to two hours, and increased the magnitude of peak particulate concentrations substantially in Los Angeles over a two-day period

(Jacobson, 1999b). Such results imply that rainfall, irrigation, and climate change all affect pollution concentrations.

6.7.3. Urban Heat Island Effect

Landcover affects ground temperature, which affects pollutant concentrations. Most of the globe is covered with water (71.3 percent) or snow/ice (3.3 percent). The remainder is covered with forests, grasslands, croplands, wetlands, barren lands, tundra, savanna, shrubland, and urban areas. Urban surfaces consist primarily of roads, walkways, rooftops, vegetation cover, and bare soil. Urban construction material surfaces increase surface temperatures due to the urban heat island effect, first recorded in 1807 by English meteorologist Luke Howard (1772–1864) (also known for classifying clouds). Howard measured temperatures at several sites within and outside London and found that temperatures within the city were consistently warmer than were those outside.

Figure 6.18 shows a satellite-derived image of daytime surface temperatures in downtown Atlanta and its environs. Road surfaces and buildings stand out as being particularly hot. Urban areas heat up during the day and night to a greater extent than do surrounding vegetated areas because urban areas generally contain less liquid water than do surrounding vegetated areas (Oke, 1988, 1999). When liquid water is absent, evaporation rates and corresponding energy losses from the surface are slower than when liquid water is present, warming surface temperatures.

Increased urban temperatures result in increased mixing depths, faster near-surface winds, and lower near-surface concentrations of pollutants. Increased urban temperatures may also be responsible for enhanced thunderstorm activity (Bornstein and Lin, 2000).

Figure 6.18. Landsat 5 satellite image with buildings superimposed showing daytime surface temperatures in Atlanta, Georgia. Temperatures from hot to cold are represented by white, red, yellow, green, and blue, respectively. Small cool areas in the middle of warm areas are often due to shadows caused by large buildings.

6.7.4. Local Winds

Another factor that affects air pollution is the local wind. Winds arise due to pressure gradients. Although large-scale pressure gradients affect winds, local pressure gradients, resulting from uneven ground heating, variable topography, and local turbulence, can modify or override large-scale winds. Important local winds include sea, lake, bay, land, valley, and mountain breezes.

6.7.4.1. Sea, Lake, and Bay Breezes

Sea, lake, and bay breezes form during the day between oceans, lakes, or bays, respectively, and land. Figure 6.19 illustrates a basic sea-breeze circulation. During the day, land heats up relative to water because land has a lower specific heat than does water. Rising air over land forces air aloft to diverge horizontally, decreasing surface-air pressures (setting up a shallow thermal low-pressure system) over land. As a result of the pressure gradient between land and water, air moves from the water, where the pressure is now relatively high, toward the land. In the case of ocean water meeting land, the movement of near-surface air is the **sea breeze**. Although the ACoF acts on the sea-breeze air, the distance traveled by the sea-breeze is too short (a few tens of kilometers) for the Coriolis force to turn the air noticeably.

Meanwhile, some of the diverging air aloft over land returns toward the water. The convergence of air aloft over water increases surface air pressure over water, prompting a stronger flow of surface air from the water to the land, completing the basic sea-breeze circulation cell. At night, land cools to a greater extent than does water, and all the pressures and flow directions in Fig. 6.19 reverse themselves, creating a **land breeze**, a near-surface flow of air from land to water.

Figure 6.19 illustrates that a basic sea-breeze circulation cell can be embedded in a large-scale sea-breeze cell. The Los Angeles Basin, for example, is bordered on its southwestern side by the Pacific Ocean and on its eastern side by the San Bernardino Mountains. The Mojave Desert lies to the east of the mountains. The desert heats up more than does land near the coast during the day, creating a thermal low over the desert, drawing air in from the coast, creating the circulation pattern shown in the figure.

Figure 6.20 illustrates the variation of sea- and land-breeze wind speeds at Hawthorne, California, near the coast in the Los Angeles Basin, over a three-day period. Sea-breeze wind speeds peak in the afternoon, when land–ocean temperature

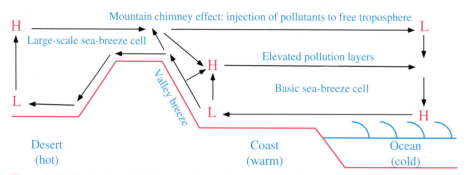

Figure 6.19. Illustration of a large-scale sea-breeze circulation cell, basic sea-breeze circulation cell, valley breeze, the chimney effect, and the formation of elevated pollution layers, as described in the text. Pressures shown (L and H) are relative to other pressures in the horizontal.

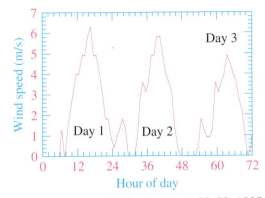

Figure 6.20. Wind speeds at Hawthorne, California, August 26–28, 1987.

differences peak. Similarly, sea-breeze winds peak in summer and are minimum in winter. Land breezes are relatively weak in comparison with sea breezes because at night, the land–ocean temperature difference is small.

6.7.4.2. Valley and Mountain Breezes

A **valley breeze** is a wind that blows from a valley up a mountain slope and results from the heating of the mountain slope during the day. Heating causes air on the mountain slope to rise, drawing air up from the valley to replace the rising air. Figure 6.19 illustrates that in the case of a mountain near the coast, a valley breeze can become integrated into a large-scale sea-breeze cell. The opposite of a valley breeze is a **mountain breeze**, which originates from a mountain slope and travels downward. Mountain breezes typically occur at night, when mountain faces cool rapidly. As a mountain face cools, air above the face also cools and drains downslope.

6.7.4.3. Effects of Sea and Valley Breezes on Pollution

In the Los Angeles Basin, sea and valley breezes are instrumental in transferring primary pollutants, emitted mainly from source regions on the west side of the basin, to receptor regions on the east side of the basin, where they arrive as secondary pollutants. Figure 4.10 showed an example of the daily variation of $NO(g)$, $NO_2(g)$, and $O_3(g)$ at a source and receptor region in Los Angeles. The valley breeze, in particular, moves ozone from Fontana, Riverside, and San Bernardino, in the eastern Los Angeles Basin, up the San Bernardino Mountains to Crestline, one of the most polluted locations in the United States in terms of ozone.

6.7.4.4. Chimney Effect and Elevated Pollution Layers

Pollutants in an enclosed basin, such as the Los Angeles Basin, can escape the basin through mountain passes and over mountain ridges. A third mechanism of escape is through the **mountain chimney effect** (Lu and Turco, 1995). Through this effect, a mountain slope is heated, causing air containing pollutants to rise, injecting the pollutants from within the mixed layer into the free troposphere. The mountain chimney effect is illustrated in Fig. 6.19.

Instead of dispersing, gases and particles may build up in **elevated pollution layers**. Pollutant concentrations in these layers often exceed those near the ground. Fig. 6.21 shows a brilliant sunset through an elevated pollution layer in Los Angeles. Elevated layers form in one of at least four ways.

Figure 6.21. Brilliant sunset through elevated pollution layer over Los Angeles, California, May 1972. Photo by Gene Daniels, U.S. EPA, available from the Still Pictures Branch, U.S. National Archives.

Figure 6.22. Elevated pollution layer formed over Long Beach, California, on July 22, 2000, by a sea-breeze circulation.

Figure 6.23. Elevated layer of smoke trapped in an inversion layer following a greenhouse fire in Menlo Park, California in June 2001.

First, sea-breeze circulations themselves can form elevated layers by lifting and injecting polluted air into the inversion layer during its return flow to the ocean. Elevated pollution layers formed by this mechanism have been reported in Tokyo (Wakamatsu et al., 1983), Athens (Lalas et al., 1983), and near Lake Michigan (Fitzner et al., 1989; Lyons et al., 1991). Figure 6.22 shows an example of an elevated pollution layer caused by a sea-breeze circulation over Long Beach, California.

Second, some of the air forced up a mountain slope by winds may rise into and spread horizontally in an inversion layer. Of the air that continues up the mountain slope past the inversion, some may circulate back down into the inversion (Lu and Turco, 1995). Elevated pollution layers formed by this mechanism have been observed adjacent to the San Bernardino and San Gabriel Mountain Ranges in Los Angeles (Wakimoto and McElroy, 1986).

Third, emissions from a smokestack or fire can rise buoyantly into an inversion layer. The inversion limits the height to which the plume can rise and forces the plume to spread horizontally. Figure 6.23 shows an example of a pollution layer formed by this mechanism.

Fourth, elevated ozone layers in the boundary layer may form by the destruction of surface ozone. During the afternoon, ozone dilutes uniformly throughout a mixing depth.

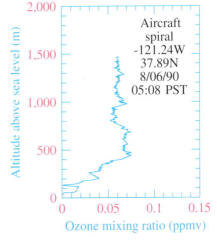

Figure 6.24. Vertical profile of ozone mixing ratio over Stockton, California, on August 6, 1990, at 05:08 Pacific Standard Time (PST) showing a nighttime elevated ozone layer that formed by the destruction of surface ozone. Data from the SARMAP field campaign (see Solomon and Thuillier, 1995).

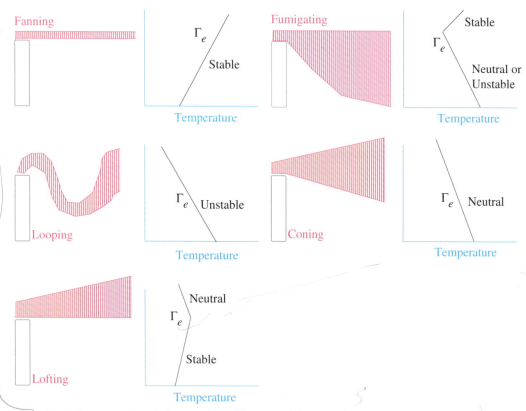

Figure 6.25. Cross-sections of plumes under different stability conditions.

In the evening, cooling of the ground stabilizes the air near the surface without affecting the stability aloft. In regions of nighttime NO(g) emissions, the NO(g) destroys near-surface ozone. Because nighttime air near the surface is stable, ozone aloft does not mix downward to replenish the lost surface ozone. Figure 6.24 shows an example of an elevated ozone layer formed by this process. The next day, the mixing depth increases, recapturing the elevated ozone and mixing it downward. McElroy and Smith (1992) estimated that daytime downmixing of elevated ozone enhanced surface ozone in certain areas of Los Angeles by 30 to 40 ppbv above what they would otherwise have been due to photochemistry alone. When few nighttime sources of NO(g) exist, elevated ozone layers do not form by this mechanism.

Figure 6.26. Lofting of pollution from a power plant using coal to generate electricity. Photo by National Renewable Energy Laboratory Staff.

6.7.5. Plume Dispersion

Another effect of local meteorology on air pollution is the effect of stability on plume dispersion from smokestacks. Figure 6.25 illustrates cross-sections of classic plume configurations that result from different stability conditions. In stable air, pollutants emitted from stacks do not rise or sink vertically; instead, they **fan** out horizontally. Viewed from above, the

Figure 6.27. Plumes rising into an inversion layer in Long Beach, California, on December 19, 2000.

pollution distribution looks like a giant fan. Fanning does not expose people to pollution immediately downwind of a stack because the fanned pollution does not mix down to the surface. Figure 6.23 shows the side view of a fanning plume originating from a fire, not a stack. The worst meteorologic condition for those living immediately downwind of a stack occurs when the atmosphere is neutral or unstable below the stack and stable above. This condition leads to **fumigation**, or the downwashing of pollutants toward the surface. **Looping** occurs when the atmosphere is unstable. Under such a condition, emissions from a stack may alternately rise and sink, depending on the extent of turbulence. **Coning** occurs when the atmosphere is neutrally stratified. Under such a condition, pollutants emitted tend to slowly disperse both upward and downward. When the atmosphere is stable below and neutral above a stack, **lofting** occurs. Lofting results in the least potential exposure to pollutants by people immediately downwind of a stack. Figure 6.26 shows an example of lofting from a smokestack. Figure 6.27 shows several plumes lofting into an inversion layer in the Los Angeles Basin.

6.8. SUMMARY

In this chapter, the forces acting on winds, the general circulation of the atmosphere, the generation of large-scale pressure systems, the effects of large-scale pressure systems on air pollution, and the effects of small-scale weather systems on air pollution were discussed. Forces examined include the pressure gradient force, apparent Coriolis force, apparent centrifugal force, and friction forces. The winds produced by these forces include the geostrophic wind, the gradient wind, and surface winds. On the global scale, the south–north and vertical components of wind flow can be described by three circulation cells – the Hadley, Ferrel, and Polar cells – in each the Northern and Southern Hemispheres. Descending and rising air in the cells result in the creation of large-scale semipermanent high-and low-pressure systems over the oceans. Pressure systems over continents (thermal pressure systems) are seasonal and controlled by ground heating and cooling. Large-scale pressure systems affect pollution through their effects on stability and winds. Surface high-pressure systems enhance the stability of the boundary layer and increase pollution buildup. Surface low-pressure systems reduce the stability of the boundary layer and decrease pollution buildup. Soil moisture affects stability, winds, and pollutant concentrations. Sea and valley breezes are important local-scale winds that enhance or clear out pollution in an area. Elevated pollution layers, often observed, are a product of the interaction of winds, stability, topography, and chemistry.

6.9. PROBLEMS

HW 6.1.   Draw a diagram showing the forces and the gradient wind around an elevated low-pressure center in the Southern Hemisphere.

HW 6.2.   Draw a diagram showing the forces and resulting winds around a surface low-pressure center in the Southern Hemisphere.

HW 6.3.   Should horizontal winds turn clockwise or counterclockwise with increasing height in the Southern Hemisphere? Demonstrate your conclusion with a force diagram at the surface and aloft.

6.4.   Characterize the stability of each of the six layers in Fig. 6.28 with one of the stability criteria given in Equation 6.4.

HW 6.5.   If the Earth did not rotate, how would you expect the current three-cell model of the general circulation of the atmosphere to change?

6.6.   In Fig. 6.28, to approximately what altitude will a parcel of pollution rise adiabatically from the surface in unsaturated air if the parcel's initial temperature is 25 °C? What if its initial temperature is 30 °C?

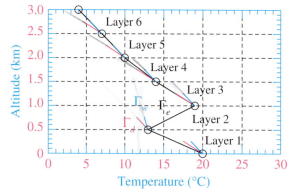

Figure 6.28. Environmental temperature profile with wet and dry adiabatic lapse rate profiles superimposed.

6.7.   In Fig. 6.28, to approximately what altitude will a parcel of pollution rise adia-
        batically from the surface in saturated air if the parcel's initial temperature is
        25 °C? What if its initial temperature is 30 °C? Assume a wet adiabatic lapse
        rate of 6 °C km^{-1}.

6.8.   Why do ozone mixing ratios peak in the afternoon, even though mixing depths
        are usually maximum in the afternoon?

6.9.   Explain why ozone mixing ratios are higher on the east side of the Los Angeles
        Basin, even though ozone precursors are emitted primarily on the west side of
        the basin.

6.10.   Summarize the meteorologic conditions that tend to enhance and reduce,
        respectively, photochemical smog buildup in an area.

6.11.   If emissions did not change with season, why might CO(g) and aerosol particle
        concentrations be higher in winter than in summer?

6.12.   If a diesel-engine truck is driving in front of you, and much of the exhaust is
        entering your car and not rising in the air, how would you define the plume-
        dispersion shape?

6.13.   From Fig. 6.28, determine the following quantities:

        (a) inversion base height
        (b) inversion top height
        (c) inversion base temperature
        (d) inversion top temperature
        (e) inversion strength
        (f) inversion thickness
        (g) mixing depth.

EFFECTS OF POLLUTION ON VISIBILITY, ULTRAVIOLET RADIATION, AND ATMOSPHERIC OPTICS

isibility, ultraviolet (UV) radiation intensity, and optical phenomena are affected by gases, aerosol particles, and hydrometeor particles interacting with solar radiation. In clean air, gases and particles affect how far we can see along the horizon and the colors of the sky, clouds, and rainbows. In polluted air, gases and aerosol particles affect visibility, optical phenomena, and UV radiation intensity. In this chapter, visibility, optics, and UV transmission in clean and polluted atmospheres are discussed. An understanding of these phenomena requires a study of the interaction of solar radiation with gases, aerosol particles, and hydrometeor particles through several optical processes, including reflection, refraction, diffraction, dispersion, scattering, absorption, and transmission. These processes are described next.

7.1 PROCESSES AFFECTING SOLAR RADIATION IN THE ATMOSPHERE

The solar spectrum is divided into **UV** (0.01 to 0.38 μm), **visible** (0.38 to 0.75 μm), and **near-infrared (IR)** (0.75 to 4.0 μm) wavelength ranges (Section 2.2).

In 1666, **Sir Isaac Newton** (1642–1727; Fig. 7.1), an English physicist and mathematician, showed that when white, visible light passed through a glass prism, each wavelength of the light bent to a different degree, resulting in the separation of the white light into a variety of colors that he called the **light spectrum**. Although the spectrum is continuous (the eye can distinguish 10 million colors), Newton discretized the spectrum into seven colors: red, orange, yellow, green, blue, indigo, and violet, to correspond to the seven notes on a musical scale. When the colors of the spectrum were recombined, they reproduced white light.

Newton also defined **primary colors** as blue, green, and red. When pairs of primary colors are added together, they produce new colors. For example, equal amounts of red plus green produce yellow; equal amounts of red plus blue produce magenta; and equal amounts of green plus blue produce cyan. For simplicity, the visible spectrum here is divided into wavelengths of the primary colors blue (0.38 to 0.5 μm), green (0.5 to 0.6 μm), and red (0.6 to 0.75 μm).

Figure 7.1. Sir Isaac Newton (1642–1727).

The visible spectrum provides most of the energy that keeps the Earth warm. The visible spectrum also affects the distance we can see and colors in the atmosphere. It is no coincidence that the acuity (keenness) of our vision peaks at 0.55 μm, in the green part of the visible spectrum, which is near the wavelength of the sun's peak radiation intensity. Our eyes have evolved to take advantage of the peak intensity in this part of the visible spectrum.

When radiation passes through the Earth's atmosphere, it is attenuated or redirected by absorption and scattering by gases, aerosol particles, and hydrometeor particles. These processes are discussed next.

Table 7.1. Wavelengths of Absorption in the Visible and UV Spectra by Some Atmospheric Gases

Gas Name	Chemical Formula	Absorption Wavelengths (μm)
Visible/Near-UV/Far-UV absorbers		
Ozone	$O_3(g)$	<0.35,0.45–0.75
Nitrate radical	$NO_3(g)$	<0.67
Nitrogen dioxide	$NO_2(g)$	<0.71
Near-UV/Far-UV absorbers		
Formaldehyde	$HCHO(g)$	<0.36
Nitric acid	$HNO_3(g)$	<0.33
Far-UV absorbers		
Molecular oxygen	$O_2(g)$	<0.245
Carbon dioxide	$CO_2(g)$	<0.21
Water	$H_2O(g)$	<0.21
Molecular nitrogen	$N_2(g)$	<0.1

7.1.1. Gas Absorption

Absorption occurs when radiative energy (such as from the sun or Earth) enters a substance and is converted to internal energy, increasing the temperature of the substance. Absorption removes radiation from an incident beam, reducing the amount of radiation received past the point of absorption. Next, absorption by gases is discussed with respect to the solar spectrum.

7.1.1.1. Gas Absorption at Ultraviolet and Visible Wavelengths

Gases selectively absorb radiation in different portions of the electromagnetic spectrum. Table 7.1 lists selected gases that absorb visible or UV radiation. Of the gases, all absorb UV, but only a few absorb visible radiation. Gases that absorb visible or UV radiation are often, but not always, photolyzed by this radiation into simpler products.

The gases that affect UV radiation the most are molecular oxygen [$O_2(g)$], molecular nitrogen [$N_2(g)$], and ozone [$O_3(g)$]. Oxygen absorbs wavelengths shorter than 0.245 μm and nitrogen absorbs wavelengths shorter than 0.1 μm. Oxygen and nitrogen prevent nearly all solar wavelengths shorter than 0.245 μm from reaching the troposphere.

In 1880, M. J. Chappuis found that ozone absorbed visible radiation between the wavelengths of 0.45 and 0.75 μm. The absorption bands in this wavelength region are now called Chappuis bands. In 1881, John Hartley first suggested that ozone was present in the upper atmosphere, and hypothesized that the reason that the Earth's surface received little radiation shorter than 0.31 μm was because ozone absorbed these wavelength. The absorption bands of ozone below 0.31 μm are now called the Hartley bands. In 1916, English physicists Alfred Fowler (1868–1940) and Robert John Strutt (1875–1947, the son of Lord Rayleigh) showed that ozone also weakly absorbed between 0.31 and 0.35 μm wavelengths. The bands of absorption in this region are now called the Huggins bands. Ozone absorption of UV and solar radiation, followed by reemission of thermal-IR radiation, heats the stratosphere, causing the stratospheric temperature inversion (Fig. 3.3). Ozone absorption also protects the surface of the Earth by preventing nearly all UV wavelengths 0.245 to 0.29 μm and most wavelengths 0.29 to 0.32 μm from reaching the troposphere.

Although the gases in Table 7.1, aside from $O_3(g)$, $O_2(g)$, and $N_2(g)$, absorb UV radiation, their mixing ratios are too low to have much effect on such radiation. For instance, stratospheric mixing ratios of **water vapor** (0 to 6 ppmv) are much lower than are those of $O_2(g)$, which absorbs many of the same wavelengths as does water vapor. Thus, water vapor has little effect on UV attenuation in the stratosphere. Similarly, stratospheric mixing ratios of **carbon dioxide** (370 ppmv) are much lower than are those of $O_2(g)$ or $N_2(g)$, both of which absorb the same wavelengths as does carbon dioxide.

The only gas that absorbs visible radiation sufficiently to affect visibility is **nitrogen dioxide** [$NO_2(g)$], but its effect is important only in polluted air where its mixing ratios are sufficiently high. Mixing ratios of the **nitrate radical** [$NO_3(g)$], which also absorbs visible radiation, are low, except at night or in the early morning. Thus, $NO_3(g)$ does not affect visibility. Although ozone mixing ratios can be high, ozone is a relatively weak absorber of visible light.

7.1.1.2. Gas Absorption Extinction Coefficient

The determination of visibility depends on all processes that attenuate or enhance radiation. In this subsection, attenuation by gas absorption is briefly discussed.

Figure 7.2 illustrates how radiation passing through a gas is reduced by absorption.

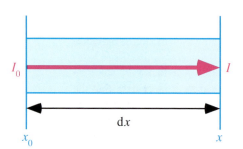

Figure 7.2. Attenuation of incident radiance I_0 due to absorption in a column of gas.

Suppose incident radiation of intensity I_0 travels a distance $dx = x - x_0$ through a uniformly mixed absorbing gas q of number concentration N_q (molecules per cubic centimeter of air). As the radiation passes through the gas, gas molecules intercept it. If the molecules absorb the radiation, the intensity of the radiation diminishes. The ability of a gas molecule to absorb radiation is embodied in the absorption cross section of the gas, $b_{a,g,q}$ (cm^2 per molecule), where the subscripts a, g, and q mean absorption, by a gas, and by gas q, respectively. Wavelength subscripts were omitted. An **absorption cross section** of a gas is an effective cross section that results in radiation reduction by absorption. Its size is on the order of, but usually not equal to, the real cross-sectional area of a gas molecule.

The product of the number concentration and absorption cross section of a gas is called an *absorption extinction coefficient*. An **extinction coefficient** measures the loss of electromagnetic radiation due to a specific process per unit distance. Extinction coefficients, symbolized with σ, have units of inverse distance (cm^{-1}, m^{-1}, or km^{-1}) and vary with wavelength. The **absorption extinction coefficient** (cm^{-1}) of gas q is

$$\sigma_{a,g,q} = N_q b_{a,g,q} \qquad (7.1)$$

The gas absorption extinction coefficient due to the sum of all gases ($\sigma_{a,g}$) is the sum of Equation 7.1 over all absorbing gases. The greater the absorption extinction coefficient of a gas in the visible spectrum, the more the gas reduces visibility. Figure 7.3 shows absorption extinction coefficients of both $NO_2(g)$ and $O_3(g)$, at two mixing ratios. The figure indicates that nitrogen dioxide affects extinction (and therefore visibility)

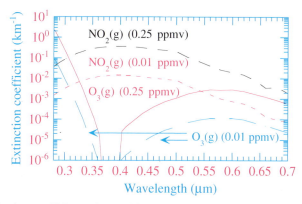

Figure 7.3. Extinction coefficients due to $NO_2(g)$ and $O_3(g)$ absorption when $T = 298$ K and $p_a = 1013$ mb.

primarily at high mixing ratios and at wavelengths below about 0.5 μm. In polluted air, such as in Los Angeles, $NO_2(g)$ mixing ratios typically range from 0.01 to 0.1 ppmv and peak near 0.15 ppmv during the morning. A typical value is 0.05 ppmv. $O_3(g)$ has a larger effect on extinction than does $NO_2(g)$ at wavelengths below about 0.32 μm. $O_3(g)$ mixing ratios in polluted air usually peak between 0.05 and 0.25 ppmv.

The reduction in radiation intensity (where I denotes intensity) at a given wavelength with distance through an absorbing gas can now be defined as

$$I = I_0 e^{-\sigma_{a,g,q}(x-x_0)} = I_0 e^{-N_q b_{a,g,q}(x-x_0)} \tag{7.2}$$

This equation states that incident radiation I_0 is reduced by a factor $e^{-\sigma_{a,g,q}(x-x_0)}$ between points x_0 and x by gas absorption. From this equation, it is evident that the absorption cross section of a gas can be determined experimentally by measuring the attenuation of radiation through a homogenous gas column, such as the one shown in Fig. 7.2, of known path length $N_q(x - x_0)$ (molecules cm^{-2}), then extracting the cross section from Equation 7.2.

EXAMPLE 7.1.

Find the fraction of incident radiation intensity (the transmission) passing through a uniform gas column of length 1 km when the number concentration of the gas is $N_q = 10^{12}$ molecules cm^{-3} and the absorption cross section of a gas molecule is $b_{a,g,q} = 10^{-19}$ cm^2 per molecule.

Solution

From Equation 7.1, the extinction coefficient through the gas is $\sigma_{a,g,q} = N_q b_{a,g,q} = 10^{-7}$ cm^{-1}. From Equation 7.2, the transmission is $I/I_0 = e^{-10^{-7} \times 10^5} = 0.99$. Thus 99 percent of incident radiation traveling through this column reaches the other side.

7.1.2. Gas Scattering

Gas scattering is the redirection of radiation by a gas molecule without a net transfer of energy to the molecule. When a gas molecule scatters, incident radiation is redirected symmetrically in the forward and backward direction and somewhat off to the side, as shown in Fig. 7.4. The figure shows the probability distribution of incident energy scattered by a gas molecule.

The scattering of radiation by gas molecules or by aerosol particles much smaller than the wavelength of light is called **Rayleigh scattering**. Because gas molecules are on the order of 0.0005 μm in diameter and a typical wavelength of visible light is 0.5 μm, all gas molecules are Rayleigh scatterers. Because molecular nitrogen [$N_2(g)$], molecular oxygen [$O_2(g)$], argon [$Ar(g)$], and water vapor [$H_2O(g)$] are the most abundant gases in the air, these are the most important Rayleigh scatterers. Rayleigh scattering is named after **Lord Baron Rayleigh**, born John William Strutt (1842–1919; Fig. 7.5), who also discovered argon gas with Sir William Ramsay in 1894 (Chapter 1). Rayleigh's theoretical work on gas scattering was published in 1871, (Rayleigh, 1871) 23 years before his discovery of argon.

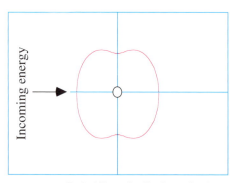

Figure 7.4. Probability distribution of where incident energy is scattered by a gas molecule (located at the center of the diagram).

Figure 7.5. Lord Baron Rayleigh (John William Strutt) (1842–1919).

7.1.2.1. Gas-Scattering Extinction Coefficient

The **gas scattering extinction coefficient** (cm^{-1}) of total air (Rayleigh scattering extinction coefficient) is analogous to that of the gas absorption extinction coefficient, and is defined as

$$\sigma_{s,g} = N_a b_{s,g} \tag{7.3}$$

where N_a is the number concentration of air molecules (molecules cm^{-3}) at a given altitude and $b_{s,g}$ is the scattering cross section (cm^2 molecule^{-1}) of a typical air molecule. The **scattering cross section** of an air molecule is an effective cross section that results in radiation reduction by scattering, and is proportional to the inverse of the fourth power of the wavelength. This means that gas molecules in the air scatter short (blue) wavelengths preferentially over long (red) wavelengths.

7.1.2.2. Colors of the Sky and Sun

The variation of the Rayleigh scattering cross section with wavelength explains why the sun appears white at noon, yellow in the afternoon, and red at sunset and why the sky is blue. White sunlight that enters the Earth's atmosphere travels a shorter distance before reaching a viewer's eye at noon than at any other time during the day, as illustrated in Fig. 7.6. During white light's travel through the atmosphere, blue wavelengths are preferentially scattered out of the direct beam by air molecules, but not enough blue light is scattered for a person looking at the sun to notice that the incident beam has changed its color from white. Thus, a person looking at the sun at noon often sees a **white sun**. The blue light that scatters out of the direct beam is scattered by gas molecules multiple times, and some of it eventually enters the viewer's eye when the viewer looks away from the sun. As such, a viewer looking away from the sun sees a **blue sky**.

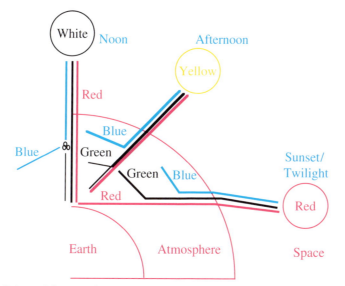

Figure 7.6. Colors of the sun. At noon, the sun appears white because red, green, and some blue light transmit to a viewer's eye. In the afternoon, sunlight traverses a longer path in the atmosphere, removing more blue. At sunset, most green is removed from the line of sight, leaving a red sun. After sunset, the sky appears red due to refraction between space and the atmosphere.

Although Lord Rayleigh formalized the theory of scattering by gases and aerosol particles much smaller than the wavelength of light, it may have been **Leonardo DaVinci** (1452–1519) who first suggested that blue colors in the sky were due to interactions of sunlight with atmospheric constituents. He wrote (c. 1500),

> I say that the blueness we see in the atmosphere is not intrinsic color, but is caused by warm vapor evaporated in minute and insensible atoms on which the solar rays fall, rendering them luminous against the infinite darkness of the fiery sphere which lies beyond and includes it. . . . If you produce a small quantity of smoke from dry wood and the rays of the sun fall on this smoke and if you place (behind it) a piece of black velvet on which the sun does not fall, you will see that the black stuff will appear of a beautiful blue color. . . . Water violently ejected in a fine spray and in a dark chamber where the sunbeams are admitted produces then blue rays. . . . Thus it follows, as I say, that the atmosphere assumes this azure hue by reason of the particles of moisture which catch the rays of the sun.

What DaVinci saw, however, was not scattering by gas molecules but scattering by small aerosol particles, namely, dry wood-smoke particles and small liquid-water drops. Because such particles are smaller than the wavelength of light, they preferentially scatter blue light as do gas molecules. Small aerosol particles are well known to scatter blue light preferentially. Preferential scattering of blue light by small aerosol particles in the air is responsible for the blue appearance of the **Blue Ridge Mountains** in Virginia and the **Blue Mountains** in Australia. Occasionally after a forest fire, a **blue moon** appears due to the scattering of blue light by small organic aerosol particles. Similarly, after a heavy rain, the relative humidity is low, and aerosol particles lose much of their water, shrinking their size enough to preferentially scatter blue light and giving the sky a deep blue appearance.

Figure 7.7. Yellow sun in the early evening off the island of Maui in the South Pacific Ocean.

In the afternoon, light takes a longer path through the air than it does at noon; thus, more blue and some green light is scattered out of the direct solar beam in the afternoon than at noon. Although a single gas molecule is less likely to scatter a green than a blue wavelength, the number of gas molecules along a viewer's line of sight is so large in the afternoon that the probability of green light scattering is sizable. Nearly all red and some green are still transmitted to the viewer's eye in the afternoon, causing the sun to appear yellow. In clean air, the sun can remain yellow until just before it reaches the horizon, as shown in Fig. 7.7.

When the sun reaches the horizon at sunset, sunlight traverses its longest distance through the atmosphere, and all blue and green and some red wavelengths are scattered out of the sun's direct beam. Only some direct red light transmits, and a viewer sees a **red sun**. Sunlight can be seen after sunset because sunlight refracts as it enters Earth's atmosphere, as illustrated in Fig. 7.6. **Refraction** is the bending of light as it passes from a medium of one density to a medium of a different density, and is discussed in Section 7.1.4.2.

During and after sunset, the whole horizon often appears red (Fig. 7.8) due to the presence of aerosol particles along the horizon. Aerosol particles scatter blue, green, and red wavelengths of direct sunlight, sending such light in all directions along the horizon. At all points along the horizon, additional particles re-scatter some of the scattered blue, green, and red light toward the viewer's eye. As these wavelengths pass through the air, blue and green are scattered preferentially out of the viewer's line of site by gas molecules, whereas most red wavelengths are transmitted, causing the horizon to appear red. Red horizons are common over the ocean, where sea-spray particles are present, as well as over land, where soil and pollution particles are present.

The time after sunset during which the sky is still illuminated due to refraction is called **twilight**. Twilight also occurs before sunrise. The length of twilight increases

Figure 7.8. Red horizon after the sun dips below a deck of stratus clouds over the Pacific Ocean, signifying the beginning of twilight.

with increasing latitude. At midlatitudes in the summers, the length of twilight is about one-half hour. At higher latitudes in summers, twilight in the morning may merge with that in the evening, creating a twilight that lasts all night, called a **white night**.

7.1.3. Aerosol and Hydrometeor Particle Absorption

All aerosol and hydrometeor particles absorb thermal-IR and near-IR radiation, but only a few absorb visible and UV radiation. Next, visible-and UV-absorption properties of aerosol particles and the effect of aerosol particle absorption on UV radiation and pollution are discussed.

7.1.3.1. Important Absorbers of Visible and UV Radiation

The strongest aerosol particle absorber in the visible spectrum is black carbon. Other visible absorbers include **hematite** [$Fe_2O_3(s)$] and **aluminum oxide** [alumina-$Al_2O_3(s)$]. Hematite is found in many soil-dust particles. Some forms of aluminum oxide are found in soil-dust particles and others are found in combustion particles.

Aerosol particle absorbers in the UV spectrum include black carbon, hematite, aluminum oxide, and certain organic compounds. The organics absorb UV radiation but little visible radiation. The strongest near-UV absorbing organics include certain **nitrated aromatics**, **polycyclic aromatic hydrocarbons** (PAHs), benzaldehydes, benzoic acids, aromatic polycarboxylic acids, and phenols (Jacobson, 1999c).

Most particulate components are weak absorbers of visible and UV radiation. **Silicon dioxide** [$SiO_2(s)$, silica], which is the white, colorless, crystalline compound found in quartz, sand, and other minerals, is an example. **Sodium chloride** [$NaCl(s)$], **ammonium sulfate** [$(NH_4)_2SO_4(s)$], and **sulfuric acid** [$H_2SO_4(aq)$] are also weak absorbers of visible and UV radiation. Soil-dust particles, which contain $SiO_2(s)$, $Al_2O_3(s)$, $Fe_2O_3(s)$, $CaCO_3(s)$, $MgCO_3(s)$, and other substances, are moderate absorbers of such radiation. Their absorptivity increases from the visible to the UV spectra (e.g., Gillette et al., 1993; Sokolik et al. 1993).

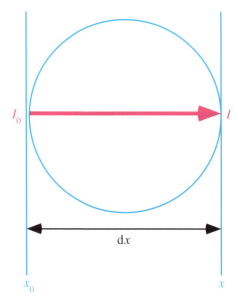

I_0

I

dx

x_0

x

Figure 7.9. Attenuation of incident radiation I_0 due to absorption by an aerosol particle.

7.1.3.2. The Imaginary Refractive Index

Figure 7.9 shows a possible path of radiation through a single spherical aerosol-particle.

The attenuation due to absorption of incident radiation of wavelength λ as it travels through a single particle is

$$I = I_0 e^{-4\pi\kappa\,(x-x_0)/\lambda} \tag{7.4}$$

where κ is the imaginary index of refraction, $x - x_0$ is the distance through the particle, and I_0 is the initial radiation intensity. The **imaginary index of refraction** is a measure of the extent to which a particle absorbs radiation. The term $4\pi\kappa/\lambda$ is an absorption extinction coefficient for a single aerosol-particle. Table 7.2 gives imaginary refractive indices for some substances at wavelengths of 0.50 and 10 μm. Black carbon has the largest imaginary indices of refraction among the substances shown.

EXAMPLE 7.2

Find the fraction of incident radiation intensity that transmits through a uniform particle of diameter 0.1 μm at a wavelength of 0.5 μm when the particle is composed of (a) black carbon and (b) water.

Solution

From Table 7.2, the imaginary refractive indices of black carbon and water at $\lambda = 0.5$ μm are $\kappa = 0.74$ and 10^{-9}, respectively. From Equation 7.4, the transmission of light through (a) black carbon is $I/I_0 = e^{-4\pi \times 0.74 \times 0.1/0.5} = 0.16$ and that through (b) liquid water is 0.999999997. Thus, a 0.1 μm black carbon particle absorbs 84 percent of incident radiation, whereas a 0.1 μm water particle absorbs only 0.0000003 percent of incident visible radiation passing through it. As such, black carbon is a strong absorber of visible light, but liquid water is not.

Table 7.2. Real and Imaginary Indices of Refraction for Some Substances at $\lambda = 0.50$ and 10.0 μm

	$\lambda = 0.5$ μm		$\lambda = 10$ μm	
	Real (n)	Imaginary (κ)	Real (n)	Imaginary (κ)
$H_2O(aq)$[a]	1.34	1.0×10^{-9}	1.22	0.05
Black Carbon (s)[b]	1.82	0.74	2.40	1.0
Organic Matter(aq,s)[b]	1.45	0.001	1.77	0.12
$H_2SO_4(aq)$[b]	1.43	1.0×10^{-8}	1.89	0.46

[a]Hale and Querry (1973).
[b]Krekov (1993).

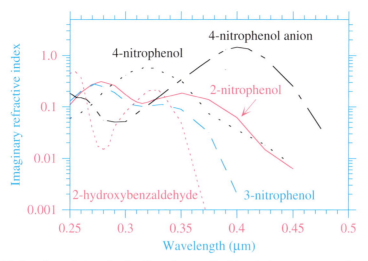

Figure 7.10. Imaginary index of refraction of some liquid organics versus wavelength. From Jacobson (1999c).

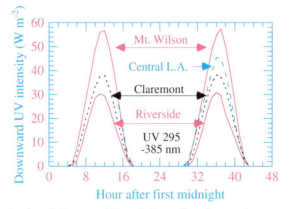

Figure 7.11. Effect of pollution on UV radiation. The figure shows measurements of downward UV radiation intensity at four sites in Los Angeles, California, on August 27–28, 1987. The three nonmountain sites were progressively further inland. The elevations of the four sites were as follows: Mt. Wilson, 1.7 km; Central Los Angeles, 87 m; Claremont, 364 m; Riverside, 249 m. UV radiation between the mountain and surface was reduced the most at Riverside, where most pollution occurred. Measurements in Central Los Angeles were available only on the second day. Data were provided by the California Air Resources Board.

7.1.3.3. Effects of Aerosol-Particle Absorption on UV Radiation

Figure 7.10 shows the imaginary refractive indices of several liquid organic compounds versus wavelength. The figure indicates that these organics preferentially absorb UV radiation.

Absorption of UV radiation by organics in aerosol particles affects the amount of UV radiation reaching the ground in smog. Measurements in the Los Angeles Basin in 1987, shown in Fig. 7.11, indicate that UV radiation 0.295 to 0.385 μm was reduced by 22 percent in central Los Angeles (near the coast), 33 percent at Claremont (further inland), and 48 percent at Riverside (much further inland) in comparison with UV radiation at Mount Wilson, 1.7 km above the basin. Measurements also indicated that total solar radiation 0.285 to 2.8 μm was reduced by only 8 to 14 percent between Riverside and Mount Wilson. Some components of photochemical smog appeared to have preferentially reduced UV radiation over total solar radiation.

Whereas some of the preferential UV reductions were due to gas absorption, Rayleigh scattering, and aerosol-particle scattering, the sum of these effects could not account for the up to 50 percent observed decreases in UV radiation at Riverside. Thus, aerosol particle absorption most likely played a role in the reductions. Because black carbon is not a preferential absorber of UV radiation (it absorbs UV and visible wavelengths relatively equally), and because soil-dust concentrations were relatively low, these UV absorbers did not account for the remainder of preferential UV reductions. Nitrated and aromatic aerosol components, which preferentially absorb UV (Fig. 7.10) may have accounted for a portion of the remainder of UV absorption. Figure 5.14 shows that nitrate concentrations at Riverside, where the largest UV reductions were occurring, were high (up to 60 μg m^{-3}); thus, it is plausible that organics at Riverside were heavily nitrated and absorbed UV radiation.

7.1.3.4. Effects of UV-Radiation Reductions on Ozone

The effect of UV radiation loss on ozone in Los Angeles cannot be measured but it can be examined by a model. A modeling study of 1987 air pollution in Los Angeles found that decreases in UV radiation due to smog decreased photolysis rates, decreasing near-surface ozone mixing ratios by an average of 5 to 8 percent (Jacobson, 1998, 1997b). The study also found that

- in regions of the boundary layer where absorption of UV radiation by aerosol particles was strong, photolysis of UV-absorbing gases decreased and ozone decreased
- in regions of the boundary layer where UV scattering dominated UV absorption by aerosol particles, photolysis of UV-absorbing gases increased and ozone increased

In a study of relatively nonaborbing aerosols in Maryland, Dickerson et al. (1997) found that highly scattering aerosol particles increased ozone, consistent with the second result.

Although reduced UV radiation and ozone may appear to be ironic benefits of certain smogs, the cause of UV reductions is heavy particle loadings. Particles, particularly small ones, cause harmful health effects that far outweigh the benefits of reduced UV radiation or the small level of reduced ozone that they might trigger. In addition, although ozone mixing ratios slightly decrease in the presence of absorbing particles, the mixing ratios of other pollutant gases increase.

7.1.4. Aerosol and Hydrometeor Particle Scattering

Particle scattering is the redirection of incident energy by a particle without a loss of energy to the particle. Particle scattering is really the combination of several processes, including reflection, refraction, and diffraction. These processes are discussed next.

7.1.4.1. Reflection

Reflection occurs when radiation bounces off an object at an angle equal to the angle of incidence. No energy is lost during reflection. Figure 7.12 shows an example of reflection. Radiation can reflect off of aerosol particles, cloud drops, or other surfaces. The colors of most objects that we see are due to preferential reflection of certain wavelengths by the object. For example, an apple appears red because the apple's skin absorbs blue and green wavelengths and reflects red wavelengths to our eye.

7.1.4.2. Refraction

Refraction occurs when a wave or photon leaves a medium of one density and enters a medium of another density. In such a case, the speed of the wave changes, changing the angle of the incident wave relative to a surface normal, as shown in Fig. 7.12. If a wave travels from a medium of one density to a medium of a higher density, it bends (refracts) toward the surface normal (the vertical line in Fig. 7.12). The angle of refraction is related to the angle of incidence by **Snell's law**,

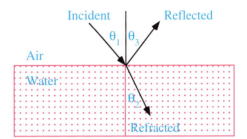

Figure 7.12. Examples of reflection and refraction. During reflection, the angle of incidence (θ_1) equals the angle of reflection (θ_3). During refraction, the angles of incidence and refraction are related by Snell's law. The line perpendicular to the air–water interface is the surface normal.

$$\frac{n_2}{n_1} = \frac{\sin\theta_1}{\sin\theta_2} \qquad (7.5)$$

In this equation, n is the real index of refraction (dimensionless), θ is the angle of incidence or refraction, and subscripts 1 and 2 refer to media 1 and 2, respectively. The **real index of refraction** is the ratio of the **speed of light in a vacuum** ($c = 2.99792 \times 10^8$ m s^{-1}) to that in a different medium (c_1, m s^{-1}). Thus,

$$n_1 = \frac{c}{c_1} \qquad (7.6)$$

Because light cannot travel faster than its speed in a vacuum, the real index of refraction of a medium other than a vacuum must exceed unity. The refractive index of air at a wavelength of 0.5 μm is 1.000279. Real refractive indices of some liquids and solids were given in Table 7.2.

EXAMPLE 7.3

(a) Suppose light at a wavelength of 0.5 μm travels between the atmosphere (medium 1) and liquid water (medium 2). Suppose also that the angle between the incident light and the surface normal is $\theta_1 = 45°$. By how many degrees is the light bent toward the surface normal when it enters medium 2? (b) Do the same calculation for light traveling between outer space (medium 1) and the atmosphere (medium 2).

Solution

(a) At a wavelength of 0.5 μm, the real index of refraction of air is $n_1 = 1.000279$ and that of liquid water is $n_2 = 1.335$. From Equation 7.5, the angle between the light and the surface normal in medium 2 is $\theta_2 = 32°$; thus, the light is bent by 13° toward the surface normal when it enters water from the atmosphere.

(b) At a wavelength of 0.5 μm, the real index of refraction of a vacuum is $n_1 = 1.0$. From Equation 7.5, the angle between the light and the surface normal in medium 2 is $\theta_2 = 44.984°$; thus, the light is bent by 0.016° toward the surface normal when it enters the atmosphere from space.

In sum, the angle of refraction between the atmosphere and water is much greater than that between space and the atmosphere.

Because the real index of refraction is wavelength dependent, different wavelengths are refracted by different angles when they pass from one medium to another. For instance, when visible light passes from air to liquid water at an incident angle of $\theta_1 = 45°$, wavelengths of 0.4 and 0.7 μm are bent by 13.11° and

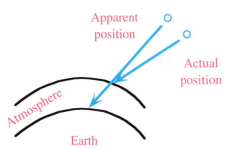

Figure 7.13. Refraction of starlight by the atmosphere makes stars appear to be where they are not.

12.90°, respectively. Thus, refraction bends short (blue) wavelengths of visible light more than it bends long (red) wavelengths. Separation of white visible light into individual colors by this selective refraction is called **dispersion** (or **dispersive refraction**). When Sir Isaac Newton separated white light into multiple colors by passing it through a glass prism (Section 7.1), he discovered dispersive refraction.

As shown in Figs. 7.6 and 7.8, refraction between space and the atmosphere is responsible for twilight, which is the sunlight we see after the sun sets and before the sun rises. Such refraction also causes stars to appear positioned where they are not, as shown in Fig. 7.13. Layers of air at different densities in the Earth's atmosphere cause starlight to refract multiple times and, thus, **flicker**, **twinkle**, or **scintillate**.

7.1.4.3. Diffraction

Diffraction is a process by which the direction of propagation of a wave changes when the wave encounters an obstruction. In terms of visible wavelengths, it is the bending of light as it passes by the edge of an obstruction. In the air, waves diffract as they pass by the surface of an aerosol particle, cloud drop, or raindrop.

Diffraction can be explained in terms of **Huygens's principle**, which states that each point of an advancing wavefront may be considered the source of a new series of secondary waves. If a stone is dropped in a tank of water, waves move out horizontally in all directions, and wavefronts are seen as concentric circles around the stone. If a point source emits waves in three dimensions, wavefronts are concentric spherical surfaces. When a wavefront encounters the edge of an obstacle, such as in Fig. 7.14, waves appear to bend (diffract) around the obstacle because a series of secondary concentric waves is emitted at the edge of the obstacle.

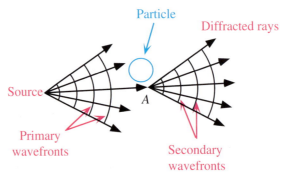

Figure 7.14. Diffraction around a spherical particle. Any point along a wavefront may be taken as the source of a new series of secondary waves. Rays emitted from point A appear to cause waves from the original source to bend around the particle.

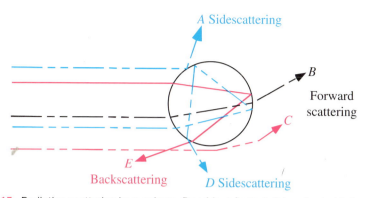

Figure 7.15. Radiative scattering by a sphere. Ray *A* is reflected; *B* is refracted twice; *C* is diffracted; *D* is refracted, internally reflected twice, then refracted; and *E* is refracted, reflected once, then refracted. Rays *A, B, C,* and *D* scatter in the forward or sideward direction; *E* scatters in the backward direction.

7.1.4.4. Summary of Particle Scattering

Particle scattering is the combination of the effects of reflection, refraction, and diffraction. When a wave approaches a spherical particle, such as a cloud drop, it can reflect off the particle, diffract around the edge of the particle, or refract into the particle. Once in the particle, the wave can be absorbed, transmit through the particle and refract out, or reflect internally one or more times and then refract out. Figure 7.15 illustrates these processes, except for absorption, which is not a scattering process. The processes that affect particle scattering the most are diffraction and double refraction, identified by rays C and B, respectively. Thus, particles scatter light primarily in the forward direction. They also scatter some light to the side and in the backward direction. **Backscattered** light results primarily from a single internal reflection (ray E). The light rays seen in Fig. 7.16 are the result of light scattering off of cloud drops in the forward and sideward directions.

7.1.4.5. Rainbows

A rainbow results from two light-scattering processes, dispersive refraction and reflection, and can be seen only if the sun is at the viewer's back and raindrops are falling in front of the viewer. The seven most prominent colors in a rainbow are red, orange, yellow, green, blue, indigo, and violet. In a **primary rainbow**, red appears on the top and violet appears on the bottom. Figure 7.17 shows an example of a primary rainbow. In a **secondary rainbow**, sometimes seen faintly above a primary rainbow, violet appears on top and red appears on the bottom. For convenience, the discussion of rainbows below considers only red, green, and blue.

Figure 7.18 shows how light interacts with raindrops to form a primary rainbow. As a beam of visible light enters a raindrop, all wavelengths bend toward the surface normal due to refraction. Blue light bends the most as a result of dispersive refraction. When light hits the back of the drop, much of it reflects internally. When reflected light reaches the front edge of the drop, it leaves the drop and refracts away from the surface normal. The angles of the blue and red wavelengths that reach a viewer's eye are 40° and 42°, respectively, in relation to the incident beam. Only one wavelength from each raindrop impinges on a viewer's eye. Thus, a rainbow appears when individual waves from many raindrops hit the viewer's eye. As seen in

Figure 7.16. Forward- and side-scattering of sunlight by a cloud. The thickness of the cloud prevents most sunlight from being transmitted.

Figure 7.17. Rainbow over an Alaskan lake in September 1992. Photo by Commander John Bortniak, NOAA Corps, available from the NOAA Central Library.

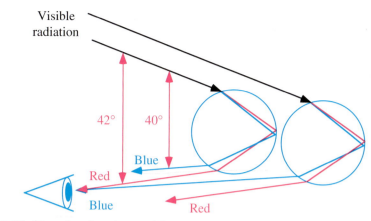

Figure 7.18. Geometry of a primary rainbow.

Fig. 7.18, red appears on the top of a primary rainbow. A secondary rainbow occurs if a second reflection occurs inside each raindrop.

Because winds at midlatitudes originate from the west or southwest (Fig. 6.5) and a rainbow appears only when the sun is at a viewer's back, sailors at midlatitudes knew that if they saw a rainbow in the morning, the rainbow was to the west and the winds were driving the storm creating the rainbow toward them. If they saw a rainbow in the evening, the rainbow was to the east and the winds were driving the storm creating the rainbow away from them. These factors led to the rhyme,

> Rainbow in the morning, sailors take warning
> Rainbow at night, sailor's delight.

7.1.5. Particle Scattering and Absorption Extinction Coefficients

The quantification of particle scattering and absorption is more complex than is that of gas scattering or absorption due to the variety of sizes and compositions of aerosol particles. Aerosol particle absorption and scattering extinction coefficients (cm^{-1}) at a given wavelength can be estimated with

$$\sigma_{a,p} = \sum_{i=1}^{N_B} n_i \pi r_i^2 Q_{a,i} \qquad \sigma_{s,p} = \sum_{i=1}^{N_B} n_i \pi r_i^2 Q_{s,i} \qquad (7.7)$$

respectively, where the summations are over N_B particle sizes, n_i is the number concentration (particles per cubic centimeter of air) of particles of radius r_i (cm), πr_i^2 is the actual cross section of a particle assuming it is spherical (cm^2 per particle), and $Q_{a,i}$ and $Q_{s,i}$ are single-particle absorption and scattering efficiencies (dimensionless), respectively.

A **single-particle scattering efficiency** is the ratio of the effective scattering cross section of a particle to its actual cross section. The scattering efficiency can exceed unity because a portion of the radiation diffracting around a particle can be intercepted and scattered by the particle. Scattering efficiencies above unity account for this additional scattering.

A **single-particle absorption efficiency** is the ratio of the effective absorption cross section of a particle to its actual cross section. The absorption efficiency can exceed unity because a portion of the radiation diffracting around a particle can be intercepted and absorbed by the particle. Absorption efficiencies above unity account for this additional absorption. The larger the imaginary index of refraction of a particle, the greater its absorption efficiency.

Single-particle absorption and scattering efficiencies vary with particle size, radiation wavelength, and refractive indices. Figures 7.19 and 7.20 show $Q_{a,i}$ and $Q_{s,i}$ for black carbon and liquid water, respectively, at a wavelength of 0.5 μm. The figures also show the **single-particle forward scattering efficiency** $Q_{f,i}$, which is the efficiency with which a particle scatters light in the forward direction. The forward scattering efficiency is always less than is the total scattering efficiency. The proximity of $Q_{f,i}$ to $Q_{s,i}$ in both figures indicates that aerosol particles scatter strongly in the forward direction. The difference between $Q_{s,i}$ and $Q_{f,i}$ is the scattering efficiency in the backward direction.

When a particle's diameter (D) is much smaller than the wavelength of light (λ) (e.g., when (D/λ < 0.03), the particle is in the **Rayleigh regime** and is called a **Tyndall absorber or scatterer**. **John Tyndall** (1820–1893) was an English experimental physicist who demonstrated experimentally that the sky's blue color results from scattering of visible light by gas molecules and that a similar effect occurs with small particles.

When a particle's diameter is near the wavelength of light ($0.03 \leq D/\lambda < 32$), the particle is in the **Mie regime**. **Gustav Mie** (1868–1957) was a German physicist who derived equations describing the scattering of radiation by particles in this regime (Mie, 1908).

When a particle has diameter much larger than the wavelength of light ($D/\lambda \geq 32$), the particle is in the **geometric regime**. Figures 7.19 and 7.20 show the diameters corresponding to these regimes for a wavelength of 0.5 μm.

Figure 7.19 shows that visible-light absorption efficiencies of black carbon particles peak when the particles are 0.2 to 0.4 μm in diameter. Such particles are in the accumulation mode with respect to particle size and in the Mie regime with respect to the ratio of particle size to the wavelength of light. Figure 7.20 shows that water particles 0.3 to 2.0 μm in diameter scatter visible light more efficiently than do smaller or larger particles. These particles are also in the accumulation mode with respect to particle size and in the Mie regime with respect to the ratio of particle size to the wavelength of light. Because the accumulation mode contains a

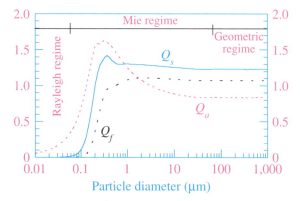

Figure 7.19. Single-particle absorption (Q_a), total scattering (Q_s), and forward scattering (Q_f) efficiencies of black carbon particles of different sizes at λ = 0.50 μm (n = 1.94, κ = 0.66).

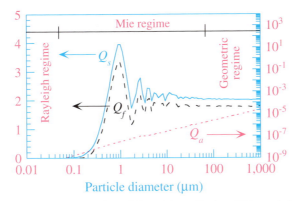

Figure 7.20. Single-particle absorption (Q_a), total scattering (Q_s), and forward scattering (Q_f) efficiencies of liquid water drops of different sizes at $\lambda = 0.50$ μm ($n = 1.335$, $\kappa = 1.0 \times 10^{-9}$).

relatively high particle number concentration, and because particles in this mode have high scattering and absorption (with respect to black carbon) efficiencies, the accumulation mode almost always causes more light reduction than do the nucleation or coarse particle modes (Waggoner et al., 1981). In many urban regions, 20 to 50 percent of the accumulation mode mass is sulfate. Thus, sulfate is correlated with particle scattering more closely than is any other particulate species, aside from liquid water.

Figure 7.20 shows that liquid water hardly absorbs visible light until particles are raindrop-size (larger than 1,000 μm in diameter). The absorptivity of rain drops causes the bottoms of precipitating clouds to appear gray or black.

7.2. VISIBILITY

Visibility is a measure of how far we can see through the air. Even in the cleanest air, our ability to see along the Earth's horizon is limited to a few hundred kilometers by background gases and aerosol particles. If we look up through the sky at night, however, we can discern light from stars that are millions of kilometers away. The difference between looking horizontally and vertically is that more gas molecules and aerosol particles lie in front of us in the horizontal than in the vertical.

Several terms describe maximum visibility. Two subjective terms are *visual range* and *prevailing visibility*. **Visual range** is the actual distance at which a person can discern an ideal dark object against the horizon sky. **Prevailing visibility** is the greatest visual range a person can see along 50 percent or more of the horizon circle (360°), but not necessarily in continuous sectors around the circle. It is determined by a person who identifies landmarks known distances away in a full 360° circle around an observation point. The greatest visual range observed over 180° or more of the circle (not necessarily in continuous sectors) is the prevailing visibility. Thus, half the area around an observation point may have visibility worse than the prevailing visibility, which is important, because most prevailing visibility observations are made at airports. If the visual range in a sector is significantly different from the prevailing visibility, the observer at an airport usually denotes this information in the observation record.

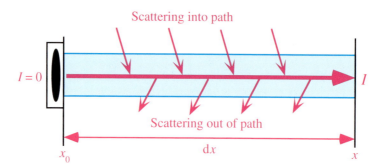

Figure 7.21. Change of radiation intensity along a beam. A radiation beam originating from a dark object has intensity $I = 0$ at point x_0. Over a distance dx, the beam's intensity increases due to scattering of background light into the beam. This added intensity is diminished somewhat by absorption along the beam and scattering out of the beam. At point x, the net intensity of the beam has increased close to that of the background intensity.

A less subjective and, now, a regulatory definition of visibility is the meteorological range. **Meteorological range** can be explained in terms of the following example. Suppose a perfectly absorbing dark object lies against a white background at a point x_0, as shown in Fig. 7.21. Because the object is perfectly absorbing, it reflects and emits no visible radiation; thus, its visible radiation intensity (I) at point x_0 is zero, and it appears black. As a viewer backs away from the object, background white light of intensity I_B scatters into the field of view, increasing the intensity of light in the viewer's line of site. Although some of the added background light is scattered out of or absorbed along the field of view by gases and aerosol-particles, at some distance away from the object, so much background light has entered the path between the viewer and the object that the viewer can barely discern the black object against the background light.

The meteorological range is a function of the **contrast ratio**, defined as

$$C_{ratio} = \frac{I_B - I}{I_B} \tag{7.8}$$

The contrast ratio gives the difference between the background intensity and the intensity in the viewer's line of site, all relative to the background intensity. If the contrast ratio is unity, then an object is perfectly visible. If it is zero, then the object cannot be differentiated from background light.

The meteorological range is the distance from an object at which the contrast ratio equals the liminal contrast ratio of 0.02 (2 percent). The **liminal or threshold contrast ratio** is the lowest visually perceptible brightness contrast a person can see. It varies from individual to individual. Koschmieder (1924) selected a value of 0.02. Middleton (1952) tested 1,000 people and found a threshold contrast range of between 0.01 and 0.20, with the mode of the sample between 0.02 and 0.03. Campbell and Maffel (1974) found a liminal contrast of 0.003 in laboratory studies of monocular vision. Nevertheless, 0.02 has become an accepted liminal contrast value for meteorological range calculations. In sum, the meteorological range is the distance from an ideal dark object at which the object has a 0.02 liminal contrast ratio against a white background.

The meteorological range can be derived from the equation for the change in object intensity along the path described in Fig. 7.21. This equation is

$$\frac{dI}{dx} = \sigma_t(I_B - I) \tag{7.9}$$

where all wavelength subscripts have been removed, σ_t is the total extinction coefficient, $\sigma_t I_B$ accounts for the scattering of background light radiation into the path, and $-\sigma_t I$ accounts for the attenuation of radiation along the path due to scattering out of the path and absorption along the path. A **total extinction coefficient** is the sum of extinction coefficients due to scattering and absorption by gases and particles. Thus,

$$\sigma_t = \sigma_{a,g} + \sigma_{s,g} + \sigma_{a,p} + \sigma_{s,p} \tag{7.10}$$

Integrating Equation 7.9 from $I = 0$ at point $x_0 = 0$ to I at point x with constant σ_t yields the equation for the contrast ratio,

$$C_{ratio} = \frac{I_B - I}{I_B} = e^{-\sigma_t x} \tag{7.11}$$

When $C_{ratio} = 0.02$ at a wavelength of 0.55 μm, the resulting distance x is the meteorological range (also called the **Koschmieder equation**).

$$x = \frac{3.912}{\sigma_t} \tag{7.12}$$

In polluted tropospheric air, the only important gas-phase visible light attenuation processes are Rayleigh scattering and absorption by $NO_2(g)$. Table 7.3 shows the meteorological ranges derived from calculated extinction coefficients, resulting from these two processes in isolation. For $NO_2(g)$, the table shows values at two mixing ratios, representing clean and polluted air respectively. For Rayleigh scattering, one meteorological range is shown because Rayleigh scattering is dominated by molecular nitrogen and oxygen, whose mixing ratios do not change much between clean and polluted air.

At a wavelength of 0.55 μm, the meteorological range due to Rayleigh scattering alone was 334 km, indicating that, in the absence of all pollutants, the furthest one can see along the horizon, assuming a liminal contrast ratio of 2 percent, is near 350 km. Waggoner et al. (1981) reported a total extinction coefficient at Bryce Canyon, Utah, corresponding to a meteorological range of less than 400 km, suggesting that visibility was limited by gas scattering.

Table 7.3. Meteorological Ranges (km) Resulting from Rayleigh Scattering and $NO_2(g)$ Absorption at Selected Wavelengths

λ (μm)	Rayleigh Scattering	$NO_2(g)$ Absorption	
		0.01 ppmv $NO_2(g)$	0.25 ppmv $NO_2(g)$
0.42	112	296	11.8
0.50	227	641	25.6
0.55	334	1,590	63.6
0.65	664	13,000	520

In Table 7.3, the meteorological range due to $NO_2(g)$ absorption decreased from 1,590 to 63.6 km when the $NO_2(g)$ mixing ratio increased from 0.01 to 0.25 ppmv. Thus, $NO_2(g)$ absorption reduced visibility more than did Rayleigh scattering when the $NO_2(g)$ mixing ratio was high. Results from a project studying **Denver's brown cloud** showed that $NO_2(g)$ accounted for about 7.6 percent of the total reduction in visibility, averaged over all sampling periods, and 37 percent of the total reduction during periods of maximum $NO_2(g)$. Scattering and absorption by aerosol particles caused most remaining extinction (Groblicki et al. 1981). In sum, $NO_2(g)$ attenuates visibility in urban air when its mixing ratios are high.

Although the effects of Rayleigh scattering and $NO_2(g)$ absorption on visibility are nonnegligible in polluted air, they are less important than are scattering and absorption by aerosol particles. Scattering by aerosol particles causes between 60 and 95 percent of visibility reductions. Absorption by aerosol particles causes between 5 and 40 percent of reductions (Cass, 1979; Tang et al., 1981; Waggoner et al., 1981).

Table 7.4 shows meteorological ranges derived from extinction coefficient measurements for a polluted and less polluted day in Los Angeles. Particle scattering dominated light extinction on both days. On the less-polluted day, gas absorption, particle absorption, and gas scattering all had similar small effects. On the polluted day, the most important visibility reducing processes were particle scattering, particle absorption, gas absorption, and gas scattering, in that order.

Equation 7.12 relates a theoretical quantity, meteorological range, to a measured extinction coefficient. When prevailing visibility, a subjective quantity, and the extinction coefficient are measured simultaneously, they usually do not satisfy Equation 7.12. Instead, the relationship between prevailing visibility and the measured extinction coefficient must be obtained empirically. Griffing (1980) studied measurements of prevailing visibility (V, in km) and extinction coefficients (km^{-1}) over a five-year period and derived the empirical relationship

$$V \approx \frac{1.9}{\sigma_t} \tag{7.13}$$

Figure 7.22 shows a map of estimated extinction coefficients in the United States between 1991 and 1995 derived from prevailing visibility measurements substituted into Equation 7.13. During winters, visibility was poorest (extinction coefficients were highest) in southern and central California, Illinois, Indiana, Iowa, Kentucky, and Michigan. During summers, visibility was poorest in Los Angeles and much of the Midwest and southern United States. Winter visibility loss in central California

Table 7.4. Meteorological Ranges (km) Resulting from Gas Scattering, Gas Absorption, Particle Scattering, Particle Absorption, and All Processes at a Wavelength of 0.55 μm on a Polluted and less Polluted Day in Los Angeles

Day	Gas Scattering	Gas Absorption	Particle Scattering	Particle Absorption	All
Polluted (8/25/83)	366	130	9.6	49.7	7.42
Less Polluted (4/7/83)	352	326	151	421	67.1

Meteorological ranges derived from extinction coefficients of Larson et al. (1984).

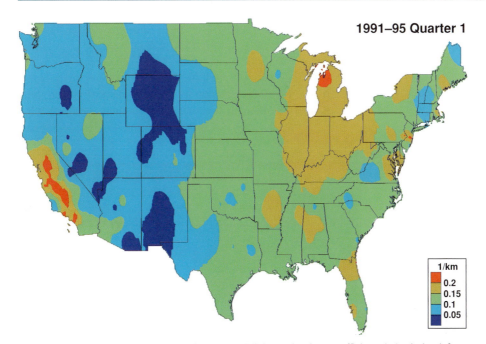

Figure 7.22. United States maps of corrected light extinction coefficient (σ_t), derived from prevailing visibility measurements (V) with the equation $\sigma_t \approx 1.9/V$ for winter (Quarter 1) and summer (Quarter 3) 1991–1995 (Schichtel et al., 2001). Large extinction coefficients correspond to poor visibility.

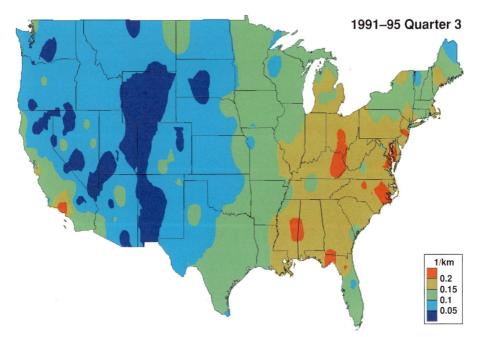

Data obtained during rain, snow, and fog were eliminated. Corrected extinction coefficients were obtained by applying relative humidity correction factors to adjust all extinction coefficients to a "dry" relative humidity of 60 percent, then selecting the "dry" extinction coefficient at each monitoring site and for each season corresponding to the 75th percentile of all extinction coefficients at the site and for the season.

was likely caused by a combination of low-lying winter inversions, high relative humidities, and heavy particle loadings. Visibility loss in Los Angeles was due to smog present all year. Poor visibility in the winter in the Midwest was due to power-plant emissions combined with high relative humidities. Summer visibility degradation in the Midwest and southern United States was due to a combination of power-plant emissions, organic particles from photochemical smog and vegetation, and high relative humidities.

7.3. COLORS IN THE ATMOSPHERE

Red sunsets and blue skies arise from preferential scattering of light by gas molecules. Rainbows arise from interactions of light with raindrops. Additional optical phenomena are discussed next.

7.3.1. White Hazes and Clouds

Figure 7.23. Haze over Los Angeles, May 1972. Photo by Gene Daniels, U.S. EPA, available from Still Pictures Branch, U.S. National Archives.

When the relative humidity is high (but still less than 100 percent), aerosol particles increase in size by absorbing liquid water. When their diameters approach the wavelength of light, the particles enters the Mie regime (Section 7.1.5) and scatter all wavelengths with equal intensity, producing a whitish haze. Hazes resulting from water growth on pollution particles, are common in urban areas under sunny skies (Fig. 7.23).

Cloud and fog drops, which contain more liquid water than do aerosol particles, appear white because they scatter all wavelengths of visible light with equal intensity. When pollution is particularly heavy and the relative humidity is high, clouds and fogs are difficult to distinguish from hazes, as shown in Fig. 7.24.

7.3.2. Reddish and Brown Colors in Smog

Reddish and brown colors in smog, such as those seen in Fig. 7.25, are due to three factors. The first is preferential absorption of blue and some green light by $NO_2(g)$, which allows most green and red to be transmitted, giving smog with $NO_2(g)$ a yellow, brown, or reddish-green color. Brown layers resulting from $NO_2(g)$ can be seen most frequently from 9 to 11 A.M. in polluted air, because these are the hours that $NO_2(g)$ mixing ratios are highest. The second is preferential absorption of blue light by nitrated aromatics and polycyclic aromatic hydrocarbons (PAHs) in aerosol particles. Figure 7.10 shows that a variety of nitrated aromatic compounds preferentially absorb not only UV, but also short visible wavelengths. PAHs are also absorbers of short visible wavelengths. The blue-light absorption and brown transmission by these compounds enhances the brownish color of smog.

Figure 7.24. Combined haze and fog over Los Angeles, near the San Gabriel Mountains, May 1972. Photo by Gene Daniels, U.S. EPA, available from Still Pictures Branch, U.S. National Archives.

Figure 7.25. Reddish and brown colors in Los Angeles smog on December 19, 2000.

Third, when soil dust concentrations are high, air can appear reddish or brown because soil particles also preferentially absorb blue and green light over red light. In the absence of a dust storm and away from desert regions, the concentration of soil-dust particles is generally low and the effect of soil dust on atmospheric optics is limited.

7.3.3. Black Colors in Smog

Black colors in smog are due primarily to absorption by black carbon, a component of soot. Soot is emitted primarily during coal, diesel fuel, jet fuel, and biomass burning. Black carbon absorbs all wavelengths of white light, transmitting none, thereby appearing black against the background sky.

7.3.4. Red Skies and Brilliant Horizons in Smog

Whereas the sun itself appears red at sunset and sunrise, and the horizon appears red during sunset and sunrise when aerosol particles are present (Fig. 7.8), the sky can also appear red in the afternoon and red horizons can become more brilliant in the presence of air pollution. As discussed in Section 7.1.2.2, red horizons occur because aerosol particles scatter the sun's blue, green, and red light, providing a source of white light to a viewer from all points along the horizon. Gas molecules scatter the blue and green wavelengths out of the viewer's line of site, allowing only the red to be transmitted from the horizon to the viewer. When heavy smog is present, particle concentrations near the surface and aloft increase sufficiently to allow this phenomenon to occur in the afternoon, as seen in Fig. 7.26. Smog particles similarly enhance the brilliance of the red horizon at sunset, as seen in Fig. 6.21.

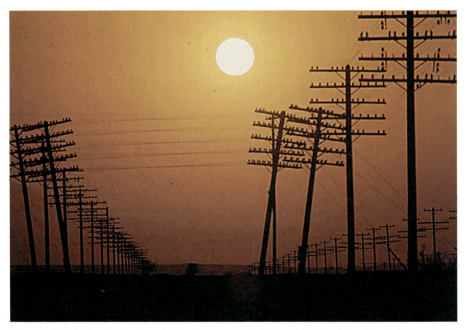

Figure 7.26. Red sky in the late afternoon over transmission lines near Salton Sea, California, May 1972. The smog causing the optical effect originated in Los Angeles. Photo by Charles O'Rear, U.S. EPA, available from Still Pictures Branch, U.S. National Archives.

Figure 7.27. Purple sunset taken from Palos Verdes, California, following the 1982 El Chichon volcanic eruption. Photo by Jeffrey Lew, UCLA.

7.3.5. Purple Glow in the Stratosphere

After a strong volcanic eruption, a purple glow may appear in the stratosphere, as shown in Fig. 7.27. This glow results because volcanos emit sulfur dioxide gas, which converts to sulfuric acid–water particles, many of which reach the stratosphere. Such particles scatter light through the stratospheric ozone layer. Ozone weakly absorbs green and some red wavelengths (Fig. 7.3), transmitting blue and some red, which combine to form purple. The purple light is scattered back to a viewer's eye by the enhanced stratospheric particle layer, causing the stratosphere to appear purple.

7.4. SUMMARY

In this chapter, processes that affect radiation were discussed with an emphasis on the effects of gases and aerosol particles on visibility degradation and UV light reduction. Processes affecting radiation through the air include absorption and scattering. Gas absorption of UV radiation is responsible for the stratospheric inversion and most photolysis reactions. Gas absorption of visible light is important only when $NO_2(g)$ mixing ratios are high. Gas (Rayleigh) scattering affects only UV and visible wavelengths. The most important gas scatterers are molecular oxygen and molecular nitrogen. All aerosol particles absorb near-and thermal-IR radiation, but only a few, including black carbon and certain soil components, absorb visible (including UV) radiation. Several organic aerosol components, particularly nitrated and aromatic compounds, preferentially absorb UV radiation, causing UV and ozone reductions in polluted air. All aerosol components scatter radiation. The most important visibility degrading processes in polluted air are particle scattering, particle absorption, gas absorption, and gas scattering, in that order. In clean air, visibility is limited in the horizontal to a few hundred kilometers by gas scattering.

7.5. PROBLEMS

7.1.   Calculate the fractional transmission of radiation through a swimming pool 2.0 m deep at a wavelength of 0.5 μm when only absorption of light by liquid water is considered. At what depth in the water do you expect the incident light to be reduced by 99 percent due to the absorption?

7.2.   If the angle of incidence, relative to the surface normal, of 0.5 μm wavelength light entering water from the air is 30°, determine the angle of refraction in the water.

7.3.   If 0.5 μm wavelength light travels from water to air, what angle of incidence (with respect to the surface normal in water) is necessary for the angle of refraction to be 90°? This angle of incidence is referred to as the critical angle. What happens if the angle of incidence in water exceeds the critical angle?

7.4.   If gas molecules in the atmosphere preferentially scattered longer wavelengths over shorter wavelengths, what colors would you expect the sky to be and the sun to be at noon, afternoon, and sunset?

7.5.   Explain why nitrogen dioxide gas causes more visibility reduction than does ozone gas.

7.6.   Calculate the meteorological range at 0.55 μm resulting from aerosol particle

   (a)   scattering alone
   (b)   absorption alone
   (c)   absorption plus scattering

   assuming the atmosphere contains 10^4 particles cm^{-3} of 0.1 μm diameter spherical particles made of black carbon. Use Fig. 7.19 to obtain single-particle absorption and scattering efficiencies and assume efficiencies at 0.5 μm are similar to those at 0.55 μm.

7.7.   If a person's liminal contrast ratio is 0.01 instead of 0.02, what would be the expression for their meteorological range? For a given total extinction coefficient, how much further (in percent) can the person see when their liminal contrast ratio is 0.01 instead of 0.02?

7.8.   Assume that the atmosphere contains 10,000 μg m^{-3} of liquid water, and the density of liquid water is 1.0 g cm^{-3}. For which size spherical particles is visibility reduced the most at a wavelength of 0.55 μm:

(a)   0.5 μm diameter particles
(b)   10.0 μm diameter particles
(c)   100.0 μm diameter particles?

Assume single-particle scattering efficiencies in Fig. 7.20 can be used for wavelengths of 0.55 μm. Show calculations.

7.9.   Explain why soil often appears brown. What is the difference between soil that appears brown and soil that appears red?

7.6. PROJECT

Find an outside, elevated location where, on a clear day, it is possible to see far in at least some directions without obstruction by buildings or mountains. Predetermine the distances to a set of landmarks that cover at least 180° along the horizon circle, but not necessarily in continuous sectors around the circle. On three different days and at two times during each day (pick a morning and an afternoon time, and use the same two times on each day),

(a)   estimate the prevailing visibility
(b)   make a list of the colors you see in the sky along the horizon and where you see them

After all data are gathered, discuss your findings in terms of differences among days and between times of day. Comment on how the relative humidity, wind, and large-scale pressure system may have affected results if that information is available. Which of these factors do you think had the greatest effect on visibility?

INTERNATIONAL REGULATION OF URBAN SMOG SINCE THE 1940s

Until the 1940s, efforts to control air pollution in the United States were limited to a few municipal ordinances and state laws regulating smoke (Chapter 4). In the United Kingdom, regulations were limited to the Alkali Act of 1863 and some relatively weak Public Health Bills designed to abate smoke. Regulations in other countries were similarly weak or nonexistent. The main reason for the lack of regulation in polluted cities was that the coal, oil, chemical, and auto industries, the ultimate sources of much of the pollution, had political power and used it to resist efforts of government intervention (e.g., Section 4.1). Because the long-term health effects of pollutants in outdoor concentrations were not well known at the time, it was also difficult for public health agencies to recommend the banning of a pollutant, particularly in the face of political pressure from industry and arguments that such a ban would hurt economic growth (e.g., Midgley's defense of tetraethyl lead in Section 3.6.9). In the late 1940s through mid-1950s, damage due to photochemical smog in the United States was sufficiently apparent that the federal government decided to take steps to address the problem. Similarly, deadly London-type smog events in the United Kingdom spurred government legislation in the 1950s. Today, many countries have instituted air pollution regulations. Nevertheless, regulations in most countries are still weak, resulting in severe pollution problems. Many countries, for instance, continue to allow tetraethyl lead in their gasoline and do not require catalytic converters in automobiles. All still promote the use of diesel fuel, which results in soot emissions. In this chapter, air pollution regulations since the 1950s are discussed. The chapter discusses regulation of outdoor pollution and air quality trends in the United States, and regulations and air quality trends in many other countries. International regulations involving transboundary air pollution are discussed in Section 10.7.

8.1. REGULATION IN THE UNITED STATES

In the 1940s, photochemical smog in Los Angeles became a cause for concern and led to the formation of the Los Angeles Air Pollution Control District in 1947. In 1949, the first National Air Pollution Symposium was held in Los Angeles. In 1951, the state of Oregon set up an agency to oversee and regulate air pollution. Other states followed suit, so that by 1960, seventeen statewide air pollution agencies existed. In 1948, a heavy air pollution episode in Donora, Pennsylvania, killed 20 people, and in 1948, 1952, and 1956, air pollution episodes in London killed a total of nearly 5,000. Pollution in Los Angeles reached its peak in severity in the mid-1950s, when ozone levels as high as 0.68 ppmv were recorded. In 1952, Arie Haagen-Smit isolated the mechanism of ozone formation in photochemical smog. The combination of a better understanding of air pollution formation and experiences with the effects of air pollution contributed to the first U.S. federal legislation concerning air pollution in 1955.

8.1.1. Air Pollution Control Act of 1955

Because of the accelerating number of problems associated with air pollution, in 1955, President Dwight D. Eisenhower asked the U.S. Congress to consider legislation addressing the issue. Until this time, state and local governments had received no federal guidance in combating air pollution. On July 14, 1955, the U.S. Congress passed the first of several major pieces of air pollution regulation, the **Air Pollution Control Act of 1955** (Public Law 84-159).

The Air Pollution Control Act of 1955 granted $3 million per year for five years to the Public Health Service (PHS), in the Department of Health, Education, and Welfare, to study air pollution. The act directed the PHS to provide technical assistance to states for combating air pollution, train individuals in the area of air pollution, and perform more research on air pollution control.

In 1959, the act was extended by four years, with funding of $5 million per year. In 1960, it was amended (Public Law 86-493, June 6, 1960) to authorize the U.S. Surgeon General to study the health effects of automobile exhaust. In 1962, it was amended again (Public Law 87-761, October 9, 1962). Although the 1955 law raised air pollution issues to a federal level, it did not impose any federal regulations on air pollution and delegated air pollution control and prevention to state and local levels.

8.1.2. Clean Air Act of 1963

By 1963, public concern about air pollution increased sufficiently for the U.S. Congress to regulate smokestack, but not automobile, emissions. In December 1963, Congress enacted the Clean Air Act of 1963 (CAA63, Public Law 88-206). The purpose of this act was "to improve, strengthen, and accelerate programs for the prevention and abatement of air pollution." CAA63 contained provisions giving the federal government authority to reduce interstate air pollution. Such reductions would be obtained by specifying emission standards for stationary pollution sources, including power plants and steel mills, and by encouraging the use of technologies to remove sulfur from coal and oil.

CAA63 did not specify controls for automobiles, the most serious source of air pollution. At the state level, some regulations were already in place. In the 1950s, new cars typically emitted about 13 grams per mile (g/mi) of gas-phase hydrocarbons (HCs), 87 g/mi of carbon monoxide [$CO(g)$], and 3.6 g/mi of nitrogen oxides [$NO_x(g)$]. In 1959, the California state legislature created the Motor Vehicle Pollution Control Board, which set the first automobile emission standards in the world. The first requirement the board implemented was to reduce crankcase emissions of unburned hydrocarbons, beginning with 1963 model cars. At the time, crankcase emissions were responsible for about 20 percent of HC emissions from automobiles. Fuel tank evaporation (9 percent), carburetor evaporation (9 percent), and tailpipe exhaust (62 percent) accounted for the remainder of HC emissions. All $CO(g)$ and $NO_x(g)$ emissions were thought to originate from tailpipes. Per the new regulations, new cars were required to reroute crankcase HC emissions back to the manifold, where they were reburned instead of emitted into the air.

8.1.3. Motor Vehicle Air Pollution Control Act of 1965

Whereas the federal government was not ready to set automobile standards in 1963, CAA63 initiated a process for reviewing the status of automobile emissions by setting up a technical committee consisting of representatives from the Department of Health, Education, and Welfare and the automobile, control device, and oil industries. Investigations by this committee and subsequent hearings in 1964 by the Senate Public Works Subcommittee on Air and Water Pollution led to the first amendment of CAA63, the Motor Vehicle Air Pollution Control Act of 1965 (Public Law 89-272, October 20, 1965). By this act, the federal government set its first federal automobile emission standards for HCs and $CO(g)$. Table 8.1 lists federal light-duty vehicle

Table 8.1. Federal Emission Standards for Light-Duty Vehicles

Year*	HCs (g/mi)	CO(g) (g/mi)	NO$_x$(g) (g/mi)	Pb(s) (g/gal)	PM (g/mi)
1968–70	3.2	33	—	—	—
1971–2	4.6	47	4.0	—	—
1972	3.4	39	—	—	—
1973–4	3.4	39	3.0	—	—
1975–6	1.5	15	3.1	—	—
1977–9	1.5	15	2.0	0.8	—
1980	0.41	7.0	2.0	0.5	—
1981	0.41	3.4	1.0	0.5	—
1982–6	0.41	3.4	1.0	0.5	0.6[a]
1987–92	0.41	3.4	1.0	0.5	0.2[a]
Tier 1: Intermediate Useful Life Standards					
1993–	0.41[a,b,e,f] 0.25[a,c,d, e,g]	3.4	0.4[c,e] 1.0[a]	0.5	0.08
Tier 2: Full-Life Standards					
1993–	0.31[a,c,d,e,g]	4.2	0.6[c,e] 1.25[a]	0.5	0.10

*Test method changed in 1971.
[a]Diesel.
[b]LVW≥3751 lbs.
[c]Gasoline.
[d]Nonmethane hydrocarbons.
[e]Methanol.
[f]Organic matter hydrocarbons.
[g]Organic matter nonmethane hydrocarbons equivalent.

Adapted from Wark et al., 1998.

emission standards from 1968 to the present. The standards for 1968 were applicable to 1968 model cars and were patterned after California state standards developed for 1966 model cars sold in California. The federal standards were intended to reduce emissions to 72 percent of their 1963 values for tailpipe HCs, 56 percent for tailpipe CO(g), and 100 percent for crankcase HCs. Despite the good intentions of the 1965 Act, more than half of all 1968 and 1969 model cars failed to meet the new emission standards. In 1966, Congress passed a second amendment to the 1963 Clean Air Act to expand local air pollution control programs.

8.1.4. Air Quality Act of 1967

In 1967, a third amendment to the 1963 Act, the **Air Quality Act of 1967** (Public Law 90-148, November 21, 1967), was passed. This act divided the United States into several inter- or intrastate **Air Quality Control Regions** (AQCRs). Officials in each AQCR were required to conduct studies to determine the extent of air pollution in the region, to collect ambient air quality data, and to develop emission inventories. The 1967 Act also specified that the Department of Health, Education, and Welfare develop and publish **Air Quality Criteria** (AQC) reports, which were science-based reports containing information about the effects, as a function of concentration, of pollutants that damage human health and welfare. The reports also contained suggestions about acceptable levels of pollution. Each state was then required to set and enforce its own air quality standards based on the acceptable levels suggested in the AQC reports. The standards had to be the same as or more stringent than those suggested in the reports. Each state was required to submit a State Implementation Plan (SIP) to the federal government discussing how the state intended to meet its air quality standards. A SIP consisted of a list of regulations the state would implement to clean a polluted region. If the SIP was not approved by the federal government or if the state did not enforce its regulations, the federal government had the authority to bring suit against the state. If the SIP was approved, the state was delegated federal authority to regulate air pollution in the state.

The 1967 Act also recommended the publication of air pollution control methods through control technology documents, provided funds to states for motor vehicle inspection programs, and allowed California to set its own automobile emission standards. In 1969, the CAA63 was amended again to authorize research on low-emission fuels and automobiles.

8.1.5. Clean Air Act Amendments of 1970

In late 1970, President Richard Nixon combined several existing federal air pollution programs to form the U.S. Environmental Protection Agency (U.S. EPA), whose purpose was to enforce federal air pollution regulations. On December 31, 1970, Congress passed the Clean Air Act Amendments of 1970 (CAAA70, Public Law 91-604). The purpose of CAAA70 was "to amend the Clean Air Act to provide for a more effective program to improve the quality of the Nation's air." CAAA70 resulted in the transfer of administrative duties for air pollution regulation from the Department of Health, Education, and Welfare to the U.S. EPA. CAAA70 specified that the U.S. EPA design National Ambient Air Quality Standards (NAAQSs) for criteria air pollutants, so-called because their permissible levels were based on health guidelines, or criteria, obtained from AQC reports.

NAAQSs were divided into primary standards, designed to protect the public health (particularly of people most susceptible to respiratory problems, such as asthmatics, the elderly, and infants), and secondary standards, designed to protect the public welfare (visibility, buildings, statues, crops, vegetation, water, animals, transportation, other economic assets, personal comfort, and personal well-being). The U.S. EPA was required to set primary standards based on health considerations alone, not on the cost of or technology available for attaining the standard. Regions in which primary standards for criteria pollutants were met were called attainment areas, and those in which primary standards were not met were called nonattainment areas.

The six original criteria pollutants specified by CAAA70 were $CO(g)$, $NO_2(g)$, $SO_2(g)$, total suspended particulates (TSP), HCs, and photochemical oxidants. Particulate lead [Pb(s)] was added to the list in 1976; ozone [$O_3(g)$] replaced photochemical oxidants in 1979; HCs were removed from the list in 1983; and TSPs were changed to include only particulates with diameter ≤ 10 μm and called PM_{10}, and a $PM_{2.5}$ standard was added in 1997. PM_{10} and $PM_{2.5}$ are, more precisely, the concentration of aerosol particles that pass through a size-selective inlet with a 50 percent efficiency cutoff at 10 μm and 2.5 μm aerodynamic diameter, respectively. Table 8.2 lists criteria pollutants for which NAAQS primary and, in most cases, secondary

Table 8.2. Ambient California State and Primary and Secondary Federal Air Quality Standards

Pollutant	California Standard	Federal Primary Standard (NAAQS)	Federal Secondary Standard (NAAQS)
Ozone $[O_3(g)]^a$			
1-hour average	0.09 ppmv (180 μg m^{-3})	0.12 ppmv (235 μg m$^{-3})^b$	Same as primary
8-hour average	—	0.08 ppmv (160 μg m$^{-3})^c$	Same as primary
Carbon monoxide $[CO(g)]^a$			
8-hour average	9.0 ppmv (10 mg m^{-3})	9.5 ppmv (10.5 mg m^{-3})	—
1-hour average	20 ppmv (23 mg m^{-3})	35 ppmv (40 mg m^{-3})	—
Nitrogen dioxide $[NO_2(g)]^a$			
Annual average	—	0.053 ppmv (100 μg m^{-3})	Same as primary
1-hour average	0.25 ppmv (470 μg m^{-3})	—	—
Sulfur dioxide $[SO_2(g)]^a$			
Annual average	—	0.03 ppmv (80 μg m^{-3})	—
24 hours	0.05 ppmv (131 μg m^{-3})	0.14 ppmv (365 μg m^{-3})	—
3 hours	—	—	0.5 ppmv (1300 μg m^{-3})
1 hour	0.25 ppmv (655 μg m^{-3})	—	—
Particulate matter ≤ 10 μm in diameter $(PM_{10})^a$			
Annual geometric mean	30 μg m^{-3}	—	—
24-hour average	50 μg m^{-3}	150 μg m^{-3d}	Same as primary
Annual arithmetic mean	—	50 μg m^{-3e}	Same as primary
Particulate matter ≤ 2.5 μm in diameter $(PM_{2.5})^a$			
24-hour average	—	65 μg m^{-3f}	Same as primary
Annual arithmetic mean	—	15 μg m^{-3g}	Same as primary
Lead $[Pb(s)]^a$			
30-day average	1.5 μg m^{-3}	—	—
Calendar quarter	—	1.5 μg m^{-3}	Same as primary
Particulate sulfates			
24-hour average	25 μg m^{-3}	—	—
Hydrogen sulfide $[H_2S(g)]$			
1-hour average	0.03 ppmv (42 μg m^{-3})	—	—
Vinyl chloride			
24-hour average	0.01 ppmv (26 μg m^{-3})	—	—

^{a}Criteria air pollutants.

^{b}Standard exceeded if the daily maximum 1-hour average concentration exceeds 0.12 ppmv more than once per year, averaged over 3 consecutive years. (This standard was in effect prior to 1997. It now applies only to areas designated as nonattainment areas prior to the 1997 revision.)

^{c}Standard exceeded if the 3-year average of the fourth-highest daily maximum 8-hour average ozone mixing ratio exceeds 0.08 ppmv. (New standard from 1997 Clean Air Act revision.)

^{d}Standard exceeded if the 99th percentile of the distribution of the 24-hour concentrations for a period of 1 year, averaged over 3 years, exceeds 150 μg m^{-3} at each monitor within an area. (New wording from 1997 Clean Air Act revision.)

^{e}Standard exceeded if the 24-hour measured concentration of PM_{10}, arithmetically averaged over a period of 1 year, exceeds 50 μg m^{-3} on average for 3 consecutive years.

^{f}Standard exceeded if the 98th percentile of the distribution of 24-hour concentrations measured for 1 year, averaged over 3 years, exceeds 65 μg m^{-3} at each monitor within an area. (New Standard from 1997 Clean Air Act revision.)

^{g}Standard exceeded if the 3-year average of the annual arithmetic mean of 24-hour measured concentrations exceeds 15.0 μg m^{-3}. (New standard from 1997 Clean Air Act revision.)

standards have been set. Secondary standards are the same as primary standards for most criteria air pollutants. One exception is $CO(g)$, for which no secondary standard has been set because its major impact is on human health. A second exception is $SO_2(g)$, for which a separate secondary standard has been set. The table also shows California state standards. For all pollutants, California state standards are stricter than are NAAQS.

CAAA70 further specified that the U.S. EPA design **New Source Performance Standards** (NSPS) for new stationary sources to limit emissions from such sources. Each state was required to inspect new stationary sources and certify that pollution controls indeed worked and would remain working for the lifetime of the source. CAAA70 also required **National Emission Standards for Hazardous Air Pollutants** (NESHAPS). Hazardous pollutants were defined as pollutants

> to which no ambient air standard is applicable and that . . . . causes, or contributes to air pollution which may be anticipated to result in an increase in mortality or an increase in serious irreversible, or incapacitating reversible illness.

In 1973, the list of hazardous pollutants included asbestos, beryllium, and mercury. By 1984, the list was expanded to include benzene, arsenic, coke-oven emissions, vinyl chloride, and radionuclides. CAAA70 also specified that

> each state shall have the primary responsibility for assuring air quality within the entire geographic area comprising each state by submitting an implementation plan for such state which shall specify the manner in which national primary and secondary ambient air quality standards will be achieved and maintained within each air quality control region in each state.

Thus, the use of SIPs, which originated with the Air Quality Control Act of 1967, continued under CAAA70. CAAA70 required that SIPs address primary and secondary standards. Through an SIP, each state was required to set ambient air quality standards at least as stringent as federal standards, evaluate air quality in each air quality control region within the state, and establish methods and timetables for improving air quality in each AQCR to meet state standards. The SIP was required to address approval procedures for new pollution sources and methods of reducing pollution from existing sources. Once submitted, an SIP required U.S. EPA approval; otherwise, the U.S. EPA had the power to take control of the state's air pollution program.

CAAA70 further required that the U.S. EPA develop aircraft emission standards, expand the number of air quality control regions, and establish tough fines and criminal penalties for violations of SIPs, emission standards, and performance standards. It also permitted citizen's suits "against any person, including the United States, alleged to be in violation of emission standards or an order issued by the administrator." The national aircraft emission standards set by CAAA70 were enacted after California enacted a state standard in 1969.

CAAA70 required that new automobiles emit 90 percent less HCs and $CO(g)$ in 1975 than in 1970 and 90 percent less $NO_x(g)$ in 1976 than in 1971. Thus Congress, and not the U.S. EPA, set the automobile emission standards, but the U.S. EPA was authorized to extend deadlines for auto-emission reductions. The U.S. EPA extended the deadline in 1975 for all reductions by one year in 1973, by a second year in 1974, and by a third year for HCs and $CO(g)$.

8.1.6. Catalytic Converters

In 1975, an important automobile emission control technology, the catalytic converter, was developed directly in response to automobile emission regulations instigated by CAAA70. Automobile engines produce incompletely combusted $NO_x(g)$, $CO(g)$, and HCs and completely combusted $CO_2(g)$ and $H_2O(g)$. Catalytic converters convert $NO_x(g)$ to $N_2(g)$, $CO(g)$ to $CO_2(g)$, and unreacted HCs to $CO_2(g)$ and $H_2O(g)$. Since 1975, three types of catalytic converters have been developed: (1) single-bed catalysts, (2) dual-bed catalysts, and (3) single-bed three-way catalysts. In all cases, exhaust gases travel through the converter with a residence time of about 50 milliseconds at a temperature of 250 to 600 °C. Single-bed converters typically use a combination of the metals platinum (Pt) and palladium (Pd) in the ratio 2:1 as the catalysts. These catalysts are applied as a coating over porous alumina spherical pellets, ceramic honeycombs, or metallic honeycombs within the converter to increase the surface area contacting the exhaust. The use of noble-metal catalysts requires the use of unleaded fuel because lead deactivates the catalyst through chemical reaction. Thus, the implementation of the catalytic converter conveniently resulted in the forced reduction in emissions of another pollutant, lead.

Single-bed catalysts convert $CO(g)$ and unreacted HCs to $CO_2(g)$, but do not control $NO_x(g)$. A second bed was developed to convert $NO_x(g)$ to $N_2(g)$ or $N_2O(g)$. The catalyst in the second converter is typically rhodium (Rh), ruthenium (Ru), Pt, or Pd.

The three-way catalyst, developed in 1979, allowed for the simultaneous oxidation of unreacted HCs and $CO(g)$ and reduction of $NO_x(g)$ in a single bed. The use of this catalyst requires a specific input air to fuel ratio of 14.8:1 to 14.9:1 and a temperature range of 350 to 600 °C for it to remain effective in converting all three groups of compounds. At higher ratios, $CO(g)$ and HCs are converted efficiently, but $NO_x(g)$ is not. At lower ratios, the reverse is true. At temperatures below 35 °C, conversion efficiency falls off fast. At 25 °C, it is near zero. The catalysts in the three-way structures are usually platinum and rhodium at a ratio of Pt:Rh = 5:1

8.1.7. Clean Air Act Amendments of 1977

In 1977, Congress passed the Clean Air Act Amendments of 1977 (CAAA77, Public Law 95-95, August 7, 1977). CAAA77 extended the date for mandated automobile emission reductions of HCs and $CO(g)$ to 1980. The $NO_x(g)$ standard was relaxed from 0.4 g/mi to 1.0 g/mi and the deadline for compliance extended to 1981. A 0.8 grams per gallon (g/gal) standard was also introduced for lead. The lead standard was tightened in 1980 to 0.5 g/gal. In 1980, the U.S. EPA also set limits on diesel fuel particulate emissions and required $CO(g)$ emissions from heavy-duty trucks to be reduced by 90 percent by 1984.

In 1977, most U.S. states had nonattainment areas where at least one NAAQS had not been achieved. CAAA77 required states that had at least one nonattainment area to describe, in a revised SIP, how they would achieve attainment by December 31, 1982.

CAAA77 also formalized a permitting program, initiated by the U.S. EPA in 1974, to prevent significant deterioration (PSD) of air quality in regions that were already in attainment of NAAQSs. Under the program, three classes of regions, Class I, II, and III regions, were designated. Class I regions included pristine areas, such as national and international parks and national wilderness areas, where no new sources of pollution were allowed. Class II regions included areas where moderate changes in air quality were allowed, but where stringent regulations were desired. Class III regions included areas where major growth and industrialization were allowed so long as pollutant levels did not exceed NAAQS. Before a new pollution source can be built or an existing source can be modified to increase pollution in a PSD region that allows growth, a PSD permit must be obtained. To obtain a permit, the polluter proposing the new source or change must ensure that the **best available control technology** (BACT) will be installed and the resulting pollution will not lead to a violation of an NAAQS. A BACT is a pollution control technology that results in the removal of the greatest amount of emissions from a particular industry or process.

CAAA77 also mandated that computer modeling be performed to check whether each proposed new source of pollution could result in an exceedence of emission limits or in a violation of an NAAQS. CAAA77 contained the first regulations in which the federal government attempted to control the emissions of chlorofluorocarbons, precursors to the destruction of stratospheric ozone.

8.1.8. Clean Air Act Amendments of 1990

On November 15, 1990, Congress passed the Clean Air Act Amendments of 1990 (CAAA90, Public Law 101-549), which was "An Act to amend the Clean Air Act to provide for attainment and maintenance of health protective national ambient air quality standards, and for other purposes." The primary goals of CAAA90 were to ameliorate problems related to urban air pollution, air toxics, acid deposition, and stratospheric ozone reduction. CAAA90 was motivated in part by the fact that CAAA70 had not eliminated air pollution problems in the United States. In 1990, 96 U.S. cities were still in violation of the NAAQS for ozone, 41 cities were in violation of the NAAQS for carbon monoxide, and 70 cities were in violation of the NAAQS for PM_{10}. Only seven air toxics had been regulated with NESHAPS between 1970 and 1990, although many more had been identified.

CAAA90 was similar to CAAA70 with respect to NAAQSs, NSPSs, and PSD standards. Major changes were enacted with respect to other issues. In particular, nonattainment areas for $O_3(g)$, $CO(g)$, and PM_{10} were divided into six classifications, depending on the severity of nonattainment, and each district in which nonattainment occurred was given a different deadline for reaching attainment. For ozone, only Los Angeles was designated as "extreme" and given until 2010 to reach attainment of the NAAQS. Baltimore and New York were designated as "severe" and given until 2007. Chicago, Houston, Milwaukee, Muskegan, Philadelphia, and San Diego were also designated as "severe" but slightly less so, and given until 2005 to reach attainment.

For attainment to be achieved, all new pollution sources in nonattainment areas, regardless of size, were required to obtain their **lowest achievable emissions rate** (LAER), which is the lowest emissions rate achieved for a specific pollutant by a similar source in any region. LAERs were required to be less stringent than NSPS for the source. To achieve LAERs, states or ACQRs were required to adopt **reasonably achievable control technologies** (RACTs) for all existing major emission sources. RACTs are control technologies that are reasonably available and technologically and economically feasible. They are usually applied to existing sources in nonattainment areas.

Per CAAA90, state or local air quality districts overseeing a nonattainment area were required to develop emission inventories for ROGs, $NO_x(g)$, and $CO(g)$. Emissions from mobile, stationary point, area, and biogenic sources were to be included in the inventories. The 1990 Act also mandated that computer modeling be carried out with current and projected future inventories to demonstrate that attainment could be obtained under proposed reductions in emissions.

CAAA90 created a list of 189 hazardous air pollutants (HAPs) from hundreds of source categories. Under CAAA90, the U.S. EPA was required to develop emission standards for each source category under a timetable. For each new or existing source anticipated to emit more than 10 short tons per year of one HAP or 25 short tons per year of a combination of HAPs, the U.S. EPA was required to establish a maximum achievable control technology (MACT) to reduce hazardous pollution from the source. In selecting MACTS, the U.S. EPA was permitted to consider cost, non-air quality health and environmental impacts, and energy requirements. Because the use of a MACT does not necessarily mean that hazardous pollutant concentrations will be reduced to a safe level, the U.S. EPA was also required to consult with the Surgeon General to evaluate the risk resulting from the implementation of each MACT.

The control of toxic air pollutants under CAAA90 differed from the control of criteria air pollutants. In the former case, the U.S. EPA was required to develop a program to control toxic emissions; in the latter, states were required to develop programs to control criteria air pollutant emissions and their ambient concentrations.

CAAA90 also tightened emission standards for automobiles and trucks, required additional reductions in emissions of acid-deposition precursors, established a federal permitting program for point sources of pollution (previously, states were responsible for permits), and mandated reductions in the emission of chlorofluorocarbons (CFCs) to combat stratospheric ozone reduction. With respect to acid deposition, CAAA90 required a 10 million ton reduction in sulfur dioxide emissions from 1980 levels and a 2 million ton reduction in nitrogen oxide emissions from 1980 levels. With respect to CFCs, CAAA90 required a complete phase-out of CFCs, Halons (synthetic bromine containing compounds), and carbon tetrachloride by 2000 and methyl chloroform by 2002. CAAA90 also required the U.S. EPA to publish a list of safe and unsafe substitutes for these compounds and a list of the ozone-depletion potential, atmospheric lifetimes, and global-warming potentials of all regulated substances suspected of damaging stratospheric ozone. CAAA90 banned the production of nonessential products releasing ozone-depleting chemicals. Such products included aerosol spray cans releasing CFCs and certain noninsulating foam.

8.1.9. Clean Air Act Revision of 1997

In 1997, Congress passed the 1997 Clean Air Act Revision (CAAR97), by which it modified the NAAQSs for ozone and particulate matter (see Table 8.2). The new federal ozone standard is based on an 8-hour average rather than a 1-hour average. The new averaging time is intended to protect those who spend time working or playing outdoors, the people most vulnerable to the effects of ozone.

The particulate standard was modified to include a new $PM_{2.5}$ standard because studies have shown that aerosol particles ≤ 2.5 μm in diameter have more effect on respiratory illness, premature death, and visibility than do larger aerosol particles (Section 5.6). The purpose of the new $PM_{2.5}$ standard was to reduce risks associated with disease and early death associated with $PM_{2.5}$.

8.1.10. Regulation of Interstate and Transboundary Air Pollution in the United States

As described in Section 6.6.2.4, long-range transport of air pollution is a problem that affects most countries. International efforts to control transboundary air pollution are discussed with respect to acid deposition in Section 10.7. Here, control of long-range transport in and between the United States and its neighbors is discussed.

Interstate transport of air pollution in the United States is recognized by the government through Section 110 of the Clean Air Act Amendments of 1970, which requires that SIPs be submitted to the U.S. EPA by each state and contain provisions to address problems of emissions transported to downwind states. Section 126 of CAAA70 allows downwind states to file petitions with the U.S. EPA to take action to reduce emissions in upwind states when such emissions make it difficult for the downwind state to meet federal air quality standards.

In 1994, the U.S. EPA itself recognized that it was difficult for some states to meet federal standards merely by reducing emissions in their own state. As a result, the Ozone Transport Assessment Group (OTAG), a partnership between the U.S. EPA, the Environmental Council of the States (ECOS) (37 easternmost states and the District of Columbia), and several industry and environmental groups, was set up in 1995 (and concluded in 1997) to develop a mechanism to reduce ozone buildup due to interstate ozone transport, particularly in the northeast United States. In 1997, during the period of OTAG, eight northeastern states filed petitions under Section 126 of the Clean Air Act Amendments requesting the U.S. EPA to take action against 22 upwind states (Alabama, Connecticut, Delaware, Georgia, Illinois, Indiana, Kentucky, Massachusetts, Maryland, Michigan, Missouri, North Carolina, New Jersey, New York, Ohio, Pennsylvania, Rhode Island, South Carolina, Tennessee, Virginia, Wisconsin, and West Virginia) and the District of Columbia for emitting excess oxides of nitrogen [$NO_x(g)$]. Such emissions were suspected of exacerbating ozone in the petitioning states. In September 1998, the U.S. EPA implemented a rule [known as the $NO_x(g)$ SIP Call] that required these 22 states and the District of Columbia to submit SIPs that addressed the regional transport of ozone. The rule required reductions of $NO_x(g)$ but not reactive organic gases (ROGs) in these states by 2003.

Transboundary pollution between the United States and Canada is formally recognized through the Canada/U.S. Air Quality Agreement. Article V of the agreement states that Canada must notify the United States of any proposed projects within 100 km of the border that would be likely to emit more than 90 tons/year of $SO_2(g)$, $NO_x(g)$, $CO(g)$, TSP, or ROGs. On a smaller scale, the United States and Mexico implemented a cooperative study, the Big Bend Regional Preliminary Visibility Study, to examine causes of visibility reduction in Big Bend National Park. In 1998, this preliminary study concluded that sources in both the United States and Mexico degraded visibility in the park, depending on the wind conditions.

8.1.11. Smog Alerts

Because ozone is a criteria pollutant and its mixing ratios exceed federal and state standards more than do those of any other pollutant and because high levels of ozone are often good indicators of the severity of photochemical smog problems, many cities in the U.S. and now worldwide issue smog alerts when ozone levels reach certain plateaus. Smog alert levels exist for pollutants aside from ozone as well. Table 8.3

Table 8.3. Mixing Ratios Required before the Given 1-Hour Standard Is Exceeded

Health Standard Level	Ozone 1-Hour Average Mixing Ratio (ppmv)	Carbon Monoxide 1-Hour Average Mixing Ratio (ppmv)	Nitrogen Dioxide 1-Hour Average Mixing Ratio (ppmv)
California standard	0.09	20	0.25
Federal standard (NAAQS)	0.12[a]	35	—
Health advisory	0.15[b]	—	—
Stage 1 smog alert	0.20[b]	40[b]	0.60[b]
Stage 2 smog alert	0.35[b]	75[b]	1.20[b]
Stage 3 smog alert	0.50[b]	100[b]	1.60[b]

[a]Prior to 1997.
[b]Applies to California.

identifies the mixing ratios of ozone, carbon monoxide, and nitrogen dioxide required to trigger a California standard violation, a federal standard violation, a California health advisory, and a California Stage 1, 2, and 3 smog alert.

Smog alerts in Los Angeles have been in place since the 1950s. Ozone levels required for a Stage 1, 2, or 3 alert are based on the relative health risk associated with each level. When an ozone smog alert occurs, individuals are advised to limit outdoor activity to morning or early evening hours because ozone levels peak in the mid-afternoon. Outdoor activities in schools are usually eliminated during a smog alert. Health advisories for ozone were initiated in California in 1990 by the California Air Resources Board because evidence indicated that ozone levels above the federal standard and below a Stage 1 smog alert may affect healthy, exercising adults. When a health advisory occurs, all individuals, including children and healthy adults, are advised to limit prolonged and vigorous outdoor exercise. Individuals with heart or lung disease are advised to avoid outdoor activity until the advisory is no longer in effect.

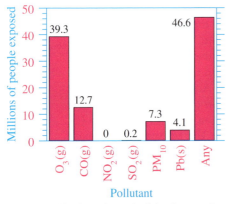

Figure 8.1. Number of people living in counties exposed to pollutants above federal primary air quality standards (NAAQS) in 1996.

Source: U.S. EPA Office of Air and Radiation.

8.1.12. U.S. Air Quality Trends from the 1970s to Present

In 1996, more than 46 million people lived in areas of the United States in which at least one NAAQS was violated, as shown in Fig. 8.1. Most people exposed to high pollutant levels were exposed to ozone.

The U.S. districts in which air quality problems exist are in southern and central California, the Boston through Washington (BoWash) corridor, the Milwaukee through Chicago Great Lakes Region, Phoenix, El Paso, Dallas–Ft. Worth, Houston, Baton Rouge, and Atlanta.

Although pollution levels are far from ideal in the United States, (e.g., Fig. 8.1), they have improved in many urban U.S. cities since the 1950s. Much of the improvement can be attributed to regulations arising at the district (AQCR), state, and federal levels. Arguably, the Los Angeles Air Pollution Control District, formed in 1947 and now called the South Coast Air Quality Management District, has been at the forefront in initiating regulations that were ultimately adopted at the federal level. Improvements in air

Table 8.4. Percentage Changes in the Number of Federal Primary Standard (NAAQS) Exceedences and Tons of Emissions in the United States from 1988 to 1997 for Several Pollutants

Pollutant	Percentage Change in Number of NAAQS Exceedences 1988–1997	Percentage Change in Tons of Emissions 1988–1997
Ozone [$O_3(g)$]	−19	—
Carbon monoxide [$CO(g)$]	−38	−25
Nitrogen dioxide [$NO_2(g)$]	−14	−1
Sulfur dioxide [$SO_2(g)$]	−39	−12
Particulate matter $\leq 10\mu m$ (PM_{10})	−26	−12
Lead [$Pb(s)$]	−67	−44
Reactive organic gases (ROGs)	—	−20

Source: U.S. EPA Office of Air and Radiation.

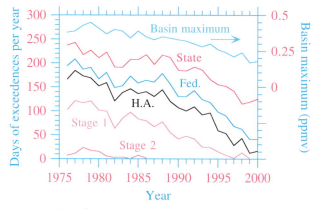

Figure 8.2. Days per year that the ozone mixing ratio in the Los Angeles Basin exceeded the California state standard (State), the NAAQS (Fed.), the California health advisory (H.A.) level, the Stage 1 smog alert level, and the Stage 2 smog alert level from 1976 to 2000. Also shown is the maximum ozone mixing ratio each year in the basin.

Source: South Coast Air Quality Management District.

quality resulting from regulation are phenomenal, particularly in Los Angeles. Figure 8.2 shows the number of days per year that state standards, federal standards, health advisory levels, Stage 1 alerts, and Stage 2 alerts were exceeded between 1975 and 2000 in the Los Angeles Basin. The figure shows that the California state ozone standard, exceeded 237 days per year in 1976, was exceeded only 125 days per year in 2000. Between 1976 and 2000, NAAQS exceedences were reduced from 194 to 40 days per year, and Stage 2 and Stage 1 alerts were eliminated. Stage 3 alerts, frequent in the 1950s, when peak ozone mixing ratios reached 0.68 ppmv, have not occurred in Los Angeles since prior to 1976. Figure 8.2 shows that, between 1976 and 2000, the highest ozone level recorded in the Los Angeles Basin was 0.45 ppmv (1979).

Nationwide, pollutant levels have likewise decreased since the late 1980s. Table 8.4 shows the percentage changes in (a) the number of federal primary standard exceedences and (b) emissions of criteria air pollutants and volatile organic compounds between 1988 and 1997. The table indicates that the number of exceedences has dropped for all criteria pollutants, and the number of emitted tons has decreased for all pollutants shown, even though the number of automobiles and other pollutant sources

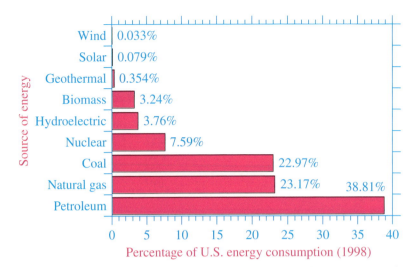

Figure 8.3. Percentage U.S. energy consumption by source in 1998. Total consumption was 94.2 quadrillion Btu.

Source: U.S. Energy Information Administration (EIA).

increased during this period. The U.S. EPA Office of Air and Radiation reports that between 1970 and 1996, the U.S. population increased 29 percent, vehicle miles traveled increased 121 percent, and gross domestic product increased 104 percent, but total emissions of the six major emitted pollutants decreased 32 percent. The reason is due to the widespread use of emission reduction technologies invented following CAAA regulations of emissions and outdoor concentrations of air pollutants.

The improvement in air quality in the United States since 1970 and the corresponding improvement in the economy attests to the fact that air pollution regulations do not damage an economy. Instead, air pollution regulations lead to inventions and new or expanded industries. Areas of invention include air pollution control technologies, engine technologies, renewable energy technologies, and improved fuel technologies. New or expanded industries include the renewable energy industries, the pollution control device industry, the pollution measurement device industry, the pollution remediation industry, the pollution software industry, and the pollution modeling industry. Regulations have also resulted in the employment of public- and educational-sector workers in the area of pollution/climate regulation, policy, science, and engineering and have led to the expansion of the supercomputer industry to satiate the demand for the researchers devoted to studying and mitigating air pollution and climate problems.

One of the industries that has expanded as a result of air pollution regulation is the renewable energy industry. Between 1988 and 1998, consumption of renewable energy (hydroelectric, biomass, geothermal, solar, and wind energy) in the United States increased by 19.9 percent, compared with an increase in comsumption of fossil-fuel energy (petroleum, natural gas, and coal) of 12.3 percent. During the same period, consumption of nuclear energy, a non-air polluting energy (except in the case of a radioactive leak), also increased by 26.5 percent. Although alternative energy consumption has increased in the United States, it represented only 7.4 percent of total consumption in 1998. Fossil fuels were the source of 85 percent of total energy consumed (Fig. 8.3). Figures 8.4 to 8.6 show examples of renewable energy sources.

Figure 8.4. Reflectors focusing solar energy onto a 10-megawatt receiver power tower, called Solar One, in Barstow, California. Photo by Sandia National Laboratory Staff, available from the National Renewable Energy Laboratory.

Figure 8.5. Wind machines at dusk at a wind farm in Palm Springs, California. Photo by Warren Gretz, available from the National Renewable Energy Laboratory.

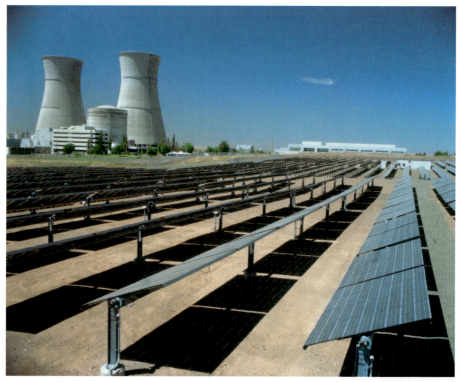

Figure 8.6. Solar photovoltaic system colocated with a nuclear power plant at the Sacramento Municipal Utility District's Rancho Seco facility. The photovoltaic system (2 megawatts), spread over 8,094 m², produces power for 660 homes. Photo by Warren Gretz, available from the National Renewable Energy Laboratory.

Table 8.5. Number of Days per Year That California State Visibility Standard Was Exceeded

Location	1990	1994
Azusa	No data	91
Burbank	180	No data
Lancaster	14	No data
Long Beach	155	No data
Los Angeles	154	No data
Ontario	250	No data
Riverside	200	No data
San Bernardino	200	176

Source: South Coast Air Quality Management District.

8.1.13. Visibility Regulations and Trends

Currently, no U.S. national standard exists for visibility degradation. California first set a visibility standard in 1959 and modified it in 1969. The 1969 standard required that the prevailing visibility outside of Lake Tahoe exceed 10 miles (16.09 km) when the relative humidity was less than 70 percent. In Lake Tahoe, the minimum allowable visibility was set to 30 miles (48.3 km). Measurements were made by a person looking

Table 8.6. Percentage Contributions to Visibility Reduction in the Western and Eastern United States as a Result of Different Aerosol Particle Components

Substance	West	East
Sulfates	25–65	>60
Organic matter	15–35	10–15
Nitrates	5–45	10–15
Black carbon	15–25	10–15
Soil dust	10–20	10–15

Source: U.S. EPA Office of Air Quality Planning and Standards.

for landmarks a known distance away. The furthest landmark that could be seen along 180° or more of the horizon circle defined the prevailing visibility.

Because prevailing visibility is a subjective measure of visibility, the California Air Resources Board changed the California visibility standard in 1991 to one based on the use of the meteorological range. The new standard required that the meteorological range at a wavelength of 0.55 μm not be less than 10 miles (16.09 km) outside of Lake Tahoe and 30 miles (48.3 km) in Lake Tahoe when the relative humidity was less than 70 percent.

Table 8.5 shows the number of days per year that the visibility standard was exceeded at different locations in Los Angeles in 1990 and 1994. The 1990 data were based on the prevailing visibility standard and the 1994 data were based on the meteorological range standard. The data indicate that in the eastern Los Angeles Basin, visibility was less than the state limit on 50 percent or more of the days of the year in 1990 and 1994. Such visibility degradation was due primarily to aerosol particle buildup in the eastern basin.

Although no visibility regulations exist on a national level, prevailing visibility data have been collected at 280 monitoring stations at airports in the United States since 1960. These data have been used primarily for air-traffic control purposes. Statistical analysis of the data by the U.S. EPA Office of Air and Radiation indicate that visibility deteriorated in the eastern United States between 1970 and 1980, but improved slightly between 1980 and 1990. Visibility changes during these two decades correlate with similar trends in emissions of sulfur dioxide. Table 8.6 shows an estimated percentage contribution to visibility reduction in the western and eastern United States due to different pollutants. The largest contributor to visibility impairment in the east and west was sulfates. Visibility impairment in the east was generally worse than that in the west because higher relative humidities in the east resulted in aerosol particles with higher liquid water contents and greater size than in the west. Figure 7.22 showed maps of summer versus winter visibility in the United States between 1991 and 1995.

8.2. POLLUTION TRENDS AND REGULATIONS OUTSIDE THE UNITED STATES

Whereas the United States has made an effort to reduce urban smog since 1955, other countries have also made efforts, some over a period of decades and others over a period of a few years. In the 1950s, Los Angeles, California, may have been the most polluted city in the world in terms of ozone, and London may have been the most polluted in terms of particulates. Today, Mexico City, New Delhi, Beijing, Cairo, Sao

Paolo, Shanghai, Jakarta, Bangkok, Tehran, and Calcutta are among the most polluted. The World Health Organization (WHO) calculates that twelve of the fifteen cities with the highest particulate levels and six of the fifteen cities with the highest sulfur dioxide levels are in Asia. The WHO also calculates that about 2.7 million people die each year from air pollution: 900,000 in cities and 1.8 million in rural areas. The largest source of mortality from air pollution is indoor burning of biomass and coal (WHO, 2000). Whereas health problems from lead have declined in the United States, western Europe, and Japan, they have increased in many other countries. Lead levels found in children in Cairo, Jakarta, Mexico City, and many cities in Africa and China are still high. In this section, air quality trends and regulations in several countries are discussed.

8.2.1. Canada

Canadian air today is relatively clean in comparison with that in many other countries. Canada's yearly averaged concentrations of sulfur dioxide, nitrogen dioxide, and particulates are less than standards for these pollutants set by the WHO. Nevertheless, pollution is still a problem in many Canadian cities, and acid deposition affects its lakes and forests.

Vehicles are the largest source of air pollution in Canada, responsible for 60 percent of emissions. An increase in emissions of approximately 45 percent since 1990 is attributable to an increase in the number of vehicles, particularly sport utility vehicles (SUVs). Although emission controls have improved in vehicles during the same period, they have not compensated for the increase in SUV sales (EIA, 2000).

Canada has been concerned with air pollution since at least the late 1960s. In 1969, it initiated an air pollution monitoring network called the National Air Pollution Surveillance Network (NAPS). This network contains nearly 240 air quality monitoring stations, located mostly in urban areas. More recently, in 1999, Canada enacted a new Canadian Environmental Protection Act (CEPA), under which the government obtained new powers to control emissions, particularly of toxic substances. PM_{10} was declared toxic, allowing it to be regulated under the act. The act reduced the allowable level of sulfur in gasoline and diesel, reduced the allowable level of benzene in gasoline, and doubled funding for outdoor air pollution monitoring (Environment Canada, 2000a).

Ontario is the most polluted province in Canada. Environment Canada, the Canadian national environmental agency, ordered refineries in Ontario to reduce sulfur dioxide emissions by 2005. The government of Ontario also requires the measurement of all harmful emissions from power plants.

A long-term goal of the Canadian government is to implement a Canada-wide 65 ppbv ozone standard by 2010. To obtain this goal, Canadians must reduce $NO_x(g)$ and VOC emissions by about 35 percent by 2010. Canada also plans to increase reliance on renewable energy sources. In 1998, about 29 percent of its electricity came from hydroelectric power. The government intends to increase this source by 2 percent in the coming years (EIA, 2000).

8.2.2. Mexico

In the 1940s, Mexico City's air was relatively clean. Since 1940, its human population has grown from 1.8 million to 20 million (making it the most populated city in the world), its vehicle population has increased from tens of thousands to 3.5 million, and its factory population has skyrocketed. Today, Mexico City's air is among the dirtiest in the world.

Like Los Angeles, Mexico City sits in a basin surrounded by mountains, and pollution is frequently trapped beneath the Pacific high-pressure system. Unlike Los Angeles, Mexico City is not bounded by an ocean on one side.

Air pollution regulations in Mexico City were first enacted in 1971, with a Decree to Prevent and Control Environmental Pollution, under which monitors were to be set up to measure outdoor pollutant mixing ratios. A full set of monitors was implemented only by 1986, at which time ozone was found to be the main pollutant. Ambient standards were then set for ozone, carbon monoxide, nitrogen dioxide, sulfur dioxide, total suspended particulates, and lead. In 1986, low-leaded gasoline was introduced, but by 1996, 44 percent of gasoline sold was still leaded (EIA, 2000). Other steps the government took to reduce pollution were to provide incentives to relocate industry out of Mexico City (1978) and to require vehicle inspections (1989).

In 1989, Mexico City enacted a program by which only cars with odd-numbered license plates could be driven on a given day and cars with even-numbered plates could be driven on the next. The program did not reduce pollution because many people violated its spirit by switching license plates or cars. In other cases, trips were postponed by a day, but not canceled. More recently, a system was enacted that allows only cars and trucks that use alternative fuels to be used seven days a week.

Ozone levels in Mexico City exceeded national standards 324 days in 1995. Also in 1995, Mexico initiated a five-year National Environmental Program (1996–2000) designed to spend $13.3 million to clean up air pollution in Mexico City. The government also passed a tax-incentive program for the purchase of pollution-control equipment.

Although Mexico City was named by the World Resources Institute as the most dangerous city in the world for children in terms of air pollution in 1999, the pollution that year was the city's lowest in a decade. Still, in August 2000, a smog event required factories to shut down and students to stay indoors.

8.2.3. Brazil

Most of Brazil is covered with the Amazon rainforest, which makes up 30 percent of the world's remaining tropical forests. Sao Paolo, the capital of Brazil, is the second-most populated city in the world after Mexico City. Air pollution in Sao Paolo results from a large number of high-emitting automobiles, poor road systems, and low fuel prices. Road-construction and railway projects in the city are expected to alleviate traffic by 2002. Another city, Curitiba, was the fastest growing city in Brazil in the 1970s. Curitiba has reduced its air pollution, despite population growth, by designing an efficient road network and increasing the use of public transportation.

Brazil has an alcohol-fuel program, the Brazilian National Alcohol Program, which was started in 1975 to reduce Brazil's reliance on imported fuel following the worldwide spike in oil prices in 1973. The program ensured that all gasoline sold in Brazil contained 22 percent anhydrous ethanol and that the price of the ethanol–gasoline blend would remain similar to that of gasoline alone. Ethanol for the project was produced primarily from sugarcane. Because the market price of an ethanol–gasoline blend was higher than that of pure gasoline, the program required government subsidies. When gasoline prices fell and sugarcane prices increased in the late 1980s, ethanol prices rose sharply, and the program disintegrated. In 1997, only 1 percent of new cars sold in Brazil used ethanol. Nevertheless, 41 percent of Brazil's vehicles still run on ethanol. In 1999, the Brazilian government revived the alcohol-fuel program by encouraging the replacement of taxis and government vehicles with new vehicles that run on 100 percent ethanol. The government also hopes to increase the percentage of ethanol in ethanol–gasoline blends from 24 its 2000 level, to 26, but the high price of ethanol may prevent this (EIA, 2000).

Although the use of ethanol fuel instead of gasoline reduces the emission of aromatic hydrocarbons in smog, it increases the production of acetaldehyde (Section 4.3.7), leading to ozone and PAN formation. Mixing ratios of PAN, an eye irritant in smog, are high in several Brazilian cities.

Brazil has a strong renewable-energy program. About 95 percent of its electricity is obtained from hydropower, derived mostly from the Amazon River. About 20 million Brazilians live without electricity. The government plans to provide electricity to these people through renewable-energy sources, such as solar and wind energy.

8.2.4. Chile

The air in Santiago, Chile, ranks among the most polluted in the world, especially in terms of particulates. The pollution is caused by vehicles, industrial emissions, unpaved road dust, and dust from eroded land. Between 1980 and 1997, Santiago's population increased from 4.3 to 5.8 million, its number of private cars increased from 312,000 to 736,000, and its number of taxis increased from 16,500 to 54,000 (Jorquera et al., 2000). The city, nestled in a valley, frequently experiences a large-scale subsidence inversion. As such, winds are often light and rainfall is scarce, exacerbating pollution problems.

Pollution regulations in Santiago were first enacted in 1987. That year, a law was passed allowing only cars with specified last digits on their license plates to operate within the city limits on a given workday. Starting in 1989, inspections of carbon monoxide, particulate matter, and hydrocarbon emissions from vehicles were mandated. In 1992, the Chilean government outlined a long-term plan to combat air pollution. The two main control strategies were regulation of sources and economic incentives. This plan culminated in the 1994 "Basic Law about the Environment," which set up a framework for environmental management. In the meantime, in 1993, a law was passed requiring catalytic converters on all new cars. Such cars were not subject to traffic bans. The same year, emissions from stationary industrial sources were regulated, and open-burning wood stoves without particulate abatement equipment were banned. In 1995, a measure requiring stationary sources to shut down when particulate concentrations rose above a critical level was passed. In 1997, laws controlling emissions from nonindustrial stationary sources and residential heating boilers were passed. Another method of reducing pollution in Santiago was to eliminate old buses. Between 1980 and 1990, the number of buses in the city increased from 9,500 to 13,000, but by 1997, the number had decreased to 9,000 (Jorquera et al., 2000).

Pollution abatement in Santiago has been somewhat effective. Ozone levels decreased between 1989 and 1995, but they did not change from 1995 to 1998. Between 1989 and 1998, particulate levels decreased by 1.5 to 7 percent per year. Monthly average $PM_{2.5}$ levels still range from 100 to 150 μg m^{-3} (Jorquera et al., 2000). Black carbon and organic matter levels in Santiago are still more than seven times those in Los Angeles (Didyk et al., 2000).

8.2.5. European Union

The European Union (E.U.; previously the European Community) today consists of fifteen member countries: Austria, Belgium, Denmark, Finland, France, Germany,

Greece, Ireland, Italy, Luxembourg, the Netherlands, Portugal, Spain, Sweden, and the United Kingdom. The European Community was established in 1957 (originally with six member countries) under the Treaty of Rome to integrate the countries of Europe economically and politically. More recently, the E.U. has taken action to improve air quality across Europe.

Since 1970, the E.U. has enacted more than 700 environmentally friendly laws in the form of directives, regulations, and decisions. Most laws are **directives**, which are binding on all member nations but take into account particular conditions in each nation. For example, a directive for emission reductions from a power plant may allow each member nation to reduce emissions by a different amount. About 10 percent of E.U. laws are **regulations**, which are laws applied uniformly to all member nations. An example of a regulation is the appointment of a qualified person to inspect the use of a dangerous chemical. **Decisions**, which are the least common form of environmental law, are requirements that may be directed at specific member nations or may be a modification of a regulation.

Some directives that have been passed in the E.U. include those limiting the allowable amount of ozone, carbon monoxide, sulfur dioxide, nitrogen dioxide, oxides of nitrogen, benzene, particulate matter, and lead in ambient air, and those limiting emissions from point and mobile sources. Although E.U. directives have strengthened controls on emissions from gasoline and diesel fuel, about 25 percent of passenger cars and nearly all trucks in Europe still run on diesel fuel, which produces soot, an unhealthy aerosol particle component that enhances global warming. A directive in 1989 required that all new cars in the E.U. run on unleaded fuel, and leaded fuel was all but eliminated from the E.U. automobile fleet by January 2000.

8.2.6. United Kingdom

In January 1956, soon after the U.S. Air Pollution Control Act was passed, the **United Kingdom (U.K.) Clean Air Act** also passed. The U.K. Act was a response to the devastating smog event in London that killed 4,000 people in 1952. It controlled both household and industrial emissions of pollution for the first time in the United Kingdom, but it dealt only with smoke, particularly black and dark smoke. It did not deal with sulfur dioxide, although presumably reductions in smoke would also reduce sulfur dioxide. The act resulted in 90 percent of London being controlled by smoke regulations. The act resulted in smokeless zones in London and the relocation of many power plants to rural areas.

The next major piece of air pollution legislation in the U.K. was the Clean Air Act of 1968, which required that industries burning fossil fuels (solid, liquid, or gas) build tall chimneys so that their emissions would not deposit to the ground locally.

The U.K. joined the E.U. in 1972. Since then, E.U. directives on the controls of pollution have been followed in the U.K. Some of these directives were formalized in the 1990 U.K. Environmental Protection Act, which permitted local governments to control emissions from small industrial processes and set up a statutory system for local air quality management. The national government still controlled emissions from large sources under the 1956 and 1968 Clean Air Acts. In 1992, U.K. vehicle emission standards were strengthened in accordance with an E.U. directive. In 1993, catalytic converters were required in all new gasoline-powered vehicles, again in accordance with an E.U. directive. In 1995, the U.K. passed the Environment Act, which outlined the need for scientific studies of the effects of air pollution and the need for new health-based air quality standards by 2005. The Act required the publication of a plan to combat air pollution. In 1997, a plan called the *United Kingdom National Air Quality Strategy* was published. Under this strategy, ambient limits for ozone, nitrogen dioxide, sulfur dioxide, carbon monoxide, benzene, 1-,3-butadiene, lead, and particulates were set.

Prior to the 1980s, most energy in the U.K. was obtained from coal, a dirty fuel that caused much of the smoke problems experienced in London and other big cities. Between 1980 and 1998, coal consumption decreased by 57 percent, a factor that enabled air quality to improve. By 1998, the U.K. was obtaining 37 percent of its energy from oil, 34 percent from natural gas, 14 percent from coal, 12 percent from nuclear power, 1 percent from hydroelectric power, and 2 percent from other renewable sources (EIA, 2000). Although there are fewer vehicles in the U.K. than in many other countries of the E.U., those in the U.K. are driven more miles than are those in other E.U. countries, resulting in vehicles being one of the largest sources of pollution in the U.K. today. About 25 percent of vehicles in the U.K. run on diesel fuel, which produces soot.

The U.K. instituted an energy tax effective April 2001. Renewable energy sources are exempt. By 2010, reliance on renewable energy is expected to increase from 2 to 10 percent. The U.K. has also agreed to E.U. directives to improve energy efficiency of new cars 25 percent by 2008.

8.2.7. France

Because it lacked domestic fossil-fuel energy sources and desired independence from reliance on imported fuel, France developed a large nuclear-power industry in the 1970s. The use of nuclear energy has reduced France's reliance on fossil fuels in its energy sector. In 1998, 38 percent of France's energy was from nuclear power, 42 percent was from oil, and 14 percent was from natural gas (EIA, 2000). Vehicles are now the major source of air pollution in France; Paris alone is clogged with 3 million vehicles, leading to smog events and health problems. Carbon dioxide emissions from vehicles increased dramatically in the 1990s, but due to improved vehicle emission standards, emissions of carbon monoxide, reactive organic gases, oxides of nitrogen, and particulate matter decreased.

Modern air pollution monitoring in Paris began in 1956 when the Laboratoire d'Hygiène installed an outdoor surveillance network. In 1972, a monitoring network that measured pollution from automobiles near roads was introduced, and in 1973, a 25-station monitoring network that measured pollution near electric utilities was funded. In 1979, the French Ministry of the Environment created AIRPARIF, a government agency responsible for measuring pollution and assessing its impacts in the Paris area. The agency combined existing with new monitoring stations to produce a network of 75 stations. Pollutants measured include nitrogen dioxide, carbon monoxide, ozone, sulfur dioxide, particulates, and lead.

Since December 1990, the French Environment and Energy Control Agency, a public industrial and commercial corporation under the supervision of the Ministries of the Environment, Industry, and Research, has overseen air pollution regulation in France. In 1992, this agency set up a national air quality index, "ATMO." The index, which ranges from 1 (cleanest air) to 10, describes overall air quality of a region during a day. In 1994, the agency set up a nationwide system of alerting the public to high levels of pollution. In 1997, a system by which Parisians would be asked to reduce driving or take alternate routes if heavy smog was present or impending was designed. The agency has also encouraged the use of public transport and clean fuels. France passed an ecological tax that penalizes polluters and reduces taxes for those whose conduct is "ecologically useful." Although France leads the E.U. in nonemitting energy consumption, further improvement in its renewable energy capacity is limited by subsidies that favor nonrenewable energy.

8.2.8. Spain

In the late 1970s and early 1980s, two of the largest cities in Spain, Madrid and Bilbao, had some of the country's most serious air quality problems. Particulate levels in Madrid, for example, reached an average of 200 μg m^{-3}, compared with the government's recommendation of 80 μg m^{-3}. Bilboa's carbon monoxide levels were higher than in most other western European cities. Heavy pollution was due to the large number of automobiles. In the late 1970s, Spain had more registered vehicles than all but seven countries of the world. Today, the ratio of automobiles to people is 1:2. Additional pollution sources include oil-fired space heating and industry. Starting in the late 1980s, improvements in environmental protection alleviated some of the pollution in Madrid and Bilboa, but pollution in smaller industrial cities remained an issue, particularly because of the rapid increases in population in these cities. Today, air quality throughout Spain is better than it was in the 1970s and 1980s. Improvements are the result of E.U. and internal regulations focusing on reducing emissions.

Since the mid-1970s, Spain's energy demand has increased by about 75 percent. Of the energy consumed in Spain in 1998, 57 percent was from oil, 10 percent was from natural gas, 13 percent was from coal, 11 percent was from nuclear power, 7 percent was from hydroelectric power, and 2 percent was from other sources (EIA, 2000). Almost all of Spain's oil is imported. Spain has the fourth-largest installed wind capacity in the world after Germany, the United States, and Denmark.

8.2.9. Germany

From the late 1800s until the 1980s, the largest source of energy in Germany was coal. Much of the coal in Germany had a high sulfur content and originated from the **Ruhr region**. In the 1920s, soot, sulfur dioxide, and chemical pollution became so severe and industry so powerful in the Ruhr region that around 1930, the school in Solingen had to shut down for 18 months to accommodate industrial emissions (McNeill, 2000). The leveling of many factories in the Ruhr region during World War II reduced pollution temporarily. In the 1950s, industry rebounded, increasing pollution levels to new heights. In the 1960s, a law in West Germany required new smokestacks to be taller than before. This law reduced pollution locally, but increased it downwind.

In East Germany following World War II, the primary coal burned was lignite (brown) coal, a relatively dirty coal. The region between Dresden (Germany), Prague (Czech Republic), and Krakow (Poland) is rich in lignite coal and is called the **sulfur triangle**. Like in the Ruhr region, steel, coal, cement, glass, ceramic, and iron factories flourished in the sulfur triangle following World War II. The pollution may have killed up to 7,000 people per year in the mid-1970s (McNeill, 2000). After the reunification of Germany, most lignite mines and associated power plants in the former East Germany were shut down and the remaining power plants were retrofitted with pollution-control

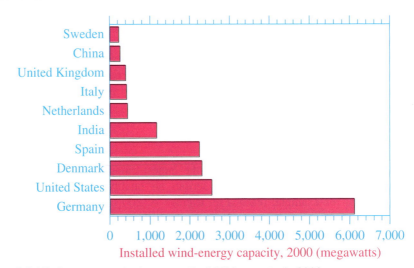

Figure 8.7. Wind-energy capacity in megawatts (MW) by country in 2000.
Source: AWEA (2001).

equipment. Nevertheless, coal is still Germany's only domestic source of nonrenewable energy, and Germany burns oil and coal at one of the highest rates in the E.U.

Today, vehicles are the largest source of pollution in unified Germany. Starting in 1993, leaded gasoline in new passenger cars was banned and catalytic converters were required, consistent with E.U. directives. Emission limits on new cars were tightened again in 1997 and 1999. High ozone levels in cities have encouraged the government to consider speed limit restrictions.

Germany has a strong environmental political party, the Greens. As a result, Germany has tougher environmental legislation than in many other E.U. nations. To promote clean air, the federal government of Germany instituted "eco-taxes" on gasoline and electricity. The first tax went into effect on April 1, 1999. Revenues from the tax are used to fund renewable energy programs. In 1999, Germany proposed that the E.U. tax fossil fuels, but the proposal was defeated. Today, Germany has the largest wind-energy capacity in the world, followed by the United States and Denmark (Fig. 8.7). By contrast, Germany's hydroelectric and solar capacity are relatively small, causing its total renewable energy capacity to lag behind those of France, Spain, and Italy in the E.U. (EIA, 2000). Thirty percent of Germany's electricity is from nuclear power.

8.2.10. Russia

Today, air pollution in Russia affects millions of people and the viability of crops. Average air pollution levels exceed Russian pollution standards in more than 1,200 towns, villages, and cities, affecting the health of 65 million people (44 percent of the population). About 14,000 deaths in 1999 were attributed directly to air pollution (EIA, 2000). Russia has also experienced deforestation over a large portion of its countryside. Pollution from factories (Fig. 8.8) and the 1986 Chernobyl nuclear disaster in nearby Ukraine have contributed to the contamination of populated regions. The Russian State Committee on Environmental Protection has called about 15 percent of Russian territory "ecologically unfavorable."

Figure 8.8. Power plant emissions into Moscow air, November 28, 1994. Photo by Roger Taylor, available from the National Renewable Energy Laboratory.

Russia regulates air pollution through national laws, but few are enforced today, particularly due to a lack of funds. Economic problems have resulted in environmental protection taking a low priority. The Prime Minister of Russia, Vladimir Putin, has stated that the national government cannot solve environmental problems until its economy has improved. As a result, cities are responsible for their own pollution cleanup, which they usually cannot afford. Some regulations have been enacted. Moscow, for example, passed regulations to close down factories and require trucks to stay out of downtown during specified hours. These measures have had little effect. One reason is that the number of vehicles in Moscow has increased in the past few years to more than the road system can handle. About 3 million of the vehicles in Moscow do not have catalytic converters and about one-third of all vehicles run on leaded gasoline. Vehicle exhaust accounts for about 90 percent of pollution in Moscow and more than 50 percent in other Russian cities (EIA, 2000).

8.2.11. Israel

Israel imports almost all of its energy resources. About 25 percent of its energy comes from coal. Most of the rest comes from oil, natural gas, wind, and solar energy. Israel expects that 25 percent of its energy will come from natural gas by 2005 (EIA, 2000). About 3.5 percent of Israel's energy originates from solar power. Vehicle traffic is a major source of pollution in Tel Aviv, Jerusalem, and Haifa. Today, Israel has about 2 million vehicles.

Air pollution in Israel was first controlled under the Abatement of Nuisances Law of 1961, which authorized the Minister of the Environment to define and regulate "unreasonable" air pollution, and also allowed the minister to issue decrees against specific emitters, require factories to comply with emission standards, and monitor air pollution. In 1994, the Ministry of the Environment used its power to design a program to control smog by reducing vehicle emissions and setting up a national air quality monitoring network. Under the program, all new cars were required to have catalytic converters, lead was gradually phased out of cars, and vehicle emission standards were required to conform with those set by the E.U. Industrial emissions of sulfur dioxide, nitrogen oxides, VOCs, heavy metals, particulates, hazardous inorganic particulate matter, and other substances were also controlled. In 1998, the Manufacturers Association of Israel signed a covenant with the Ministry of the Environment agreeing to the terms of the program.

8.2.12. Egypt

Egypt's worst air pollution is in its capital, Cairo, which houses one-fourth of Egypt's population. Cairo has about 1.2 million vehicles. Lead concentrations are up to 20 times higher than maximum levels suggested by the World Health Organization (WHO). In 1994, all children in Cairo were susceptible to lead poisoning (WHO, 2000). Total particulate concentrations in Cairo are three to four times the WHO standard. Air pollution is estimated to cause 10,000 to 25,000 deaths per year (EIA, 2000).

In 1995, a program called the Cairo Air Improvement Project (CAIP), sponsored by the Egyptian government, the Egyptian Environmental Affairs Agency, and the U.S. Agency for International Development (USAID), was initiated to improve air quality. One project of CAIP was to set up an automobile testing program that encourages motorists to tune up their cars, thereby reducing pollution. A second was to replace a lead smelter in Cairo with a new facility outside the city fitted with emission controls. Before the move, three lead smelters produced 40 to 60 percent of the lead in Cairo air. Most of the remaining lead originated from leaded gasoline. Although only unleaded fuel is sold in Cairo, many cars still use it, particularly as it can still be sold in the rest of Egypt until 2002. A third goal of CAIP was to replace public buses with new buses that run on natural gas. CAIP built 36 air pollution monitoring stations throughout the city.

8.2.13. Iran

Iran has severe air pollution problems, primarily due to its 280 percent increase in energy consumption between 1980 and 1998 (EIA, 2000). Most energy consumed has been gasoline, which is inexpensive because oil is a domestic product of Iran. Vehicles are estimated to cause 75 to 80 percent of air pollution in Tehran. About one-quarter of Tehran's 2 million vehicles are at least 20 years old and do not have catalytic converters. Many vehicles run on leaded gasoline, others have leaky engines, and still others emit clouds of smoke. The road infrastructure in Tehran was not designed for the number of vehicles currently in the city. Pollution is exacerbated by the fact that the city is bounded by mountains to the north, which slow the winds, particularly when a large-scale subsidence inversion is present. Tehran's air pollution is rivaled only by that of Mexico City, Beijing, Cairo, Sao Paolo, Shanghai, Jakarta, and Bangkok (EIA, 2000).

Smog events in Tehran have forced closures of elementary schools and the city center. They have also forced residents to wear face masks when walking outside. Longer-term measures taken in Tehran include the requirement that only vehicles with odd or even license plates can enter the city on a given day. In December 1999, the mayor of Tehran announced plans to phase out old automobiles, a measure that was expected to reduce pollution by 16 percent.

8.2.14. India

India is the second-most populous country in the world after China. The rapid population growth in India, from 300 million in 1947 to more than 1 billion today, has strained resources and increased pollution. In 1997, 53 percent of India's energy was from burning coal. Since 1980, vehicle pollution has increased by a factor of eight and industrial pollution has increased by a factor of four (EIA, 2000).

From the 1880s until recently, Calcutta, near the coal fields in West Bengal, was arguably the most polluted city in India. Following 1880, Calcutta developed iron, steel, glass, jute, chemical, and paper industries, powered by coal combustion. By the 1920s, there were 2,500 coal-fired steam boilers whose smokestacks covered the landscape (McNeill, 2000). Early smoke control was made easy by the willingness of the authoritarian colonial government of India to clamp down on smoke emissions. Regulatory controls reduced smoke emissions from the 1910s until the 1950s, when smoke emissions increased again. In the 1980s, an increase in the number of vehicles exacerbated pollution problems. Today, New Delhi and Calcutta are the most polluted cities in India. The WHO lists New Delhi as one of the ten most polluted cities in the world.

In 1976, India passed an amendment to its constitution allowing the government to intervene to protect public health, forests, and wildlife. This amendment was ineffective because another clause in it stated that the amendment was not enforceable in court. In 1984, an accidental release of methyl isocyanate from a pesticide manufacturing plant in Bhopal, India, killed 4,000 people and injured more than 200,000 others. In 1986, the Environmental Protection Act of 1986 was passed, giving the Ministry of Environment and Forests responsibility over environmental legislation. Vehicle pollution accounts for 70 percent of air pollution in India, and car ownership is on the rise. To control further increases in the number of vehicles, New Delhi has limited the sales of vehicles to 4,000 per month, a number that India's Supreme Court recently reduced to 1,500 per month. All new cars must comply with exhaust emission standards.

8.2.15. China

A 1998 WHO study of the air quality in 272 cities found that seven of the ten most polluted cities in the world were in China. China, itself, found that two-thirds of 338 cities in which the air is monitored are polluted. The two major pollutants in China are sulfur dioxide and soot, both emitted during coal burning. China is the largest producer and consumer of coal and has the third-largest coal reserves (after the United States and Russia) in the world. The high sulfur dioxide emissions from coal burning result in acid deposition, both in China and downwind, in Japan and South Korea. Large sources of pollution in China are industrial boilers and furnaces, which consume half of China's coal (EIA, 2000), and residential burning of biomass for home heating and cooking (Streets et al., 2001). Northern China also experiences severe dust storms originating from the Gobi Desert. In spring 2000, storms were more severe than in the previous 50 years, darkening the skies over northern China more than twelve times. The intensity and frequency of the dust storms were linked to soil erosion and deforestation, coupled with drought.

In 1995, China amended its 1987 Air Pollution Control law, calling for the phase out of leaded gasoline by 2000. As of January 1, 2000, all new vehicles in China were required to emit less than 0.005 grams of lead per liter of gasoline. As of July 1, 2000, sulfur and alkene contents of gasoline in Beijing, Shanghai, and Guangzhou were limited to 0.08 percent and 35 percent, respectively. These limits will apply to the rest of China in 2003.

On September 1, 2000, China again amended its Air Pollution Control law. The new law aims to reduce sulfur dioxide emissions from 18.6 million tons in 1999 to 10 million tons by 2010. It also requires the phase-out of dirty coal and provides incentives for the use of low-sulfur, low-ash coal in medium and large cities. New and expanding power plants in these cities must also install equipment that removes sulfur and particulates. Under the previous law, power-plant emitters were charged only if their emissions exceeded a certain standard. Under the new law, emitters will be charged on the basis of their total emissions. In provinces where pollution standards are not met, provincial governments may set up zones and require all factories in each zone to obtain an emission permit.

Under previous law, emissions from only industrial sources and power plants were controlled; under the new law, emissions from vehicles, ships, domestic heating and cooking, and construction are controlled as well. Initially, vehicle emission standards are being set to those enacted in the E.U. in the early 1990s. Individual cities may enact stricter standards. Existing vehicles must be retrofitted to meet the standards and must pass annual emission checks and random inspections. Additional vehicle standards, to be set in 2005, will be similar to E.U. standards of the late 1990s. Standards in 2010 will be similar to international standards set for that year.

Under the new law, cities are given incentives to replace coal-heating stoves in households with centralized heating. All new coal-burning heating boilers are banned from areas in which centralized heating is available. Restaurants in large- and medium-size cities are required to convert from coal to nature gas, liquefied petroleum gas, or electricity. Household cooking stoves will be converted from coal to gas, electricity, or coal briquettes. Construction companies are required to reduce dust emissions. Cities are given incentives to pave or grow plants over bare soil and reduce road dust.

The new law gives tax incentives to encourage the use of low-emitting vehicles and for the production and use of solar, wind, and hydroelectric energy.

Under previous law, enforcement authority resided with the State Council. Enforcement authority under the new law resides with the State Council of Environmental Protection (SCEP), a branch of the State Council. Some enforcement authority has been reserved for provincial, municipal, and county governments. Remedies against pollution violators include fines, civil penalties, and criminal penalties.

8.2.16. Japan

Some early pollution in Japan was due to mining. From the seventeenth century to 1925, the Besshi copper mine and smelter on Shikoku Island produced pollution that damaged crops, enraging farmers. In 1925, a high smokestack and desulfurization equipment reduced pollution from the smelter. The implementation of a tall stack at the Hitachi copper mine in 1905 similarly reduced local complaints about pollution.

In the early 1900s, the expansion of industry in Japan increased urban pollutant emissions. In the 1910s, pollution levels in Osaka, a city developed around coal combustion, were on par with those in other industrial cities of the world (McNeill, 2000).

In 1932, Osaka passed a law requiring industry to increase combustion efficiency and reduce smoke emissions. Instead, the lack of inspectors and an increase in the number of chemical and metallurgy plants increased pollution. During World War II, much of Osaka was leveled, decreasing pollution temporarily. Following the war, industry rebounded, and the population and number of vehicles increased. By the early 1970s, Osake, Kobe, and Kyoto had merged into a large metropolitan region with heavy smoke and vehicle pollution. Tokyo, which had also grown since the late 1800s, similarly experienced pollution problems, mostly from vehicles.

In 1960, the city of Ube, in southern Japan, passed regulations controlling soot and smoke. The success of the regulations catalyzed a national law in 1962 regulating soot and smoke emissions throughout Japan. The regulations did not solve problems immediately, as pollution in major cities continued to intensify. In 1970, following a smog episode in Tokyo, the national government passed a series of pollution-control laws and created Japan's Environment Agency. Following the passage of the U.S. Clean Air Act Amendments of 1970, Japan set its own emission standards for hydrocarbons, oxides of nitrogen, and carbon monoxide from automobiles. As in the United States, pollution problems improved in Japan following the implementation of these standards.

In 1982, the Environment Agency began to regulate hydrocarbons from stationary sources. In 1987, a program to phase out lead in gasoline was implemented. In 1992, a law was passed aimed at reducing nitrogen dioxide in Tokyo and the Kinki region to a daily average of 0.06 ppmv by March 2001. The law also required diesel-powered truck owners to switch to gasoline- or newer diesel-powered trucks by 2000. Yet, by 1998, the nitrogen dioxide standard was still exceeded at 26 percent of nonroadside air monitoring stations and 65 percent of roadside stations. Similarly, particulate matter standards were exceeded at 33 percent of nonroadside monitoring sites and 65 percent of roadside monitoring sites.

Although the highest pollution levels in Japan today are lower than they were in the 1950s through 1970s, increases in population since the 1980s have spread pollution over larger areas than they did previously. Health problems related to pollution have also been exacerbated by increases in the number of vehicles powered by diesel fuel. An estimated 4,000 people die each year in Japan of lung cancer attributable to diesel exhaust and other fine particulates (EIA, 2000).

Because its air pollution problems had not improved sufficiently since the mid-1980s, in 1999, Japan's Environment Agency began to focus on strengthening particulate emission controls from industrial sites. A revision of laws relating to air pollution is expected to address this issue.

8.2.17. South Africa

Ninety percent of South Africa's electricity is obtained from burning coal. Coal is also burned in boilers and stoves in factories and hospitals. Millions of people in townships and villages without electricity burn coal and wood for home heating. Burning coal produces sulfur dioxide and soot, and sulfur dioxide emissions exacerbate acid deposition. Coal emissions from homes and industry, combined with other pollutants from automobile exhaust, frequently create dark skies over Johannesburg. In Capetown, a major problem is nitrogen dioxide, resulting from vehicles emissions of nitric oxide. Nitrogen dioxide levels in Capetown are often higher than those in Calcutta.

Because the price of coal is so low, South Africa has little incentive to invest in alternative energy sources. However, in one recent business venture, a national power supplier and an oil company agreed to put solar panels in 50,000 homes currently without electricity. Homeowners would have to buy cards to activate the panels.

The Constitution of South Africa is somewhat unique among constitutions in that it guarantees everyone the right (a) "to an environment that is not harmful to their health or well-being" and (b) "to have the environment protected for the benefit of present and future generations through reasonable legislative and other measures." In October 1997, a White Paper on Environmental Management called for South Africa's Department of Environment Affairs and Tourism to regulate and combat air pollution, particularly from coal and fuel burning, vehicles, mining, industrial activity, and incinerators. Although emission regulations have resulted, they have not been enforced strictly to date.

8.2.18. Australia

Air pollution problems are less severe in Australia than they are in many other countries, particularly because Australian cities are less populated and have fewer sources of pollution than do more polluted cities. Australia is also surrounded by oceans, so it does not receive much transboundary air pollution. The sulfur contents of coal and oil in Australia are lower than those in other countries. Coal-fired power plants are also located away from urban areas. As a result, sulfur dioxide mixing ratios in Australian cities are relatively low. Sydney and Melbourne occasionally have bad air pollution days, particularly in summer and autumn.

Automobiles are the largest source of air pollution in Australia. Catalytic converters were required starting in 1986. Because cars that use the converter cannot run on lead, lead concentrations in the air subsequently decreased. Lead is expected to be phased out of gasoline completely by 2002, at which time automobile emission standards for particulates, nitric oxide, carbon monoxide, and hydrocarbons will be strengthened. Australia is planning to give financial incentives to industry to convert commercial vehicles to clean fuels by 2015.

8.3. SUMMARY

In this chapter, regulation of and trends in urban air pollution were discussed with respect to several countries. Until the 1950s, air pollution in the United States was not addressed at the national level. Increased concern over photochemical smog and fatal London-type smog events in the United States encouraged Congress to implement the Air Pollution Control Act of 1955 and the United Kingdom to implement the U.K. Clean Air Act of 1956. The U.S. Act provided funds to study air pollution and provided for technical assistance to states, many of which had already enacted air pollution legislation, but it did not regulate air pollution at the federal level. The U.S. Congress first regulated stationary sources of pollution through the Clean Air Act of 1963. This Act led to the Motor Vehicle Air Pollution Control Act of 1965, which was the first federal legislation controlling emission from motor vehicles. The Air Quality Act of 1967 divided the country into Air Quality Control Regions, specified the development of Air Quality Criteria for certain pollutants, and required states to submit State Implementation Plans for combating air pollution. In landmark legislation, Congress passed the Clean Air Act Amendments of 1970. This act required the newly created U.S. EPA to develop National Ambient Air Quality Standards for certain pollutants identified as being unhealthful and specified that the U.S. EPA set performance standards for new sources of air pollution. Regulations under CAAA70 led to the 1975 invention of the catalytic converter, the most important device to date for limiting emissions from automobiles. The Clean Air Act Amendments of 1990 led to controls of many hazardous air pollutants not controlled under earlier legislation. The Clean Air Act Revision of 1997 led to revised ozone and particulate standards. State and federal air pollution legislation resulted in a remarkable improvement in urban air in the United States. Strong legislation by the European Union and several industrialized countries has also led to improved air quality. Although legislation has been enacted in many developing countries, legislation in several of those countries is relatively new, has not been enforced, or has not been sufficiently strong. As such, air pollution in many cities of the world has worsened. Some of the most polluted cities in the world today include Mexico City, New Delhi, Beijing, Cairo, Sao Paolo, Shanghai, Jakarta, Bangkok, Tehran, and Calcutta.

8.4. PROBLEMS

8.1.   Suppose you drive a gasoline engine car 15,000 miles per year.

(a) How many kilograms per year of hydrocarbons and carbon monoxide would your car have emitted in 1968 versus today if your car exactly met federal emission standards in both years? (Use intermediate useful life emission standards for the present year.)

(b) If each of 10 million cars in your city were driven 15,000 miles per year, how many tons per day of carbon monoxide would have been emitted in 1968 versus today?

8.2.   Explain how the three-way catalyst works.

8.3.   The United Kingdom Alkali Act of 1863 resulted in new control technologies for converting hydrochloric acid from soda-ash factory chimneys to reusable forms of chlorine, and the Clean Air Act Amendments resulted in the development of the catalytic converter. If air pollution control regulation leads to improved technologies, why is tough air pollution legislation generally so difficult to pass?

8.4.   Identify any one-hour California state or federal standards, health advisories, or smog-alert levels exceeded by $NO_2(g)$ or $O_3(g)$ in Fig. 4.10(a) and (b).

8.5.   What are two possible explanations for the improvements in Los Angeles air pollution between 1975 and today, as shown in Fig. 8.2?

8.6.   Today, what criteria pollutants exceed U.S. federal standards the most? Which exceed the standards the least?

8.7.   Although more people in the United States are exposed to levels of $CO(g)$ than PM_{10} above the federal standard (Fig. 8.1), exposure to PM_{10} is of greater concern in terms of health effects. Why?

8.8.   Match each city below with an air pollution related characteristic of the city.

(a) Cairo

(b) Santiago

(c) Mexico City

(d) Los Angeles

(e) Beijing

(f) New Delhi

(g) Paris

(h) Sao Paolo

(i) Moscow

(j) London

(1) Heavy reliance on alcohol-fuel burning

(2) High sulfur dioxide and soot concentrations in the air

(3) Energy sector relies heavily on nuclear power

(4) Severe lead air pollution problem

(5) Experienced a devastating smog event in 1952

(6) Heavy particulate pollution, particularly from soil erosion

(7) Lack of funding limits enforcement of air pollution laws

(8) Most dangerous city (in terms of air pollution) worldwide for children, 1999

(9) Ozone levels exceeded 0.6 ppmv in the 1950s

(10) Car sales limited by law to slow rise in vehicle emissions

INDOOR AIR POLLUTION

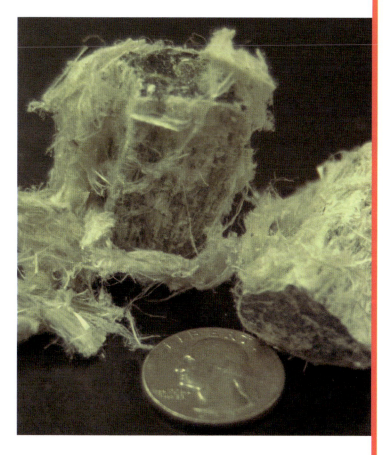

B ecause people spend most of their time indoors, it is useful to examine the composition and quality of indoor air. Because people's time is often divided between home and work, it is important to examine air quality in both residences and workplaces. Sources of indoor air pollution include outdoor air and indoor emissions. Outdoor air contains the constituents of smog, but some of these constituents dissipate quickly indoors because of the lack of ultraviolet (UV) radiation to reproduce them indoors. Major indoor sources of pollution include stoves, heaters, carpets, fireplaces, tobacco smoke, motor vehicle exhaust from garages, building materials, and insulation. Whereas indoor air pollution in the United States is regulated in workplaces, it is not regulated in residences. In this chapter, characteristics, sources, and regulation of indoor air pollution are discussed.

9.1. POLLUTANTS IN INDOOR AIR AND THEIR SOURCES

In the United States, about 89 percent of people's time is spent indoors, 6 percent is spent in vehicles, and 5 percent is spent outdoors (Robinson et al., 1991). Another study found that in nonindustrialized countries, people in urban areas spend 79 percent of their time indoors and those in rural areas spend 65 percent of their time indoors (Smith, 1993). Because people breathe indoor air more than outdoor air, an examination of indoor air is warranted.

Table 9.1 identifies major pollutants in indoor air and their primary sources. Many of the pollutant gases in indoor air are also found in outdoor air. Outdoor pollutants enter indoor air by infiltration, natural ventilation, and forced ventilation. **Infiltration**

Table 9.1. Important Indoor Air Pollutants and Their Emission Sources

Pollutant	Emission Sources
Gases	
Carbon dioxide	Metabolic activity, combustion, garage exhaust, tobacco smoke
Carbon monoxide	Boilers, gas or kerosene heaters, gas stoves, wood stoves, fireplaces, tobacco smoke, garage exhaust, outdoor air
Nitrogen dioxide	Outdoor air, garage exhaust, kerosene and gas space heaters, wood stoves, gas stoves, tobacco smoke
Ozone	Outdoor air, photocopy machines, electrostatic air cleaners
Sulfur dioxide	Outdoor air, kerosene space heaters, gas stoves, and coal appliances
Formaldehyde	Particleboard, insulation, furnishings, paneling, plywood, carpets, ceiling tile, tobacco smoke
Volatile organic compounds	Adhesives, solvents, building materials, combustion appliances, paints, varnishes, tobacco smoke, room deodorizers, cooking, carpets, furniture, draperies
Radon	Diffusion from soil
Aerosol particles	
Allergens	House dust, domestic animals, insects, pollens
Asbestos	Fire-retardant materials, insulation
Fungal spores	Soil, plants, foodstuffs, internal surfaces
Bacteria, viruses	People, animals, plants, air conditioners
Polycyclic aromatic hydrocarbons	Fuel combustion, tobacco smoke
Other	Resuspension, tobacco smoke, wood stoves, fireplaces, outdoor air

Sources: Spengler and Sexton (1983), Nagda et al. (1987).

is natural air exchange through cracks and leaks, such as through door and window frames, chimneys, exhaust vents, ducts, plumbing passages, and electrical outlets. Natural ventilation is air exchange resulting from the opening or closing of windows or doors to enhance the circulation of air. Forced ventilation is the air exchange resulting from the use of whole-house fans or blowers (Masters, 1998). Next, the pollutants in Table 9.1 are discussed briefly.

9.1.1. Carbon Dioxide

Carbon dioxide [$CO_2(g)$] is present in background air and produced indoors from breathing and combustion. It does not pose a health problem until its mixing ratio reaches 15,000 ppmv, much higher than its background mixing ratio of 370 ppmv. Mixing ratios and health effects of carbon dioxide are discussed in Section 3.6.2.

9.1.2. Carbon Monoxide

Carbon monoxide [$CO(g)$], produced outdoors by automobile and other fossil-fuel combustion sources, is emitted indoors by boilers, heaters, stoves, fireplaces, cigarettes, and in-garage cars. In the absence of indoor sources, mixing ratios of $CO(g)$ indoors are usually less than are those outdoors (Jones, 1999). In the presence of indoor sources, indoor mixing ratios of $CO(g)$ can reach a factor of four or more times those outdoors. Mixing ratios and health effects of carbon monoxide are discussed in Section 3.6.3.

9.1.3. Nitrogen Dioxide

Nitrogen dioxide [$NO_2(g)$] is produced chemically from oxidation of nitric oxide [$NO(g)$] and is emitted in small quantities indoors. Sources of $NO(g)$ and $NO_2(g)$ indoors include in-garage cars, kerosene and gas space heaters, wood stoves, gas stoves, and cigarettes. In the absence of indoor sources, indoor mixing ratios of $NO_2(g)$ are similar to those outdoors (Jones, 1999). High short-term mixing ratios of $NO_2(g)$ occur during the operation of $NO_2(g)$-producing appliances. Because UV sunlight does not penetrate indoors, the photolysis of $NO_2(g)$ and subsequent production of ozone is not a concern in indoor air. The health effects of $NO_2(g)$ are discussed in Section 3.6.8.

9.1.4. Ozone

Ozone [$O_3(g)$], produced photochemically outdoors following photolysis of nitrogen dioxide, is rarely produced indoors because UV sunlight, required for its production, is usually unavailable indoors. The major indoor source of ozone is outdoor air. Photocopy machines and electrostatic air cleaners emit sufficient UV radiation to produce ozone indoors. In residences and most workplaces, however, ozone mixing ratios indoors are almost always less than are those outdoors. The indoor to outdoor ratio of ozone ranges from 0.1:1 to 1:1, with typical values of 0.3:1 to 0.5:1 (Finlayson-Pitts and Pitts, 1999). Ozone is lost indoors by reaction with wall, floor, and ceiling surfaces; reaction with indoor gases; and deposition to floors. Health effects of ozone are discussed in Section 3.6.5.

9.1.5. Sulfur Dioxide

Sulfur dioxide [$SO_2(g)$], emitted outdoors during gasoline, diesel, and coal combustion, is emitted indoors during combustion of kerosene in space heaters. In the absence of indoor sources, indoor $SO_2(g)$ mixing ratios are typically 10 to 60 percent those of outdoor air (Jones, 1999). Once indoors, $SO_2(g)$ does not chemically degrade quickly in the gas phase because the hydroxyl radical [$OH(g)$], required to initiate its breakdown, is not produced indoors. Because the *e*-folding lifetime of the hydroxyl radical is about 1 second, $OH(g)$ brought indoors from the outside disappears quickly. Losses of $SO_2(g)$ include deposition to wall and floor surfaces, dissolution into liquid water (e.g., in bathtubs and sinks), and dissolution into aerosol particles containing liquid water. The health effects of sulfur dioxide are discussed in Section 3.6.6.

9.1.6. Formaldehyde

Formaldehyde [$HCHO(g)$], produced during biomass burning and chemical reaction outdoors, is emitted from particleboard, insulation, furnishings, paneling, plywood, carpets, ceiling tile, and tobacco smoke indoors. Formaldehyde mixing ratios indoors are usually greater than are those outdoors. In outdoor air, its major breakdown processes are photolysis by UV sunlight and reaction with $HO_2(g)$ and $OH(g)$. UV sunlight and $OH(g)$ are not present indoors, but $HO_2(g)$ sometimes is, and it is the most likely indoor chemical breakdown source of formaldehyde. Formaldehyde is also removed by deposition to the ground and reaction with wall, floor, and ceiling surfaces. Health effects of formaldehyde are discussed in Section 4.2.6.

9.1.7. Radon

Radon (Rn) is a radioactive but chemically unreactive, colorless, tasteless, and odorless gas that forms naturally in soils. Its decay products are believed to be carcinogenic and have been measured in high concentrations near uranium mines and in houses (particularly in their basements) overlying soils with uranium-rich rocks (Nazaroff and Nero, 1988). Radon plays no role in acid deposition, stratospheric ozone, or outdoor air pollution problems. Because of its indoor effects, radon is considered a hazardous pollutant in the United States under the 1990 Clean Air Act Amendments (CAAA90).

Sources and Sinks

The ultimate source of radon gas is the solid mineral uranium-238 (^{238}U), where the 238 refers to the isotope, or number of protons plus neutrons in the nucleus of an uranium atom. Of all uranium on Earth, 99.2745 percent is ^{238}U, 0.72 percent is ^{235}U, and 0.0055 percent is ^{234}U. ^{238}U has a half-life of 4.5 billion years.

Radon formation from uranium involves a long sequence of radioactive decay processes. During radioactive decay of an element, the element spontaneously emits radiation in the form of an alpha (α) particle, a beta (β) particle, or a gamma (γ) ray. An **alpha particle** is the nucleus of a helium atom, which is made of two neutrons and two protons (Fig. 1.1). It is the least penetrating form of radiation and can be stopped by a thick piece of paper. Alpha particles are not dangerous unless the emitting substance is inhaled or ingested. A **beta particle** is a high-velocity electron. Beta particles penetrate deeper than do alpha particles, but less than do other forms of radiation, such as gamma rays. A **gamma ray** is a highly energized, deeply penetrating photon emitted from the nucleus of an atom during nuclear fission (such as in the sun's core), but also sometimes during radioactive decay of an element.

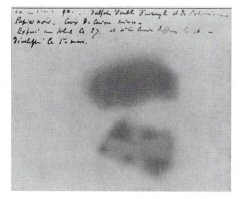

Radioactive decay was discovered on March 1, 1896, by French physicist **Antoine Henri Becquerel** (1871–1937; Fig. 9.1). To obtain his discovery, Becquerel placed a uranium-containing mineral on top of a photographic plate wrapped by thin, black paper. After letting the experiment sit in a drawer for a couple of days, he developed the plate and found that it was fogged by emissions (Fig. 9.2) that he traced to the uranium in the mineral. He referred to the emissions as **metallic phosphorescence**. What he had discovered was radioactive decay. He repeated the experiment by placing coins under the paper and found that their outlines were traced by the emissions. Two years later, British physicist **Ernest Rutherford** (1871–1937; Fig. 9.3) found that uranium emitted two types of particles, which he named *alpha* and *beta particles.* Rutherford later discovered the gamma ray as well.

Equation 9.1 summarizes the radioactive decay pathway of ^{238}U to ^{206}Pb. Numbers shown are half-lives of each decay process.

When it decays to produce radon, ^{238}U first releases an alpha particle, producing thorium-234 (^{234}Th), which decays to protactinium-234 (^{234}Pa), releasing a beta particle. ^{234}Pa has the same number of protons and neutrons in its nucleus as does ^{234}Th, but ^{234}Pa has one less electron than does ^{234}Th,

Figure 9.3. Ernest Rutherford (1871–1937).

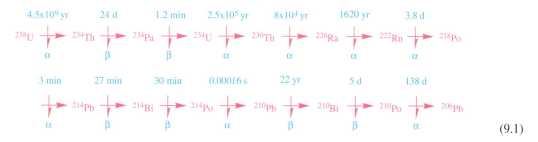

$$(9.1)$$

giving ^{234}Pa a positive charge. ^{234}Pa decays further to uranium-234 (^{234}U), then to thorium-230 (^{230}Th), then to radium-226 (^{226}Ra), then to radon-222 (^{222}Rn).

Whereas radon precursors are bound in minerals (Lyman, 1997), ^{222}Rn is a gas and can escape through soil and unsealed floors into houses, where its mixing ratio builds up in the absence of ventilation. ^{222}Rn has a half-life of 3.8 days. It decays to polonium-218 (^{218}Po), which has a half-life of three minutes and decays to lead-214 (^{214}Pb). ^{218}Po and ^{214}Pb, referred to as **radon progeny**, are electrically charged and can be inhaled or attach to particles that are inhaled (Cohen, 1998). In the lungs or in the ambient air, ^{214}Pb decays to bismuth-214 (^{214}Bi), which decays to polonium-214 (^{214}Po). ^{214}Po decays almost immediately to lead-210 (^{210}Pb), which has a lifetime of 22 years and usually settles to the ground if it has not been inhaled. It decays to bismuth-210 (^{210}Bi), then to polonium-210 (^{210}Po), then to the stable isotope, lead-206 (^{206}Pb), which does not decay further.

Concentrations

Outdoor concentrations of radon are generally low and do not pose a human health risk. Because of the lack of ventilation in many houses, indoor concentrations can become thousands of times larger than outdoor concentrations (Wanner, 1993). Indoor concentrations depend on the abundance of radon in soil and the porosity of floors. Nero et al. (1986) found mean indoor concentrations of ^{222}Rn in 552 homes as 56 Bq (becquerels) m^{-3}, where a becquerel is a the number of disintegrations of atomic nuclei per second. A variety of other studies found indoor concentrations of 46 to 116 Bq m^{-3} (Jones, 1999). Marcinowski et al. (1994) estimated that about 6 percent of U.S. homes contained radon concentrations in excess of U.S. EPA levels considered safe (148 Bq m^{-3}). Radon levels in homes can be reduced by installing check valves in drains, sealing basement walls and floors, and installing fans in crawl spaces (to speed mixing of outside air with radon-laden air under the house).

Health Effects

^{222}Rn, a gas, is not harmful itself, but its progeny, ^{218}Po and ^{214}Pb, which enter the lungs directly or on the surfaces of aerosol particles, are believed to be highly carcinogenic (Polpong and Bovornkitti, 1998). Any activity increasing the inhalation of aerosol particles enhances the risk of inhaling radon progeny; thus, the combination of radon and cigarette smoking is expected to increases lung cancer risks above the normal risks associated with smoking (Hampson et al., 1998). Whereas several studies argue that a link exists between high radon levels and enhanced cancer rates (Henshaw et al., 1990, Lagarde et al., 1997), other studies argue that no such link exists (Etherington et al., 1996).

9.1.8. Volatile Organic Compounds

Volatile organic compounds (VOCs) are organic compounds that have relatively low boiling points (50 to 260°C). Because of the their low boiling points, VOCs often evaporate from materials containing them. Sources of VOCs indoors include adhesives, solvents, building materials, combustion appliances, paints, varnishes, tobacco smoke, room deodorizers, cooking, carpets, furniture, and draperies. More than 350 VOCs with mixing ratios greater than 1 ppbv have been measured in indoor air (Brooks et al., 1991). Carpets alone emit at least 99 different VOCs (Sollinger et al., 1994). Some common VOCs in indoor air include propane, butane, pentane, hexane, *n*-decane, benzene, toluene, xylene, styrene, acetone, methylethylketone, and limonene, among many others. VOC mixing ratios indoors can easily exceed those outdoors by a factor of five (Wallace, 1991). Many VOCs are hazardous. The health effects of some VOCs are described in Table 3.16.

9.1.9. Allergens

Allergens are particles such as pollens, foods, or microorganisms that cause an allergy, which is an abnormally high sensitivity to a substance. Symptoms of an allergy include sneezing, itching, and skin rashes. Indoor sources of allergens are dust-mite feces, cats, dogs, rodents, cockroaches, and fungi. Pollens, mostly from outdoor sources, originate from trees, plants, grasses, and weeds. An omnipresent source of indoor allergens is dust mites. Dust mites grow to adult size of about 250 to 350 μm from an egg in about 25 days (D'Amado et al., 1994). Their feces range in size from 10 to 40 μm. A gram of dust may contain up to 100,000 feces. The allergen in dust-mite feces is a protein in the intestinal-enzyme coating of the feces (Jones, 1999). Allergens in airborne dust-mite feces may exacerbate symptoms in 85 percent of asthmatics (Platts-Mills and Carter, 1997). In cockroaches, the sources of allergens are body parts and feces. Cat allergens are present in the saliva, dander, and skin of the cat. Dog allergens are found in the saliva and dander of the dog. Allergens from cats, dogs, and cockroaches can trigger rapid asthmatic responses.

9.1.10. Asbestos

Asbestos is a class of natural impure hydrated silicate minerals that can be separated into flexible fibers (Fig. 9.4). Asbestos is chemically inert, does not conduct heat or electricity, and is fire resistant. Its fire-resistant properties were known by French emperor **Charlemagne** (742–814), who would amuse his guests at dinner parties by wiping his mouth with a napkin made of asbestos, throwing the napkin into the fireplace, then retrieving the unburned napkin to their surprise. Until the 1970s, asbestos was used widely in the construction industry as an electrical and thermal insulator, particularly

Figure 9.4. Unmilled Chrysotile asbestos. Photo by Robert Grieshaber.

in pipe and boiler insulation, cementboard, thermal tiles, paint, and wallpaper (Maroni et al., 1995; Jones, 1999). Today, new asbestos production is banned in many countries, but much of the asbestos installed in buildings before the bans took place still exists.

People who were most likely to be exposed to asbestos in the past were miners, insulation manufacturers, and insulation installers. From the 1950s to 1990, the town of Libby, Montana, was the center of a vermiculite (clay mineral) mining operation. Unknown to the miners, the vermiculite was laden with asbestos. Between the early 1980s and 2000, 192 people from the town of 2,700 died and 375 others were diagnosed with asbestos-related lung problems. Some miners inadvertently exposed their families to asbestos when they brought their clothes, carrying asbestos-laden dust, into their homes. Today, miners and those who remove asbestos from buildings are the people most likely to be exposed to asbestos in concentrations high enough to cause health problems. In the United States, asbestos is defined as a hazardous air pollutant under CAAA90.

Sources and Sinks

Once insulation containing asbestos has been installed, the asbestos is not expected to cause damage to humans unless the insulation is disturbed. At that time, fibers can be scattered into the air, where they can remain for minutes to days until they deposit to the ground or are inhaled. Thus, the only source of asbestos in indoor air is turbulent uplift, and the only sink is deposition.

Concentrations

Indoor concentrations of asbestos vary from building to building. Lee et al. (1992) found an average of 0.02 structures per cubic centimeter in 315 public, commercial, residential, school, and university buildings in the United States. The concentration of fibers longer than 5 μm was only 0.00013 structures per cubic centimeter.

Health Effects

The primary health effects of asbestos exposure are lung cancer, mesothelioma, and asbestosis. Mesothelioma is a cancer of the mesothelial membrane lining the lungs, and asbestosis is a slow, debilitating disease of the lungs. The most dangerous asbestos fibers are those longer than 5 to 10 μm but with a diameter less than 1 μm. Exposure to 0.0004 structures per cubic centimeter is expected to cause a lifetime cancer risk of 160 cases of mesothelioma per 1 million people (Turco, 1997). The time between first exposure to asbestos and the appearance of tumors is estimated to be 20 to 50 years (Jones, 1999). Cigarette smoking and exposure to asbestos are believed to amplify the rates of lung cancer in comparison with the rates of cancer associated with just smoking or just exposure to asbestos. Short-term acute exposure to asbestos can lead to skin irritation (Spengler and Sexton, 1983).

9.1.11. Fungal Spores, Bacteria, and Viruses

Fungal spores, bacteria, and viruses are common indoor air contaminants. Sources of fungal spores are soils, plants, foodstuff, internal surfaces, and outdoor air. Sources of bacteria and viruses are humans, animals, plants, air conditioners, and outdoor air. Fungi grow well on damp surfaces. Common fungi in buildings include *Penicillium, Cladosporium,* and *Aspergillus.* Common bacteria include *Bacillus, Staphylococcus,* and *Micrococcus* (Jones, 1999). Some diseases associated with fungi, bacteria, and viruses include rhinitis (a respiratory illness), asthma, humidifier fever, extrinsic allergic alveolitis, and atopic dermatitis. Most human illnesses due to viral and bacterial infection are due to human-to-human transmission of microorganisms rather than to building-to-human transmission (Ayars, 1997).

9.1.12. Environmental Tobacco Smoke

When a cigarette is smoked, some of the smoke is inhaled and swallowed, some is inhaled and exhaled (mainstream smoke), and the rest is emitted from the cigarette (sidestream smoke). The mainstream plus sidestream smoke is called environmental tobacco smoke (ETS), a mixture of more than 4,000 aerosol particle components and gases, at least 50 of which are known carcinogens. ETS, itself, has been classified by the U.S. EPA as a carcinogen. Also called second-hand smoke, ETS, builds up in enclosed spaces, increasing danger to others in the vicinity. Even in well-ventilated indoor areas, particle and gas concentrations associated with ETS increase. Although the cumulative effect of ETS on outdoor air pollution is relatively small compared with the effects of other sources of pollution, such as automobiles, ETS concentrations can build up outdoors in the vicinity of smokers.

Outdoor ETS emission rates are not regulated in the United States, although regulations prohibiting smoking in many public and private indoor facilities currently exist at the local, state, and federal levels. Many chemical constituents of ETS are classified as hazardous air pollutants under CAAA90, but the Act controls only sources emitting more than 10 tons of a hazardous substance per year, and individual cigarettes emit less than 1 g of all pollutants combined. The product of the number of cigarettes smoked per year multiplied by the emission rate per cigarette is much larger than 10 tons per year, but this statistic is not recognized by CAAA90.

Sources

In 1990, approximately 26 percent of the U.S. population, or 50 million people, were estimated to smoke cigarettes regularly (Rando et al., 1997). In 1986, 70 percent of all children in the United States lived in households in which at least one parent smoked (Weiss, 1986). Table 9.2 shows mainstream and sidestream emissions for some components emitted from cigarettes. The actual emission rate of a cigarette depends on the type of tobacco, the density of its packing, the type of wrapping paper, and the puffing rate of the smoker (Hines et al., 1993).

Table 9.2 indicates that emission rates of sidestream smoke are greater than are those of mainstream smoke for many pollutants. Thus, in many cases, a person standing a short distance from a cigarette is exposed to more pollution than is the smoker (Schlitt and Knöppel, 1989).

Table 9.3 compares emissions from cigarettes with those from vehicles. The table indicates that driving a vehicle that meets current U.S. EPA emission standards 1 mile results in particle emissions equivalent to emissions from about 1.4 cigarettes. Emissions of CO(g) and NO_x(g) from driving 1 mile are equivalent to those from smoking 73 and 190 cigarettes, respectively. The total emissions per day of these substances from cigarettes is much less than that from mobile sources, but ETS is often emitted in enclosed spaces, where its concentrations build up.

Concentrations

ETS contributes to the buildup of gas and particle concentrations indoors. Spengler et al. (1981) found that one pack of cigarettes per day contributes about 20 μg m^{-3} of particles over a 24-hour period. During the time a cigarette is actually smoked, particle concentrations increase to 500 to 1,000 μg m^{-3}. Leaderer et al. (1990) found that particle concentrations in homes with a cigarette smoker were up to three times those in homes without a smoker.

Health Effects

Short-term exposure to ETS results in eye, nose, and throat irritation for most individuals, and allergic skin reactions for some (Maroni et al., 1995). ETS also elevates symptoms for people who have asthma, and may induce asthma in some

Table 9.2. Mainstream and Sidestream Emission Rates per Cigarette

Substance	Mainstream Smoke (µg per Cigarette)	Sidestream Smoke (µg per Cigarette)
Carbon dioxide	10,000–80,000	81,000–640,000
Carbon monoxide	500–26,000	1,200–65,000
Nitrogen oxides	16–600	80–3,500
Ammonia	10–130	400–9,500
Hydrogen cyanide	280–550	48–203
Formaldehyde	20–90	1,000–4,600
Acrolein	10–140	100–1,700
N-nitrosodimethylamine	0.004–0.18	0.04–149
Nicotine	60–2,300	160–7,600
Total particulate	100–40,000	130–76,000
Phenol	20–150	52–390
Catechol	40–280	28–196
Naphthalene	2.8	45
Benzo(a)pyrene	0.008–0.04	0.02–0.14
Aniline	0.10–1.20	3–36
2-Naphthylamine	0.004–0.027	0.02–1.1
4-Aminobiphenyl	0.002–0.005	0.06–0.16
N-nitrosonornicotine	0.2–3.7	0.02–18

Source: Rando et al., 1997; Jones, 1999.

children (Jones, 1999). A link between exposure to ETS and lower respiratory tract illness has been found (Somerville et al., 1988). Children exposed to ETS are hospitalized more often than are children not exposed (Harlap and Davies, 1974). Several studies have linked long-term ETS exposure to lung cancer. Janerich et al. (1990) found that ETS may cause up to 17 percent of lung cancers among nonsmokers. Rando et al. (1997) found a 30 percent increase in the cancer risk to women whose husbands smoked. Ryan et al. (1992) linked brain cancer tumors to ETS.

Table 9.3. Comparison of Emissions from Cigarettes and Mobile Sources

Substance	(a) Average Cigarette Emissions (g/Cigarette)[a]	(b) Average Vehicle Emissions (g/mi)[b]	(c) Number of Cigarettes Resulting in Same Emissions as 1 Mile of Driving[c]	(d) Estimated Cigarette Emissions in the United States (tons/day)[d]	(e) Estimated Mobile-Source Direct Emissions in the United States (tons/day)[e]
Carbon monoxide	0.0464	3.4	73.3	60	189,000
Nitrogen oxides	0.0021	0.4	190.5	2.7	32,000
Particulates	0.058	0.08	1.4	75	9300

[a]Sum of average mainstream and sidestream emissions in Table 9.2.

[b]1994 federal light-duty vehicle emission standards.

[c]Column (b) divided by column (a).

[d]Assumes the current population of the United States is 250,000,000 and 26 percent of the population smokes 20 cigarettes per day.

[e]Estimated 1997 emissions from U.S. EPA (1997).

9.2. SICK BUILDING SYNDROME

In some workplaces, employees experience an unusually high rate of headaches, nausea, nasal congestion, chest congestion, eye problems, throat problems, fatigue, fever, muscle pain, dizziness, and dry skin. These symptoms, present during working hours, often improve after a person leaves work. The situation described is called sick building syndrome (SBS), the cause of which is unknown, although it may be due to certain VOCs, building ventilation systems, or the exposure to many pollutants in low doses simultaneously. SBS may also be caused by enhanced stress levels and heavy workloads or a combination of psychological and chemical factors (Jones, 1999).

9.3. REGULATION OF INDOOR AIR POLLUTION

In the United States, National Ambient Air Quality Standards (NAAQS), initiated under the Clean Air Act Amendments, control outdoor air pollutants only. No regulations control air pollution concentrations in indoor residences. Standards for pollutant concentrations in indoor workplaces are set by a government agency, the Occupational Safety and Health Administration (OSHA), which obtains recommendations for standards from another government agency, the National Institute for Occupational Safety and Health (NIOSH), and an independent professional society, the American Conference of Governmental Industrial Hygienists, Inc. (ACGIH). OSHA and NIOSH were created by the 1970 U.S. Occupational Safety and Health Act. OSHA is in the U.S. Department of Labor and is responsible for setting and enforcing workplace standards, whereas NIOSH is in the U.S. Department of Health and Human Services and is responsible for researching workplace health issues. ACGIH's primary mission is to promulgate workplace safety standards.

NIOSH recommends permissible exposure limits (PELs), short-time exposure limits (STELs), and ceiling concentrations. A PEL is the maximum allowable concentration of a pollutant in an indoor workplace over an 8-hour period during a day. An STEL is the maximum allowable concentration of a pollutant over a 15-minute period. A ceiling concentration is a concentration that may never be exceeded. ACGIH sets 8-hour time-weighted average threshold limit values (TWA-TLVs), which are similar to PELs. For most pollutants, PELs and TWA-TLVs are the same. When differences occur, they are small.

Indoor standards exist for more than 150 compounds. Table 9.4 compares indoor with outdoor standards for a few compounds. For ozone and carbon monoxide, the

Table 9.4. Comparison of Indoor Workplace Standards with Outdoor Federal and California State Standards for Selected Gases

Gas	Indoor 8-hour PEL and TWA-TLV (ppmv)[a]	Indoor 15-min STEL (ppmv)[a]	Indoor Ceiling (ppmv)[a]	Outdoor NAAQS (ppmv)	Outdoor California Standard (ppmv)
Carbon monoxide	35	—	200	9.5 (8 hour)	9 (8 hour)
Nitrogen dioxide	—	1	—	0.053 (annual)	0.25 (1 hour)
Ozone	0.1	0.3	—	0.08 (8 hour)	0.09 (1 hour)
Sulfur dioxide	2	5	—	0.14 (24 hour)	0.05 (24 hour)

[a]National Institute for Occupational Safety and Health (2000).

8-hour PEL/TWA-TLV standards are less stringent than are the outdoor standards because outdoor standards are designed to protect the entire population, particularly infants and people afflicted with disease or illness. Indoor standards are designed to protect workers, who are assumed to be healthier than is the average person. The table also indicates that the 15-minute STEL for nitrogen dioxide is four times higher (less stringent) than is the 1-hour outdoor California standard. Stringent outdoor standards for nitrogen dioxide are set because it is a precursor to photochemical smog. Because UV sunlight does not penetrate indoors, nitrogen dioxide does not produce ozone indoors, and indoor regulations of nitrogen dioxide as a smog precursor are not necessary. Indoor standards for nitrogen dioxide are based solely on health concerns.

9.4. SUMMARY

People spend most of their time indoors; thus, most of their exposure to pollution occurs indoors. Indoor air contains many of the same pollutants as does outdoor air, but pollution concentrations in indoor and outdoor air usually differ. Indoor mixing ratios of ozone and sulfur dioxide are usually less than are those outdoors, but indoor mixing ratios of formaldehyde are usually greater than are those outdoors. Indoor mixing ratios of carbon monoxide and nitrogen dioxide are generally the same as or less than are those outdoors, unless appliances or other indoor combustion sources are turned on. Indoor concentrations of radon and environmental tobacco smoke, when present, are usually greater than are those outdoors, giving rise to potentially serious health problems to people exposed to these pollutants indoors. Indoor air also contains volatile organic compounds, allergens, fungi, bacteria, and viruses. In the United States, indoor air is regulated only in the workplace and in public buildings; residential air is not regulated.

9.5. PROBLEMS

9.1.   Why are ozone mixing ratios almost always lower indoors than outdoors?

9.2.   Why are workplace standards for pollutant concentrations generally less stringent than standards for outdoor air?

9.3.   What would be the volume mixing ratio (ppmv) of carbon monoxide and the mass concentration ($\mu g\ m^{-3}$) of particulates if ten cigarettes were smoked in a 5 m $\times$ 10 m $\times$ 3 m room. How do these values compare with the California state 1-hour standard for CO(g) and the 24-hour average standard for PM_{10}? Based on the results, which pollutant do you think is more of a cause for concern with respect to indoor air quality? Assume the dry air partial pressure is 1013 mb, the temperature is 298 K, and use cigarette emission rates from Table 9.3.

9.4.   Why is radium less of a concern than radon?

9.5.   Why is removing asbestos from buildings often more dangerous than leaving it alone?

9.6.   Why does the gas-phase chemical decay of organic compounds generally take longer indoors than outdoors?

ACID DEPOSITION

10

Acid deposition occurs when sulfuric acid, nitric acid, or hydrochloric acid, emitted into or produced in the air as a gas or liquid, deposits to soils, lakes, farmland, forests, or buildings. Deposition of acid gases is **dry acid deposition**, and deposition of acid liquids is **wet acid deposition**. Wet acid deposition can be through rain **(acid rain)**, fog **(acid fog)**, or aerosol particles **(acid haze)**. On the Earth's surface, acids have a variety of environmental impacts, including damage to microorganisms, fish, forests, agriculture, and structures. In the air, acids in high concentrations are harmful to humans. Acid deposition problems have existed since coal was first combusted, but were exacerbated during the Industrial Revolution in the eighteenth century. The problem became more severe with the growth of the alkali industry in 19th-century France and the United Kingdom. In this chapter, the history, science, and regulation of acid deposition problems are discussed.

10.1. HISTORICAL ASPECTS OF ACID DEPOSITION

Acid deposition is caused by the emission or atmospheric formation of gas- or aqueous-phase **sulfuric acid** (H_2SO_4), **nitric acid** (HNO_3), or **hydrochloric acid** (HCl). Historically, coal was the first and largest source of anthropogenically produced atmospheric acids. Coal combustion produces sulfur dioxide gas [$SO_2(g)$] and hydrochloric acid gas [HCl(g)]. $SO_2(g)$ produces gas-and aqueous-phase sulfuric acid. Humans have combusted coal for thousands of years. In the 1200s,

Figure 10.1. Nicolas Leblanc (1742–1806).

sea coal was brought to London and used in lime kilns and forges (Section 4.1). It was later burned in furnaces to produce glass and bricks, in breweries to produce beer and ale, and in homes to provide heating. During the **Industrial Revolution**, which started in the eighteenth century, coal was used to provide energy for the steam engine.

In the late eighteenth century, a second major source of atmospheric acids emerged. Around 1780, the demand for **sodium carbonate** [$Na_2CO_3(s)$] (also known as **soda ash**, **washing soda**, and **salt cake**), used in the production of soaps, detergents, cleansers, glass, paper, bleaches, and dyes, increased. Although this chemical could be extracted from barilla plants, which grew near the Mediterranean Sea, and from sea kelp, a more efficient method of producing it was desired. In 1781, the French Academy of Sciences offered a prize to the person who could develop the most efficient and economic method of producing soda ash. The prize was never awarded because of the French Revolution, but **Nicolas Leblanc** (1742–1806; Fig. 10.1), encouraged by the competition, began experimenting in 1784 to find a new process. In 1789, he came up with a two-step set of reactions to produce soda ash. In the first step, he dissolved common salt into a sulfuric acid solution at a high temperature in an iron pan to produce dissolved sodium sulfate **(Glauber's salt)** and hydrochloric acid gas by

$$2NaCl(s) + H_2SO_4(aq) \xrightarrow{\text{High temperature}} Na_2SO_4(aq) + 2HCl(g) \qquad (10.1)$$

Sodium chloride (salt)   Sulfuric acid   Sodium sulfate   Hydrochloric acid

In the second step, he heated sodium sulfate together with charcoal and chalk in a kiln to form sodium carbonate by

$$Na_2SO_4(aq) + 2C + CaCO_3(s) \xrightarrow{\text{High temperature}} Na_2CO_3(aq) + CaS(s) + 2CO_2(g) \qquad (10.2)$$

Sodium sulfate   Carbon from charcoal   Calcium carbonate from chalk   Sodium carbonate   Calcium sulfide   Carbon dioxide

By-products of the second reaction, when complete, included **calcium sulfide** [CaS(s)], an odorous, yellow to light-gray powder, and carbon dioxide gas. In practice, the reaction was incomplete and produced gas-phase sulfuric acid and soot as well. The sodium carbonate residue was separated from calcium sulfide by adding water, which preferentially dissolved the sodium carbonate. The resulting solution was then dried, producing sodium carbonate crystals. Sodium carbonate was then combined with animal fat to produce **soap**.

A necessary ingredient for the production of sodium carbonate was aqueous sulfuric acid (Reaction 10.1). This was obtained by burning elemental sulfur (S) powder with saltpeter [$KNO_3(s)$], then dissolving the resulting $H_2SO_4(g)$ in water, as was done by Libavius in 1585. This process was inefficient, releasing volumes of gas-phase sulfuric acid and nitric oxide [NO(g)] (which converts to nitrogen dioxide, then to nitric acid in air).

In sum, although the production of sodium carbonate from common salt and sulfuric acid allowed the alkali industry to become self-sufficient and escape reliance on the import of natural sodium carbonate, it resulted in the release of HCl(g), $H_2SO_4(g)$, $HNO_3(g)$, and soot, causing widespread acid deposition and air pollution in France and the United Kingdom. The Leblanc process also produced large amounts of impure solid "alkali waste" containing calcium sulfide. The waste was piled near each factory. When rainwater fell on the waste, some of the calcium sulfide dissolved, producing dissolved hydrogen sulfide $H_2S(aq)$, which evaporated as a harmful, odorous, colorless gas. Much of the rest of the calcium sulfide oxidized over time to form calcium sulfate [$CaSO_4$-$2H_2O$(s), gypsum]. Piles of gypsum from soda ash factories can still be seen today.

Although Leblanc patented the soda ash technique and started a soda ash factory in 1791, he lost his patent and factory to the state, which nationalized patents and factories in 1793. Leblanc ultimately committed suicide in 1806 because of his inability to recover from his business losses.

In 1863, an estimated 1.76 million tons of raw material were burned to form soda ash, producing only 0.28 million tons of useful products. Most of the remaining 1.48 million tons was emitted as HCl(g) or other gases and produced as solid waste (Brock, 1992). Environmental damage due to the alkali industry was severe. HCl(g) and other pollutants rained down onto agricultural property and cities in France and the U.K., causing property and health damage. In the United Kingdom, St. Helens, Newcastle, and Glasgow countrysides were decimated (Brimblecombe, 1987).

Early complaints against alkali factories were in the form of civil litigation. For example, in 1838, a Liverpool landowner filed a complaint against an alkali factory, charging that it destroyed his crops and interfered with his hunting. Ultimately, HCl(g) from alkali manufacturers was regulated in France and the United Kingdom. In France, regulation took the form of planning laws that controlled the location of alkali factories. In the United Kingdom, the **1863 Alkali Act** required alkali manufacturers to reduce 95 percent of their HCl(g) emissions. The act also called for the appointment of an alkali inspector to watch over the industry.

Impetus for the 1863 Alkali Act came in the early 1860s, when William Gossage invented a technique to wash HCl(g) from waste gases before it was released from chimneys. Gossage built his own soda ash factory in Worcestershire in 1830. Spurred by his neighbors' complaints about the hydrochloric acid emitted from his factory, he worked to mitigate the problem. His solution was to convert a windmill into a tower, fill the tower with brushwood, and spray water down the top of the tower as smoke rose from the bottom. The water dissolved most of the hydrochloric acid (just as rain does) and drained it into a nearby waterway (which he was not too concerned about). Because the technique was so simple and inexpensive and because the consequences of not implementing the technique were so severe, the United Kingdom easily passed the 1863 Alkali Act shortly after Gossage's invention. Subsequent to the Act, Walter Weldon (in 1866) and Hugh Deacon (in 1868) developed processes for converting HCl(g) to chlorine that could be used for bleaching powder. These inventions allowed chlorine, which otherwise would have been wasted, to be recycled.

Despite the success of the Alkali Act at reducing HCl(g), the alkali industry continued to emit sulfuric acid, nitric acid, soot, and other pollutants in abundance. Because the act did not control emissions from factories other than those producing soda ash, pollution problems in the U.K. worsened. Although the Alkali Act was modified in 1881 to regulate other sections of the chemical industry, the number of factories and the volume of emissions from them had grown so much that the new law had little effect. In 1899, one writer described St. Helens as

> a sordid ugly town. The sky is a low-hanging roof of smeary smoke. The atmosphere is a blend of railway tunnel, hospital ward, gas works and open sewer. The features of the place are chimneys, furnaces, steam jets, smoke clouds and coal mines. (Blatchford, 1899)

The first Alkali Act inspector in the United Kingdom was Scottish chemist **Robert Angus Smith** (1817–1884; Fig. 10.2). He was charged with ensuring industry reduced HCl(g) emissions by 95 percent. Smith was also a field experimentalist. In 1872, he published *Air and Rain: The Beginnings of a Chemical Climatology,* in which he discussed results of the first monitoring network for air pollution in Great Britain. As part of the analysis, he recorded the gas-phase mixing ratios of molecular oxygen and carbon dioxide and measured the composition of chloride, sulfate, nitrate, and ammonium in rainwater in the British Isles. In his book, Smith introduced the term **acid rain** to describe the high sulfate concentrations in rain near coal-burning facilities.

Leblanc's soda ash process dominated the alkali industry until the 1880s. In 1861, Belgian chemist **Ernst Solvay** (1838–1922) developed a more efficient process of pro-

ducing soda ash by raining a salt produced from sodium chloride and ammonia gas down a tower (Solvay tower) over an upcurrent of carbon dioxide gas. The technique required no sulfuric acid or potassium nitrate (saltpeter). The disadvantages were the cost of the tower and the fact that the final chloride was tied up in a form that could not easily be reused. Only by the 1880s did the Solvay process overtake the LeBlanc process in economic efficiency and popularity.

Throughout the twentieth century, acid deposition problems continued to plague cities in the United Kingdom and municipalities in other countries where coal and chemical burning occurred. The fatal pollution episodes in London, discussed in Chapter 4, included contributions from acidic compounds in smoke.

Acid deposition became an issue of international interest in the 1950s and 1960s, when a relationship was found between sulfur emissions in continental Europe and acidification of Scandinavian lakes. These studies led to a 1972 United Nations Conference on Human Environment in Stockholm that called for an international effort to reduce acidification. Since then, numerous studies on acid

Figure 10.2. Robert Angus Smith (1817–1884).

deposition have been carried out. As a result of these studies, national governments have intervened to control acid deposition problems. Such efforts are discussed at the end of this chapter. First, the basic science of acid deposition is discussed.

10.2. CAUSES OF ACIDITY

In Chapter 5, pH was defined as

$$pH = -\log_{10}[H^+] \qquad (10.3)$$

where $[H^+]$ is the molarity (moles per liter) of H^+ in a solution containing a solvent and one or more solutes. The pH scale, shown in Fig. 10.3 for a limited range, varies from less than 0 (lots of H^+ and very acidic) to greater than 14 (very little H^+ and very basic or alkaline). Neutral pH, the pH of distilled water, is 7.0. At this pH, the molarity of H^+ is 10^{-7} mol L^{-1}. A pH of 4 means that the molarity of H^+ is 10^{-4} mol L^{-1}. Thus, water at a pH of 4 is 1,000 times more acidic (contains 1,000 times more H^+ ions) than is water at a pH of 7.

The molarity of H^+ is related to that of OH^- by the equilibrium relationship,

$$H_2O(aq) \rightleftharpoons H^+ + OH^-$$

Liquid   Hydrogen   Hydroxide

water   ion   ion   (10.4)

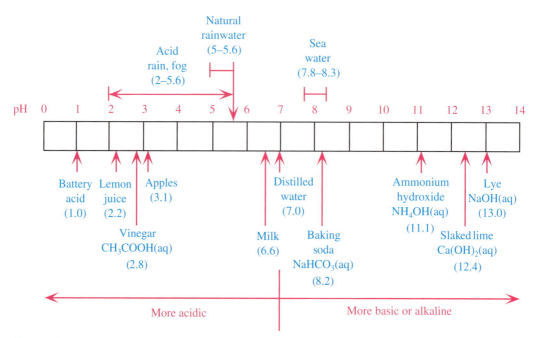

Figure 10.3. Diagram of the pH scale and the pH levels of selected solutions.

The equilibrium constant for this relationship is approximately 10^{-14} mol^2 L^{-2}, meaning that the product of [H$^+$] and [OH$^-$] must always equal approximately 10^{-14} mol^2 L^{-2} for water to be in equilibrium with H$^+$ and OH$^-$.

EXAMPLE 10.1

What is the pH of water at equilibrium when

    (a) [OH$^-$] = 10^{-7} mol L^{-1} and
    (b) [OH$^-$] = 10^{-11} mol L^{-1}?

Solution

Because [H$^+$] [OH$^-$] = 10^{-14} mol^2 L^{-2} to satisfy Equation 10.4, [H$^+$] = 10^{-7} mol L^{-1} and 10^{-3} mol L^{-1} in the two respective cases. Substituting these values into Equation 10.3 gives pHs of (a) 7 and (b) 3, corresponding to neutral and acidic conditions, respectively.

10.2.1. Carbonic Acid

Water can be acidified in one of several ways. When gas-phase carbon dioxide dissolves in water, it reacts rapidly with a water molecule to form aqueous **carbonic acid** [H$_2$CO$_3$(aq)], a weak acid, which dissociates by the reversible reactions

$$CO_2(aq) + H_2O(aq) \rightleftharpoons H_2CO_3(aq) \rightleftharpoons H^+ + HCO_3^- \rightleftharpoons 2H^+ + CO_3^{2-}$$

  Dissolved   Liquid   Dissolved   Hydrogen Bicarbonate   Hydrogen Carbonate
carbon dioxide   water   carbonic   ion   ion   ion   ion
                            acid

(10.5)

At a pH of distilled water (7), only the first dissociation (to the bicarbonate ion) occurs. The added H^+ (proton) decreases the pH of the solution, increasing its acidity. In the background air, the mixing ratio of $CO_2(g)$ is about 370 ppmv. A fraction of this $CO_2(g)$ always dissolves in rainwater. Thus, rainwater, even in the cleanest environment on Earth, is naturally acidic due to the presence of background carbonic acid in it. The **pH of rainwater affected by only carbonic acid is about 5.6**, indicating that its hydrogen ion molarity is 25 times that of distilled water.

10.2.2. Sulfuric Acid

When gas-phase sulfuric acid condenses onto rain drops, the resulting aqueous-phase **sulfuric acid** [$H_2SO_4(aq)$], a strong acid, dissociates by

$$H_2SO_4(g) \longrightarrow H_2SO_4(aq) \rightleftharpoons H^+ + HSO_4^- \rightleftharpoons 2H^+ + SO_4^{2-}$$

| Sulfuric acid gas | Dissolved sulfuric acid | Hydrogen ion | Bisulfate ion | Hydrogen ion | Sulfate ion | (10.6) |

At pH levels greater than $+2$, complete dissociation to the sulfate ion is favored, adding two protons to solution. The enhanced [H^+] decreases the pH further, increasing the acidity of rainwater. Sulfur dioxide can also dissolve in rainwater and produce the sulfate and hydrogen ions chemically, acidifying the water. This process is discussed in Section 10.3.2.

10.2.3. Nitric Acid

When gas-phase nitric acid dissolves in raindrops, it forms aqueous **nitric acid** [$HNO_3(aq)$], a strong acid that dissociates almost completely by

$$HNO_3(g) \rightleftharpoons HNO_3(aq) \rightleftharpoons H^+ + NO_3^-$$

| Nitric acid gas | Dissolved nitric acid | Hydrogen ion | Nitrate ion | (10.7) |

adding one proton to solution. As with sulfuric acid, nitric acid decreases the pH of rainwater below that of rainwater affected by only carbonic acid.

10.2.4. Hydrochloric Acid

When gas-phase hydrochloric acid dissolves in raindrops, it forms aqueous **hydrochloric acid** [$HCl(aq)$], a strong acid that dissociates almost completely by

$$HCl(g) \rightleftharpoons HCl(aq) \rightleftharpoons H^+ + Cl^-$$

| Hydrochloric acid gas | Dissolved hydrochloric acid | Hydrogen ion | Chloride ion | (10.8) |

adding one proton to solution. Hydrochloric acid also decreases the pH of rainwater below that of rainwater affected by only carbonic acid.

10.2.5. Natural and Anthropogenic Sources of Acids

Some of the enhanced acidity of rainwater from sulfuric acid, nitric acid, and hydrochloric acid is natural. Volcanos, for example, emit $SO_2(g)$, a source of sulfuric acid, and $HCl(g)$. Phytoplankton over the oceans emit dimethylsulfide [$DMS(g)$], which oxidizes to $SO_2(g)$. The main natural source of $HNO_3(g)$ is gas-phase oxidation of natural nitrogen dioxide [$NO_2(g)$]. The addition of natural acids to rainwater containing carbonic acid results in typical natural rainwater pHs of between 5.0 and 5.6, as shown in Fig. 10.3.

Acid deposition occurs when anthropogenically produced acids are deposited to the ground, plants, or lakes in dry or wet form. The two most important anthropogenically produced acids today are sulfuric and nitric acid, although hydrochloric acid can be important in some areas. In the eastern United States, about 60 to 70 percent of excess acidity of rainwater is due to sulfuric acid, whereas 30 to 40 percent is due to nitric acid (Glass et al., 1979). Thus, sulfuric acid is the predominant acid of concern. In polluted cites where fog is present, such as in Los Angeles, California, nitric acid fog is a problem. In locations where $HCl(g)$ is emitted anthropogenically, such as near wood burning or industrial processing, $HCl(aq)$ affects the acidity of rainwater. Today, however, $HCl(aq)$ contributes to less than 5 percent of total rainwater acidity by mass. Other acids that are occasionally important in rainwater include formic acid [$HCOOH(aq)$, produced from formaldehyde] and acetic acid [$CH_3COOH(aq)$, produced from acetaldehyde and the main ingrediant in vinegar].

Sulfuric acid originates from sulfur dioxide gas [$SO_2(g)$], and nitric acid originates from gas-phase oxides of nitrogen [$NO_x(g)$]. In the United States, 70 percent of $SO_2(g)$ and more than 85 percent of $NO_x(g)$ emissions are anthropogenic in origin. Thus, the excess acidification of rain in the United States is a result of primarily anthropogenic rather than natural acids.

10.2.6. Acidity of Rain-and Fogwater

Rainwater with a pH less than that of natural rainwater is acid rain. The pH of acid rain varies between 2 and 5.6, although typical values are near 4 and extreme values of less than 2 have been observed (Likens, 1976; Marsh, 1978; Graves, 1980; Graedel and Weschler, 1981). A pH of 4 corresponds to an H^+ molarity 1,000 times that of distilled water and 40 times that of natural rainwater. A pH of 2 corresponds to an H^+ molarity 100,000 times that of distilled water and 4,000 times that of natural rainwater. In Los Angeles, where fogs are common and nitric acid mixing ratios are high, fogwater pHs are typically 2.2 to 4.0 (Waldman et al., 1982; Munger et al., 1983), but levels as low as 1.7 have been recorded (Jacob et al., 1985). Nitrate ion molarities in those studies were about 2.5 times those of sulfate ions. An acidified fog is termed *acid fog*.

10.3. SULFURIC ACID DEPOSITION

Acid deposition is the deposition of acid-containing gases, aerosol particles, fog drops, or rain drops to the ground, lakes, plant leaves, tree leaves, or buildings. The most abundant acid in the air is usually sulfuric acid [$H_2SO_4(aq)$], whose source is sulfur dioxide gas [$SO_2(g)$], emitted anthropogenically from coal-fire power plants, metal-smelter operations, and other sources (Section 3.6.6).

Table 10.1. Names and Formulae of S(IV) and S(VI) Species

S(IV) Family		S(VI) Family	
Chemical Name	Chemical Formula	Chemical Name	Chemical Formula
Sulfur dioxide	$SO_2(g,aq)$		
Sulfurous acid	$H_2SO_3(aq)$	Sulfuric acid	$H_2SO_4(g,aq)$
Bisulfite ion	HSO_3^-	Bisulfate ion	HSO_4^-
Sulfite ion	SO_3^{2-}	Sulfate ion	SO_4^{2-}

Power plants usually emit $SO_2(g)$ from high stacks so that the pollutant is not easily downwashed to the surface nearby. The higher the stack, the further the wind carries the gas before it is removed from the air. The wind transports $SO_2(g)$ over long distances, sometimes hundreds to thousands of kilometers. Thus, acid deposition is often a regional and **long-range transport** problem. When acids or acid precursors are transported across political boundaries, they create **transboundary air pollution**, prevalent between the United States and Canada; among western, northern, and eastern European countries; and among several Asian countries.

Sulfur dioxide and sulfuric acid are but two of several sulfur-containing species in the air. Some additional species are listed in Table 10.1. The species in the table are conveniently divided into two families, the **S(IV) and S(VI) families**, in which the IV and the VI represent the oxidation states (+4 and +6, respectively) of the members of the respective families. Thus, S(VI) members are more oxidized than are S(IV) members. Because sulfur dioxide is in the S(IV) family and sulfuric acid, the main source of acidity in rainwater, is in the S(VI) family, the oxidation of gas-phase sulfur dioxide to aqueous-phase sulfuric acid represents a conversion from the S(IV) family to the S(VI) family. This conversion occurs along two pathways, described next.

10.3.1. Gas-Phase Oxidation of S(IV)

The fist conversion mechanism of S(IV) to S(VI) involves the following steps: (1) gas-phase oxidation of $SO_2(g)$ to $H_2SO_4(g)$; (2) condensation of $H_2SO_4(g)$ and water vapor onto aerosol particles or cloud drops to produce an $H_2SO_4(aq)$-$H_2O(aq)$ solution; and (3) dissociation of $H_2SO_4(aq)$ to SO_4^{2-} in the solution. The gas-phase chemical conversion process (Step 1) is

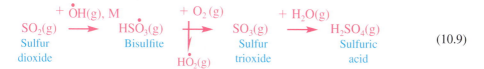

$$\tag{10.9}$$

Because sulfuric acid has a low saturation vapor pressure (SVP, Section 5.3.2.1), nearly all $H_2SO_4(g)$ produced by Reaction 10.9 condenses onto particle or drop surfaces (Step 2). At typical pHs of aerosol particles and cloud drops, nearly all condensed $H_2SO_4(aq)$ dissociates to SO_4^{2-} by Reaction 10.6 (Step 3). The dissociation releases two protons, decreasing pH and increasing acidity.

Whereas this is the dominant mechanism by which S(IV) produces S(VI) in aerosol particles, particularly when the relative humidity is below 70 percent, a second mechanism more rapidly produces S(VI) from S(IV) in cloud drops and rain drops.

10.3.2. Aqueous-Phase Oxidation of S(IV)

The second conversion process of S(IV) to S(VI) involves the following steps: (1) dissolution of $SO_2(g)$ into liquid-water drops to produce $SO_2(aq)$; (2) in-drop conversion of $SO_2(aq)$ to $H_2SO_3(aq)$ and dissociation of $H_2SO_3(aq)$ to HSO_3^- and SO_3^{2-}; and (3) in-drop oxidation of HSO_3^- and SO_3^{2-} to SO_4^{2-}. The dissolution process (Step 1) is represented by the reversible reaction

$$SO_2(g) \rightleftharpoons SO_2(aq)$$

Sulfur   Dissolved
dioxide   sulfur   (10.10)
  gas   dioxide

The formation and dissociation of sulfurous acid [$H_2SO_3(aq)$] (Step 2) occurs by

$$SO_2(aq) + H_2O(aq) \rightleftharpoons \qquad\qquad\qquad\qquad\qquad H^+ + SO_3^{2-}$$

Dissolved   Liquid   Sulfurous   Hydrogen   Bisulfite   Hydrogen Sulfite
 sulfur   water   acid   ion   ion   ion   ion
dioxide   (10.11)

Step 3 involves the irreversible conversion of the S(IV) family (primarily HSO_3^- and SO_3^{2-}) to the S(VI) family (primarily SO_4^{2-}). At pH levels of 6 or less, the most important reaction converting S(IV) to S(VI) is

$$HSO_3^- + H_2O_2(aq) + H^+ \longrightarrow SO_4^{2-} + H_2O(aq) + 2H^+$$

Bisulfite   Dissolved   Sulfate
  ion   hydrogen   ion   (10.12)
            peroxide

This reaction is written in terms of HSO_3^- and SO_4^{2-} because at pHs 2 to 6, most S(IV) exists as HSO_3^- and most S(VI) exists as SO_4^{2-}.

At pH levels greater than 6, which occur only in clouds drops that contain basic substances, such as ammonium or sodium, the most important reaction converting S(IV) to S(VI) is

$$SO_3^{2-} + O_3(aq) \longrightarrow SO_4^{2-} + O_2(aq)$$

Sulfite   Dissolved   Sulfate   Dissolved
 ion   ozone   ion   oxygen   (10.13)

This reaction is written in terms of SO_3^{2-} and SO_4^{2-} because the HSO_3^--O_3 reaction is relatively slow and at pH levels greater than 6, most S(VI) exists as SO_4^{2-}.

When $SO_2(g)$ dissolves in a drop to form $H_2SO_3(aq)$, the $H_2SO_3(aq)$ reacts to form SO_4^{2-}, forcing more $SO_2(g)$ to be drawn into the drop to replace the lost $H_2SO_3(aq)$. The more $SO_2(g)$ that dissolves and reacts, the more SO_4^{2-} that forms. In cloud drops, dissolution and aqueous reaction can convert 60 percent of $SO_2(g)$ molecules to SO_4^{2-} molecules within 20 minutes.

10.4. NITRIC ACID DEPOSITION

Nitric acid deposition occurs in and downwind of urban areas and is enhanced by the presence of clouds or fog. The origin of nitric acid [$HNO_3(g)$] is usually nitric oxide [$NO(g)$], emitted from vehicles and power plants. In the air, $NO(g)$ is oxidized to nitrogen dioxide [$NO_2(g)$], some of which is also directly emitted. $NO_2(g)$ is oxidized to nitric acid by

$$\overset{\bullet}{O}H(g) + \overset{\bullet}{N}O_2(g) \xrightarrow{M} HNO_3(g)$$
Hydroxyl   Nitrogen   Nitric   (10.14)
radical   dioxide   acid

Gas-phase nitric acid dissolves into aerosol particles or fog drops to form $HNO_3(aq)$, which dissociates to a proton [H^+] and the nitrate ion [NO_3^-] by Reaction 10.7. Thus, the addition of nitric acid to cloud water decreases the pH and increases the acidity of the water. Gas-phase nitric acid also deposits to the ground, where it can cause environmental damage.

10.5. EFFECTS OF ACID DEPOSITION

The most severe pollution episode in the twentieth century involving sulfuric acid-containing fog was probably that in London in 1952, discussed in Section 4.1.1.1. In that episode, coal burning in combination with a heavy fog resulted in more than 4,000 deaths. Although other pollutants were responsible as well, the acidified fog contributed to the disaster.

Acid deposition affects lakes, rivers, forests, agriculture, and building materials. The regions of the world that have been affected most by acid deposition include provinces of eastern Canada, the northeastern United States (particularly the Adirondack Mountain region), southern Scandinavia, middle and eastern Europe, India, Korea, Russia, China, Japan, and Thailand. Acidified forests, crops, and surface waters have also been reported in South Africa (SEI, 1998).

10.5.1. Effects on Lakes and Streams

Acids reduce the pH level in lakes and streams. Because fish and microorganisms can survive only in particular pH ranges, the changing of a lake's pH kills off many varieties of fish (including trout and salmon), invertebrates, and microorganisms. Most aquatic insects, algae, and plankton live only at pH levels above 5. The reduction of lake pH below 5 kills off these organisms, causing starvation at higher levels of the food chain. Low pH levels (less than 5.5) in lakes have also been associated with reproductive failures and mutations in fish and amphibians.

Lake acidification has particularly been a problem in Scandinavian countries. Most damage occurred in the 1950s and 1960s, during which time the average pH of Swedish lakes fell by 1 pH unit. By the end of the 1970s, about 25,000 of Sweden's 90,000 lakes were so acidified that only acid-resistant plants and animals could survive. Of the acidified lakes, about 8,000 were naturally acidic, suggesting that 17,000 had been acidified anthropogenically. Today, many lakes in Sweden and in other countries have been restored.

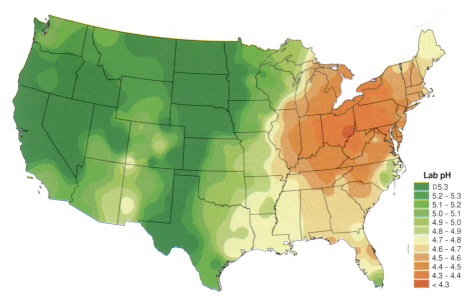

Lab pH

	≥5.3
	5.2 - 5.3
	5.1 - 5.2
	5.0 - 5.1
	4.9 - 5.0
	4.8 - 4.9
	4.7 - 4.8
	4.6 - 4.7
	4.5 - 4.6
	4.4 - 4.5
	4.3 - 4.4
	< 4.3

Figure 10.4. Map of rainwater pH in the United States in 1999 (National Atmospheric Deposition Program/National Trends Network, 2000).

Figure 10.4 is a map of rainwater acidity in the United States showing that most acidic rain occurs in Ohio, West Virginia, Pennsylvania, New York, Indiana, Michigan, Maryland, and parts of Florida.

The effects of acid rain on lakes are most pronounced after the first snowmelt of a season. Because acids accumulate in snow, runoff from melted snow can send a shock wave of acid to a lake. In some cases, the acidity of meltwater is ten times greater than that of the original rain.

10.5.2. Effects on Biomass

Acids damage plant and tree leaves and roots. When sulfuric acid deposits onto a leaf or needle, it forms a liquid film of low pH that erodes the cuticle wax, leading to the drying out (desiccation) of and injury to the leaf or needle. When acid gases, aerosol particles, or raindrops enter forest groundwater, they damage plants at their roots in two ways. First, sulfuric and nitric acid solutions dissolve and carry away important mineral nutrients, including calcium, magnesium, potassium, and sodium. Second, in acidic solutions, hydrogen ions [H^+] react with aluminum- and iron-containing minerals, such as aluminum hydroxide [$Al(OH)_3(s)$] and iron hydroxide [$Fe(OH)_3(s)$], releasing Al^{3+} and Fe^{3+}, respectively. At high enough concentrations, these metal ions are toxic to root systems (Tomlinson, 1983). As a result of acid deposition, whole forests have been decimated. In Poland and the Czech Republic, 60 to 80 percent of trees have died in recent years (SEI, 1998). Figure 10.5(a) shows a forest near the border between Germany and the Czechoslovakia in which all the lower foliage died. Figure 10.5(b) shows a forest near Most, Czechoslovakia, that was decimated by acid deposition and air pollution. Forest damage is also evident in central Europe, the United States, Canada, China, Japan, and many other countries. Acid deposition destroys crops in the same way that it destroys forests.

(a)

(b)

Figure 10.5. (a) Acidified forest, Oberwiesenthal, Germany, near the border with the Czechoslovakia, taken in 1991. The trees are of the *Picea* family. Photo by Stefan Rosengren, available from Naturbild. (b) Acidified forest in the Erzgebirge Mountains, north of the town of Most, Czechoslovakia, taken in 1987. Photo by Owen Bricker, USGS.

Figure 10.6. Sandstone figure over the portal of a castle, built in 1702, in Westphalia, Germany, photographed in 1908 (left) and in 1968 (right). The erosion of the figure is due to a combination of acid deposition and air pollution produced from the industrialized Ruhr region of Germany. Courtesy Herr Schmidt-Thomsen.

Not all damage to forests and crops is a result of acid deposition. Ozone also reacts with leaves, increasing plant and tree stress and making plants and trees more susceptible to disease, infestation, and death. Soot smothers leaves, increasing plant and tree stress. PAN discolors the leaves of plants and trees.

10.5.3. Effects on Buildings and Sculptures

Acid deposition erodes materials. In particular, acids erode sandstone, limestone, marble, copper, bronze, and brass. Of note are buildings and sculptures of historical and archeological interest, such as the Parthenon in Greece, that have decayed, partly as a result of acid deposition and partly as a result of other pollutants in the air. Ozone, for example, reduces the detail in statues. Figure 10.6 shows an example of statue erosion, due both to acid deposition and other types of air pollution.

10.6. NATURAL AND ARTIFICIAL NEUTRALIZATION OF LAKES AND SOILS

One way to reduce the effect of acid deposition on lakes is to add a **neutralizing agent** (often called a **buffer**) to the lake. Neutralizing agents increase the pH of acidified lakes. If the pH rises above 7, the lake becomes alkaline (basic). Certain chemicals also act as natural neutralizing agents in lakes and soils.

10.6.1. Ammonium Hydroxide

One anthropogenic neutralizing agent is **ammonium hydroxide** [NH₄OH(aq)], obtained by dissolving ammonia gas [NH₃(g)] in water. The net effect of adding NH₄OH(aq) to acidified water is

$$NH_4OH(aq) + H^+ \rightleftharpoons NH_4^+ + H_2O(aq)$$

Ammonium   Hydrogen   Ammonium   Liquid   (10.15)
hydroxide   ion   ion   water

which reduces the H^+ molarity, reducing acidity and increasing pH.

10.6.2. Sodium and Calcium Hydroxide

When **lye** [NaOH(aq), **sodium hydroxide**], the common component of Drano, or **slaked lime**, [Ca(OH)₂(aq), **calcium hydroxide**] is added to an acidified lake, it reacts with H^+ to form water, decreasing acidity. The net effect of adding slaked lime to acidified water is

$$Ca(OH)_2(aq) + 2H^+ \rightleftharpoons Ca^{2+} + 2H_2O(aq)$$

Calcium   Hydrogen   Calcium   Liquid   (10.16)
hydroxide   ion   ion   water

Lime is commonly added to lakes in large amounts (Fig. 10.7) to reduce the effects of acid deposition. Sweden, which has the largest liming program in the world, adds

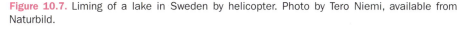

Figure 10.7. Liming of a lake in Sweden by helicopter. Photo by Tero Niemi, available from Naturbild.

200,000 tons of fine-ground limestone to lakes and watercourses each year. Since the 1970s, more than 7,000 of Sweden's 17,000 anthropogenically acidified lakes have been limed. Because lime is consumed 2 to 3 years after its application, acidified lakes need to be relimed regularly. Other countries that have large liming programs include Norway, Finland, and Canada.

10.6.3. Calcium Carbonate

Some soils that contain the minerals calcite or aragonite [$CaCO_3(s)$, **calcium carbonate**] have a natural ability to neutralize acids. When acid rain falls onto calcite-containing soils, the H^+ is removed by

$$CaCO_3(s) + 2H^+ \rightleftharpoons Ca^{2+} + CO_2(g) + H_2O(aq)$$

| Calcium carbonate | Hydrogen ion | Calcium ion | Carbon dioxide gas | Liquid water | (10.17) |

The same result occurs when soil-dust particles that contain $CaCO_3(s)$ collide with acidified raindrops. The erosion of farmland and desert borders in many locations has enhanced the quantity of soil dust in the air, inadvertently increasing the calcium carbonate content of rainwater, decreasing rainwater acidity, and increasing rainwater pH in nearby regions.

Unfortunately, the same process described by Reaction 10.17 that decreases soil acidity is partly responsible for the erosion of great statues and buildings made of or containing marble or limestone. **Marble** and **limestone**, which both contain calcite, erode when they become coated with acidified water. Coating can occur in at least two ways. The first is when acidified raindrops or aerosol particles deposit directly onto a marble or limestone surface. The second is when a gas dissolves and forms an acid in dew or rainwater that has recently coated a surface. For example, when $SO_2(g)$ dissolves in water, it oxidizes to sulfuric acid (Section 10.3.2).

When water containing sulfuric acid coats a calcite surface, the hydrogen ion dissolves the calcite by Reaction 10.17, and the sulfate ion reacts with the dissociated calcium to form the mineral **gypsum** by

$$Ca^{2+} + SO_4^{2-} + 2H_2O(aq) \rightleftharpoons CaSO_4 - 2H_2O(s)$$

| Calcium ion | Sulfate ion | Liquid water | Calcium sulfate dihydrate (gypsum) | (10.18) |

The net result is the formation of a clear-to-white gypsum crust over the marble or limestone. Bombardment by rain over time removes some of the brittle gypsum crust. Because the crust now contains part of the statue or building material (the calcium), its removal creates tiny crevices, or pits, causing erosion (Davidson et al., 1999). Because the gypsum crust and the crevices roughen the surface of a statue or building, other pollutants, such as soot, more readily bond to the surface, darkening it, as illustrated in Fig. 10.8. The figure shows photographs of the Cathedral of Learning, at the University of Pittsburgh, taken in 1930, soon after the start of its construction, and in 1934. During a four-year period, sulfate from coal smoke emitted by steel mills and locomotives (Section 4.1.6.3) roughened the limestone exterior of the building, and soot from the same smoke bonded with the roughened exterior, darkening it.

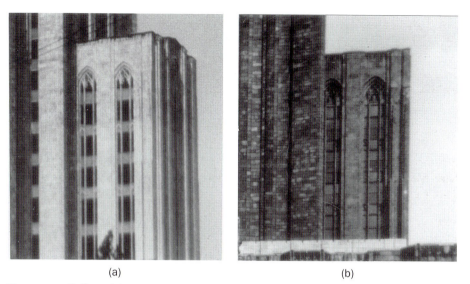

(a)   (b)

Figure 10.8. Soiling of the limestone exterior of the Cathedral of Learning at the University of Pittsburgh between (a) 1930 and (b) 1934 (Davidson et al., 1999). The building was constructed between 1929 and 1937. Sulfate and soot from coal smoke caused erosion and darkening of the building after only 4 years. Photo courtesy of the University Archives, University of Pittsburgh.

Gypsum forms not only on buildings, but also in soils and aerosol particles. When rainwater containing the sulfate ion falls on soil containing calcite, the calcite dissociates by Reaction 10.17. When the soil dries, the calcium ion reacts with the sulfate ion by Reaction 10.18, producing gypsum. The same process occurs when soil-dust particles containing calcite collide with acidified raindrops. Deposition of rain containing sulfuric acid over soils containing calcite and deposition of particles already containing gypsum have, over time, produced worldwide deposits of gypsum soil.

10.6.4. Sodium Chloride

Some acidic soils near nonpolluted coastal areas are naturally neutralized by cations originating from sea spray (e.g., Na^+, Ca^{2+}, Mg^{2+}, K^+) that have deposited onto soils over the millennia. The pH of natural seawater ranges from 7.8 to 8.3, and that of uncontaminated large sea-spray drops is similar, indicating that little H^+ exists in such drops. The deposition of sea-spray drops to coastal soils, and the subsequent desiccation of these drops produces the mineral **halite** [NaCl(s), sodium chloride or common salt]. When sulfate-containing water enters NaCl(s)-containing soils, NaCl(s) dissolves and dissociates, H^+ combines with the chloride ion [Cl^-] to form HCl(aq), and HCl(aq) evaporates to the gas phase. The net process is

$$NaCl(s) + H^+ \rightleftharpoons Na^+ + HCl(g)$$

| Sodium chloride | Hydrogen ion | Sodium ion | Hydrochloric acid | (10.19) |

which reduces H^+, increases pH, and reduces the acidity of soil water.

10.6.5. Ammonia

Ammonia gas [$NH_3(g)$] is considered an anthropogenic pollutant, but it also neutralizes raindrop and soil-water acidity. Sources of ammonia gas were discussed in Section 5.3.2.3. Once in the air, ammonia gas dissolves in water. The dissolved gas then reacts with the hydrogen ion to form the ammonium ion by

$$NH_3(aq) + H^+ \rightleftharpoons NH_4^+$$

Dissolved   Hydrogen   Ammonium   (10.20)
ammonia   ion   ion

The reduction in H^+ that results from this reaction increases pH, reducing acidity. In many cases, the pH of rainwater containing the ammonium ion exceeds 6. Aerosol particles and raindrops containing the ammonium ion deposit to soils and lakes, providing these surfaces with a neutralizing agent. Soils downwind of high ammonia gas emissions tend to have a better neutralizing capacity against acid deposition than do soils far from ammonia sources if all other conditions are the same.

10.7. RECENT REGULATORY CONTROL OF ACID DEPOSITION

The first major effort to control acid deposition was the U.K. Alkali Act of 1863, which mandated large reductions of hydrochloric acid gas emissions by soda-ash manufacturers. In more recent years, the U.S. Clean Air Act Amendments of 1970 led to lower emissions of acid-deposition precursors, namely $SO_2(g)$ and $NO_2(g)$. In 1977, the U.S. initiated the National Atmospheric Deposition Program (NADP), whose agenda was to monitor trends of acidity in precipitation. In 1980, the U.S. Congress passed the Acid Precipitation Act, which funded a program, the National Acid Precipitation Assessment Program (NAPAP). Under the program, the network of monitoring stations under NADP was enlarged to produce a National Trends Network (NTN). NAPAP reported trends observed at the monitoring sites. The U.S. Clean Air Act Amendments of 1990 mandated a 10 million ton reduction in sulfur dioxide [$SO_2(g)$] emissions from 1980 levels and a 2 million ton reduction in nitrogen oxide [$NO_x(g)$] emissions from 1980 levels by 2010. To implement these reductions, the U.S. EPA set up an emission trading system, whereby emitters could trade among themselves for limited rights to release $SO_2(g)$. Power plants were also required to install emission monitoring systems. In January 2000, the U.S. EPA issued a new rule requiring U.S. refiners to cut the sulfur content of gasoline to one-tenth its value by 2006.

Meanwhile, several studies in the 1970s concluded that winds were transporting acid-deposition precursors over long distances, often over political boundaries. Such studies culminated in the 1979 Geneva Convention on Long-Range Transboundary Air Pollution. The convention was signed by 34 governments and the European Community, and was the first agreement to deal with an international air pollution problem. As part of a 1985 amendment to the agreement (the Sulfur Protocol), member countries were required to reduce their emissions or transboundary fluxes of sulfur by 30 percent below 1980 values by 1993. Another amendment in 1988 (the Nitrogen Oxide Protocol) required countries to reduce their emissions or transboundary fluxes of nitrogen oxide to their 1987 levels by December 1994. Because the first Sulfur

Protocol did not address forest loss in central Europe sufficiently, a second Sulfur Protocol was signed in 1994 that will result in a 60 percent reduction in sulfur emissions below 1980 values by 2010.

10.7.1. Methods of Controlling Emissions

Several mechanisms are available to control emissions of acid deposition precursors. These include the mandatory use of low-sulfur coal instead of high-sulfur coal and the use of emission-control technologies. The amount of $SO_2(g)$ emitted during coal combustion depends on the sulfur content of the coal. In the United States, about 39 percent of coal is mined in the Appalachian Mountains and the rest is mined west of the Mississippi River, with Wyoming producing 31 percent of all U.S. coal, the largest percentage of any state (EIA, 2000). Coal from the Appalachian Mountains has a high sulfur content. The cost of transporting Appalachian coal to power plants, most of which are in the midwest and eastern United States, is lower than is the cost of transporting low-sulfur coal from Wyoming or other western states to these plants. As such, coal burners prefer to use high-sulfur coal. As a result of the CAAA90 requirement to reduce $SO_2(g)$ emissions, the reliance on western U.S. coal is expected to increase.

Use of low-sulfur coal is one mechanism to reduce emission of $SO_2(g)$ during coal burning. Another is to remove a certain fraction of sulfur from high-sulfur coal before burning it. A technology available for reducing $SO_2(g)$ emission from a stack is the scrubber, first developed by William Gossage to reduce $HCl(g)$ emission. A modern-day scrubber works by dissolving $SO_2(g)$ into small water drops sprayed into an exhaust stream, then removing the drops on a collecting surface, such as a bed or a wetted surface.

10.7.2. Effects of Regulation

The U.S. EPA Office of Air and Radiation estimates that sulfate concentrations in rainfall were 10 to 25 percent less in 1995 and 1996 than what they would have been if controls mandated through the Clean Air Act Amendments of 1990 had not been implemented. The largest reductions in sulfate concentrations occurred along the Ohio River Valley and states downwind. Nitrate concentrations during this period did not improve.

Reductions of sulfur dioxide emissions in Canada have reduced the acidity of some Canadian lakes and forests. For example, in the late 1960s, the Sudbury, Ontario, nickel-smelting stack (Section 6.6.2.4) was the largest individual source of $SO_2(g)$ in North America, emitting 5000 tons of $SO_2(g)$ per day, devastating nearby lakes and forests. Today, its emissions are below 500 tons of $SO_2(g)$ per day, and nearby lakes and forests have partially regenerated. Reductions in the acidity of lakes in Quebec, Atlantic Canada, and other areas of Ontario have been less dramatic (Environment Canada, 2000b). Many lakes in Sweden have been restored, but many more are still damaged by acid deposition. Acid deposition problems in eastern Europe and Asia are still severe.

10.8. SUMMARY

In this chapter, the history, science, effects, and control of acid deposition were discussed. Acidity is determined by pH, which ranges from less than 0 (very acidic) to more than 14 (very basic). The pH of distilled water is 7, of natural rainwater is 5.0 to

5.6, and of acid rain or fog is less than 5.0. Acid deposition occurs when sulfuric, nitric, or hydrochloric acid is emitted into or forms chemically in the air and is subsequently deposited as a gas or liquid to the ground, where it harms microorganisms, fish, forests, agriculture, and structures. In high concentrations in the air, acids can also harm humans. Severe acid deposition problems arose from increased coal combustion in the U.K. during the Industrial Revolution and from the growth of the alkali industry in France and the United Kingdom in the 1800s. Today, sulfuric acid is usually the most abundant acid in rainwater. Sulfuric acid is produced by gas- and aqueous-phase oxidation of sulfur dioxide. The latter process is most efficient when cloud drops are present. In polluted coastal air, nitric acid fog is often a problem. A method of ameliorating the effects of acid deposition on lakes is to add a neutralizing agent; such as slaked lime. In the United States, Canada, and western Europe, government intervention in the form of regulations limiting the emissions of acid-deposition precursors has resulted in reductions in the acidity of rainwater. Acid deposition problems in eastern Europe and Asia are still severe.

10.9. PROBLEMS

10.1.   Identify all the atmospheric acids produced by Leblanc's soda ash process.

10.2.   In terms of acid deposition precursors, what were the advantages of the Solvay versus the Leblanc soda ash process?

10.3.   Although Leblanc's process produced HCl(g), which caused widespread acid deposition problems in the early 1800s, HCl(g) was no longer the most dangerous by-product of this process in the late 1800s. Why?

10.4.   Describe the two important conversion pathways for S(IV) to S(VI). Which pathway is more important when aerosol particles are present? Why?

10.5.   What are the most important aqueous-phase oxidants of S(IV)?

10.6.   Why are nitric acid and hydrochloric acid deposition less of a problem in most parts of the world than is sulfuric acid deposition?

10.7.   How do neutralizing agents reduce the acidity of a lake?

10.8.   Suppose rainwater containing the sulfate ion enters a soil containing magnesium carbonate [$MgCO_3$(s)]. Would the magnesium carbonate act as a neutralizing agent or enhance the acidity of the water? Show the pertinent chemical process.

10.9.   Identify three products that you use or activities that you do that result in the emission of acids into atmosphere.

10.10. Identify three ways that acids or acid precursors can be controlled through legislative action.

GLOBAL STRATOSPHERIC OZONE REDUCTION

11

The stratospheric ozone layer began to form soon after the onset of oxygen-producing photosynthesis, about 2.3 billion years ago (b.y.a.). It probably did not develop fully until at least 400 million years ago (m.y.a.), when green plants evolved and molecular oxygen mixing ratios began to approach their present levels. Absorption of ultraviolet (UV) radiation by ozone is responsible for the temperature inversion that defines the present day stratosphere. This absorption is critical for preventing UV radiation from reaching the surface of the Earth, where it can harm life. The anthropogenic emission of long-lived chlorine- and bromine-containing compounds into the air since the 1930s and the slow transfer of these compounds to the stratosphere has caused a nontrivial reduction in the global stratospheric ozone layer since the 1970s. In addition, during September, October, and November each year since the early 1980s, up to 70 percent of the ozone layer has been destroyed over the Antarctic. Lesser reductions have occurred over the Arctic in March, April, and May each year. Recent international cooperation has helped reduce emissions and slow further ozone loss. In this chapter, the natural stratospheric ozone layer, global ozone reduction, and Antarctic/Arctic ozone destruction and regeneration are discussed.

11.1. STRUCTURE OF THE PRESENT-DAY OZONE LAYER

About 90 percent of all ozone molecules in the atmosphere reside in the stratosphere; most of the remaining molecules reside in the troposphere. Whereas ozone molecules near the surface harm humans, animals, plants, trees, and structures, the same ozone molecules, whether in the stratosphere or in polluted air, shield the Earth from harmful UV radiation.

A measure of the quantity of ozone in the air is the ozone column abundance, which is the sum of all ozone molecules above a square centimeter of surface between the ground and the top of the atmosphere. When this number is divided by 2.7×10^{16}, the result is the column abundance in **Dobson units** (DUs). Thus, 1 DU is equivalent to 2.7×10^{16} molecules of ozone per square centimeter of surface. The Dobson unit is named after **Gordon M. B. Dobson** (1889–1976), a researcher at Oxford University who, in the 1920s, built the first instrument, now called a *Dobson meter*, to measure total ozone column abundance from the ground. In 2000, the globally averaged column abundance of ozone from 90°S to 90°N was 293.4 DU. This column abundance contains the same number of molecules as a column of air 2.93-mm high at 1 atm of pressure and 273 K (near-surface conditions). Figure 11.1 illustrates ozone column abundance.

Figure 11.2 shows a plot of the variation in the ozone column abundance with latitude (zonally averaged – averaged over all longitudes) and month for the year 2000. The following features can be seen in the figure:

- A year-around equatorial ozone minimum due to upward motion of ozone-poor air from the troposphere that displaces ozone-rich air horizontally to higher latitudes. The column abundance over the equator is typically 250 to 290 DUs all year.
- A Northern Hemisphere (NH) spring (March–May) maximum, ranging from 350 to 460 DU, near the North Pole. The maximum is due to the northward transport of stratospheric ozone from the equator. As ozone converges at the pole, it descends, increasing the ozone column abundance. The maximum column abundance at a

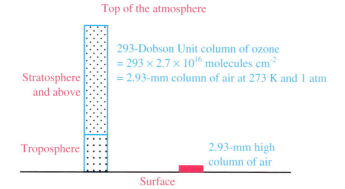

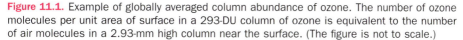

Figure 11.1. Example of globally averaged column abundance of ozone. The number of ozone molecules per unit area of surface in a 293-DU column of ozone is equivalent to the number of air molecules in a 2.93-mm high column near the surface. (The figure is not to scale.)

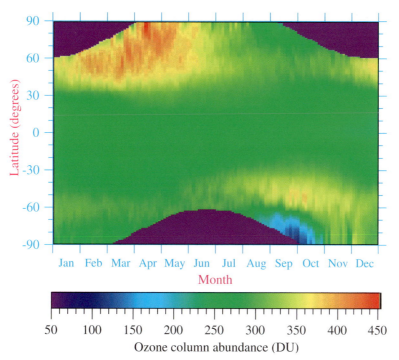

Figure 11.2. Variation of zonally averaged ozone column abundance with latitude and month during 2000. Blank regions near the poles indicate locations where data were not available. Data for the figure were obtained from the satellite-based Total Ozone Mapping Spectrometer (TOMS) and made available by NASA Goddard Space Flight Center, Greenbelt, Maryland.

specific location (not zonally averaged) in 2000 was 573 DU on January 28 at 64.5°N, 178.125°W.

- A Southern Hemisphere (SH) spring (September–November) subpolar (60 to 65°S) maximum, ranging from 350 to 420 DU. The maximum is due to the southward transport of ozone from the equator. As the ozone moves south, much of it is forced to descend in front of the **polar vortex**, a polar front jet-stream wind system that travels around the Antarctic continent in the upper troposphere and stratosphere.

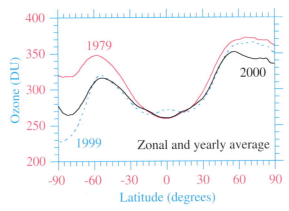

Figure 11.3. Variation of yearly and zonally averaged ozone column abundance with latitude in 1979, 1999, and 2000. Data were obtained from the satellite-based Total Ozone Mapping Spectrometer (TOMS) and made available by NASA Goddard Space Flight Center, Greenbelt, Maryland.

- A SH spring minimum of less than 150 DU over the South Pole due to chemical reactions of chlorine and bromine radicals with ozone. This minimum is called the **Antarctic ozone hole**. The minimum column abundance at a specific location (not zonally averaged) in 2000 was 94 DU on September 29 at 85.5°S, 64.375°W.

Figure 11.3 shows the variation with latitude of the yearly and zonally averaged ozone column abundance in 1979, 1999, and 2000. The figure shows that the ozone layer was thin near the equator in all three years. From 15°S to 15°N, the ozone column abundance actually increased in 1999 in comparison with 1979. Such increases, relative to 1979 ozone, occurred during about one-third of the years between 1979 and 2000.

In 1979 and earlier, the yearly and zonally averaged ozone column abundance over 60° to 90°S was greater than over the equator. Since then, the seasonal Antarctic ozone hole (Section 11.4) has caused the ozone column over 60 to 90°S to decline, even in the yearly average. Although the average column abundance over 60° to 90°S was slightly higher in 2000 than in 1999, the area of the ozone hole was larger in 2000 than in 1999.

Over 60° to 90°N, the ozone column abundance is always greater than over the equator. In most years from 1979 to 2000, the column abundance over 60° to 90°N was lower than it was in 1979. One exception was in 1999, when the column abundance was nearly the same as in 1979. In 2000, another reduction occurred. When a reduction occurs, it is called an **Arctic ozone dent** (Section 11.4).

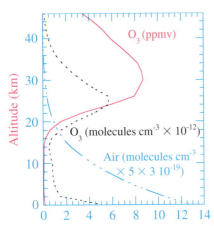

Figure 11.4. Example vertical variation in ozone mixing ratio, ozone number concentration, and air number concentration with altitude. The ozone mixing ratio at the surface is 0.20 ppmv, the level of a Stage 1 smog alert in the United States.

Figure 11.4 shows a typical variation of ozone mixing ratio, ozone number concentration, and total air number concentration with altitude. The ozone number concentration (molecules of ozone per cubic centimeter of air) in the stratosphere generally peaks at 25 to 32 km altitude. The ozone mixing ratio (number concentration of ozone divided by that of dry air) peaks at a higher altitude than does the ozone number concentration. The peak ozone number concentration in the stratosphere is close to that in polluted urban air. The peak ozone mixing ratio in the stratosphere (near 10 ppmv) is much higher than is that in polluted urban air (0.2 to 0.35 ppmv) or free-tropospheric air (0.02 to 0.04 ppmv).

11.2. RELATIONSHIP BETWEEN THE OZONE LAYER AND UV RADIATION

The ozone layer prevents damaging UV wavelengths from reaching the ground. As shown in Fig. 2.5, the UV portion of the solar spectrum is divided into far- and near-UV wavelengths. Near-UV wavelengths are further divided into UV-A, UV-B, and UV-C wavelengths. Gases, particularly ozone and oxygen, and aerosol particles absorb most solar UV radiation before it reaches the Earth's surface. Decreases in stratospheric ozone increase the transmission of UV to the surface. Enhancements in UV at the surface damage life. In this subsection, processes affecting UV radiation are summarized.

Figure 11.5 shows the intensity of downward UV and visible radiation at the top of the atmosphere (TOA) and at the ground. The figure shows that, of the incident solar radiation at the TOA, only wavelengths longer than 0.29 μm penetrate to the ground. Thus, the air filters out all far-UV and UV-C wavelengths. Of the UV that reaches the ground, about 9 percent is UV-B and the rest (91 percent) is UV-A. Of the total solar radiation reaching the surface, about 5.2 percent is UV-A/UV-B and the rest (94.8 percent) is visible/near-IR.

Table 11.1 identifies the major absorbing components responsible for reducing near- and far-UV radiation between the TOA and ground. Molecular nitrogen [$N_2(g)$] absorbs far-UV wavelengths shorter than 0.1 μm in the thermosphere and mesosphere, and molecular oxygen [$O_2(g)$] absorbs wavelengths shorter than 0.245 μm in the thermosphere, mesosphere, and stratosphere.

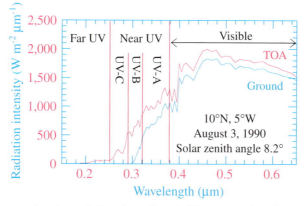

Figure 11.5. Downward solar radiation less than 0.65 μm wavelength at the top of the atmosphere (TOA) and at the ground at a location near the equator in early August. The solar zenith angle is the angle of the sun relative to a line perpendicular to the Earth's surface.

Table 11.1. Summary of Major Absorbers of UV Radiation

Name of Spectrum	Wavelengths (μm)	Dominant Absorbers	Location of Absorption
Far-UV	0.01–0.25	N_2(g)	Thermosphere, mesosphere
		O_2(g)	Thermosphere, mesosphere, stratosphere
Near-UV			
UV-C	0.25–0.29	O_3(g)	Stratosphere
UV-B	0.29–0.32	O_3(g)	Stratosphere, troposphere
		Particulate components	Polluted troposphere
UV-A	0.32–0.38	NO_2(g)	Polluted troposphere
		Particulate components	Polluted troposphere

Ozone absorbs wavelengths shorter than 0.35 μm (strongly below 0.31 μm and weakly at 0.31 to 0.35 μm). Ozone also absorbs weakly at 0.45 to 0.75 μm. Stratospheric ozone allows little UV-C, some UV-B, and most UV-A radiation to reach the troposphere. Ozone in the background troposphere absorbs some of the UV-B and UV-A not absorbed in the stratosphere. In polluted air, additional absorbers of UV-B radiation include nitrated aromatic gases and aerosol particle components, such as black carbon (BC), nitrated aromatics, polycyclic aromatic hydrocarbons (PAHs), and soil-dust components (Section 7.1.3.1).

The major UV-A-absorbing gas is nitrogen dioxide [NO_2(g)]. Its mixing ratio in clean air is too small to affect UV-A. In polluted air, its mixing ratio is high only in the morning, when UV-A intensity is low. Other UV-A absorbers in polluted air include the same aerosol particle components that absorb UV-B radiation.

Gas- and aerosol particle absorption is not the only mechanism that reduces incident downward UV radiation. Gas and aerosol particle backscattering, ground reflection, and cloud reflection return some incident UV radiation to space as well.

11.3. CHEMISTRY OF THE NATURAL OZONE LAYER

The chemistry of the natural ozone layer involves primarily oxygen-containing compounds, but the shape of the stratospheric ozone vertical profile is affected by nitrogen- and hydrogen-containing compounds as well. Next, the chemistry of the natural ozone layer is discussed.

11.3.1. The Chapman Cycle

The photochemistry of the natural stratosphere is similar to that of the free troposphere, except that stratospheric ozone is produced after photolysis of molecular oxygen, whereas tropospheric ozone is produced after photolysis of nitrogen dioxide. Next, reactions naturally producing and destroying stratospheric ozone are described.

In the stratosphere, far-UV wavelengths (shorter than 0.245 μm) break down molecular oxygen by

$$O_2(g) + h\nu \longrightarrow \bullet\overset{\bullet}{O}(^1D)(g) + \bullet\overset{\bullet}{O}(g) \qquad \lambda < 175 \text{ nm}$$

Molecular   Excited   Ground-
oxygen   atomic   state atomic
                      oxygen   oxygen

(11.1)

$$\text{O}_2(g) + h\nu \longrightarrow \bullet\overset{\bullet}{\text{O}}(g) + \bullet\overset{\bullet}{\text{O}}(g) \qquad 175 < \lambda < 245 \text{ nm}$$

Molecular   Ground-
oxygen   state atomic   (11.2)
               oxygen

The first reaction is important only at the top of the stratosphere because wavelengths shorter than 0.175 μm do not penetrate lower. Neither reaction is important in the troposphere. Excited atomic oxygen from Reaction 11.1 rapidly converts to the ground state by

$$\bullet\overset{\bullet}{\text{O}}(^1D)(g) \xrightarrow{\text{M}} \bullet\overset{\bullet}{\text{O}}(g)$$

Excited   Ground-
atomic   state atomic   (11.3)
oxygen   oxygen

Ozone then forms by

$$\bullet\overset{\bullet}{\text{O}}(g) + \text{O}_2(g) \xrightarrow{\text{M}} \text{O}_3(g)$$

Ground-   Molecular   Ozone
state atomic   oxygen
oxygen   (11.4)

This reaction also occurs in the troposphere, where the O(g) in that case originates from $\text{NO}_2(g)$ photolysis, not from $\text{O}_2(g)$ photolysis. Ozone is destroyed naturally in the stratosphere and troposphere by

$$\text{O}_3(g) + h\nu \longrightarrow \text{O}_2(g) + \bullet\overset{\bullet}{\text{O}}(^1D)(g) \qquad \lambda < 310 \text{ nm}$$

Ozone   Molecular   Excited
                 oxygen   atomic   (11.5)
                                   oxygen

$$\text{O}_3(g) + h\nu \longrightarrow \text{O}_2(g) + \bullet\overset{\bullet}{\text{O}}(g) \qquad \lambda > 310 \text{ nm}$$

Ozone   Molecular Ground-
                 oxygen   state atomic   (11.6)
                                     oxygen

Stratospheric ozone is also destroyed by

$$\bullet\overset{\bullet}{\text{O}}(g) + \text{O}_3(g) \longrightarrow 2\text{O}_2(g)$$

Ground- Ozone   Molecular
state atomic   oxygen   (11.7)
oxygen

In 1930, English physicist **Sidney Chapman** (1888–1970; Fig. 11.6) suggested that ozone in the stratosphere must be produced from UV photolysis of molecular oxygen. He further postulated that Reactions 11.2, 11.4, 11.6, 11.7, and the reaction

$$\bullet\overset{\bullet}{\text{O}}(g) + \bullet\overset{\bullet}{\text{O}}(g) \xrightarrow{\text{M}} \text{O}_2(g)$$

Ground-   Molecular
state atomic   oxygen   (11.8)
oxygen

describe the natural formation and destruction of ozone in the stratosphere (Chapman, 1930). These reactions make up the **Chapman cycle**, and they simulate the process fairly well. Some Chapman reactions are more important than are others. Reactions 11.2, 11.4, and 11.6 affect ozone the most. The non-Chapman reaction, 11.5, is also important.

Some of the Chapman cycle reactions can be used to explain why the altitudes of peak ozone concentration and mixing ratio occur where they do in Fig. 11.4. Oxygen density, like air density, decreases exponentially with increasing altitude. UV intensity decreases with decreasing altitude. Peak ozone densities occur where sufficient radiation encounters sufficient oxygen density, which is near 25 to 32 km (Fig. 11.4). At higher altitudes, the oxygen density is too low for its photolysis by Reactions 11.1 and 11.2 to produce peak ozone densities; at lower altitudes, the radiation is not intense enough for oxygen photolysis to produce peak ozone densities.

11.3.2. Effects of Nitrogen on the Natural Ozone Layer

Figure 11.6. Sidney Chapman (1888–1970).

Oxides of nitrogen [NO(g) and NO_2(g)] naturally destroy ozone, primarily in the upper stratosphere, helping shape the vertical profile of the ozone layer. In the troposphere, the major sources of nitric oxide are surface emissions and lightning. The major source of NO(g) in the stratosphere is transport from the troposphere and the breakdown of **nitrous oxide** [N_2O(g)] (laughing gas), a colorless gas emitted during denitrification by anaerobic bacteria in soils (Section 2.3.5). It is also emitted by bacteria in fertilizers, sewage, and the oceans and during biomass burning, automobile combustion, aircraft combustion, nylon manufacturing, and the use of spray cans. In the troposphere, N_2O(g) is lost by transport to the stratosphere, deposition to the surface, and chemical reaction. Because its loss rate from the troposphere is slow, nitrous oxide is long lived and well diluted in the troposphere, with an average mixing ratio of about 0.31 ppmv. The mixing ratio of N_2O(g) is relatively constant up to about 15 to 20 km, but decreases above that as a result of photolysis. Throughout the atmosphere, N_2O(g) produces nitric oxide by

$$N_2O(g) + \bullet \dot{O}(^1D)(g) \longrightarrow \dot{N}O(g) + \dot{N}O(g)$$

Nitrous   Excited   Nitric oxide
oxide   atomic
               oxygen

(11.9)

Nitric oxide naturally reduces ozone in the upper stratosphere by

$$\dot{N}O(g) + O_3(g) \longrightarrow \dot{N}O_2(g) + O_2(g)$$

Nitric   Ozone   Nitrogen   Molecular
oxide   dioxide   oxygen

(11.10)

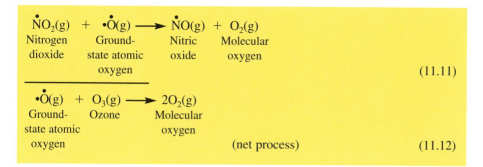

$$\overset{\bullet}{N}O_2(g) \;+\; \overset{\bullet\bullet}{\bullet O}(g) \longrightarrow \overset{\bullet}{N}O(g) \;+\; O_2(g)$$

Nitrogen   Ground-   Nitric   Molecular
dioxide   state atomic   oxide   oxygen
       oxygen

(11.11)

$$\overset{\bullet\bullet}{\bullet O}(g) \;+\; O_3(g) \longrightarrow 2O_2(g)$$

Ground-   Ozone   Molecular
state atomic   oxygen
oxygen   (net process)   (11.12)

The result of this sequence is that one molecule of ozone is destroyed, but neither $NO(g)$ nor $NO_2(g)$ is lost. This sequence is called a **catalytic ozone destruction cycle** because the species causing the $O_3(g)$ loss, $NO(g)$, is recycled. This particular cycle is the **$NO_x(g)$ catalytic ozone destruction cycle**, where $NO_x(g) = NO(g) + NO_2(g)$, and $NO(g)$ is the catalyst. The number of times the cycle is executed before $NO_x(g)$ is removed from the cycle by reaction with another gas is the **chain length**. In the upper stratosphere, the chain length of this cycle is about 10^5 (Lary, 1997). Thus, 10^5 molecules of $O_3(g)$ are destroyed before one $NO_x(g)$ molecule is removed from the cycle. In the lower stratosphere, the chain length decreases to near 10. When $NO_x(g)$ is removed from this cycle, its major loss processes are the formation of nitric acid and peroxynitric acid by the reactions

$$\overset{\bullet}{N}O_2(g) + \overset{\bullet}{O}H(g) \overset{M}{\longrightarrow} HNO_3(g)$$

Nitrogen   Hydroxyl   Nitric
dioxide   radical   acid   (11.13)

$$H\overset{\bullet}{O}_2(g) + \overset{\bullet}{N}O_2(g) \overset{M}{\longrightarrow} HO_2NO_2(g)$$

Hydroperoxy   Nitrogen   Peroxynitric
radical   dioxide   acid   (11.14)

Nitric acid and peroxynitric acid photolyze back to the reactants that formed them, but such processes are slow. Peroxynitric acid also decomposes thermally, but thermal decomposition is slow in the stratosphere because temperatures are low there.

The natural $NO_x(g)$ catalytic cycle erodes the ozone layer above ozone's peak altitude shown in Fig. 11.4. Although the $NO_x(g)$ catalytic cycle is largely natural, an unnatural source of stratospheric $NO_x(g)$ and, therefore, ozone destruction, is stratospheric aircraft emission of $NO(g)$ and $N_2O(g)$. In the 1970s and 1980s, scientists were concerned that the introduction of a fleet of supersonic transport (SST) jets into the stratosphere would enhance $N_2O(g)$ emissions sufficiently to damage the ozone layer by Reactions 11.9 through 11.12. This concern disappeared because the plan to introduce a fleet of stratospheric jets never materialized.

11.3.3. Effects of Hydrogen on the Natural Ozone Layer

Hydrogen-containing compounds, particularly the hydroxyl radical [$OH(g)$] and the hydroperoxy radical [$HO_2(g)$], are responsible for shaping the ozone profile in the lower stratosphere. The hydroxyl radical is produced in the stratosphere by one of several reactions

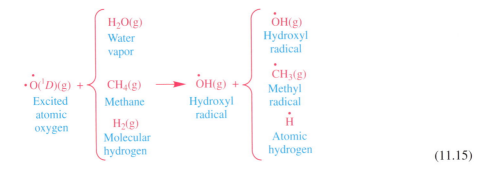

$$(11.15)$$

The hydroxyl radical participates in an **$HO_x(g)$ catalytic ozone destruction cycle**, where $HO_x(g) = OH(g) + HO_2(g)$. $HO_x(g)$ catalytic cycles are important in the lower stratosphere. The most effective $HO_x(g)$ cycle, which has a chain length in the lower stratosphere of 1 to 40 (Lary, 1997), is

$$\overset{\bullet}{O}H(g) + O_3(g) \longrightarrow H\overset{\bullet}{O}_2(g) + O_2(g)$$
Hydroxyl   Ozone   Hydroperoxy   Molecular
radical   radical   oxygen (11.16)

$$H\overset{\bullet}{O}_2(g) + O_3(g) \longrightarrow \overset{\bullet}{O}H(g) + 2O_2(g)$$
Hydroperoxy   Ozone   Hydroxyl   Molecular
radical   radical   oxygen (11.17)

$$2O_3(g) \longrightarrow 3O_2(g)$$
Ozone   Molecular
                       oxygen   (net process) (11.18)

$HO_x(g)$ species can be removed temporarily from catalytic cycles by Reactions 11.13 and 11.14 and by the reaction

$$H\overset{\bullet}{O}_2(g) + \overset{\bullet}{O}H(g) \longrightarrow H_2O(g) + O_2(g)$$
Hydroperoxy   Hydroxyl   Water   Molecular
radical   radical   vapor   oxygen (11.19)

This mechanism is particularly efficient at removing $HO_x(g)$ from catalytic cycles because it removes two $HO_x(g)$ molecules at a time.

11.3.4. Effects of Carbon on the Natural Ozone Layer

Carbon monoxide and methane produce ozone by the reaction mechanisms shown in Sections 4.2.4 and 4.2.5, respectively. The contributions of $CO(g)$ and $CH_4(g)$ to ozone production are small in the stratosphere. A by-product of methane oxidation in the stratosphere is water vapor, produced by

$$CH_4(g) + \overset{\bullet}{O}H(g) \longrightarrow \overset{\bullet}{C}H_3(g) + H_2O(g)$$
Methane   Hydroxyl   Methyl   Water
                  radical   radical   vapor (11.20)

Because water vapor mixing ratios in the stratosphere are low and transport of water vapor from the troposphere to stratosphere is slow, this reaction is a relatively important source of water vapor in the stratosphere.

11.4. RECENT CHANGES TO THE OZONE LAYER

Changes in stratospheric ozone since the early 1970s can be divided into global stratospheric changes, Antarctic stratospheric changes, and Arctic stratospheric changes.

11.4.1. Changes on a Global Scale

Figure 11.7 shows that between 1979 and 2000, the global stratospheric ozone column abundance decreased by approximately 3.5 percent (from 304.0 to 293.4 DU). Unusual decreases in global ozone occurred following the El Chichón (Mexico) volcanic eruption in April 1982, and the Mount Pinatubo (Philippines) eruption in June 1991 (Fig. 11.8). These eruption injected particles into the stratosphere. On the surfaces of these particles, chemical reactions involving chlorine took place that contributed to ozone loss. Over time, however, the concentration of these particles decreased, and the global ozone layer partially recovered. Because volcanic particles were responsible for only temporarily ozone losses, the net loss of ozone over the globe from 1979 to 2000 was still about 3.5 percent. The decrease between 60°S and 60°N was 2.5 percent (298.08 to 290.68 DU), that between 60°N and 90°N was 7.0 percent (370.35 to 344.29 DU), and that between 60°S and 90°S was 14.3 percent (335.20 to 287.23 DU).

11.4.2. Antarctic Stratospheric Changes

Between 1950 and 1980, no measurements from three ground-based stations in the Antarctic showed ozone levels less than 220 DU, a threshold for defining Antarctic ozone depletion. Every Southern Hemisphere spring (September–November) since 1980,

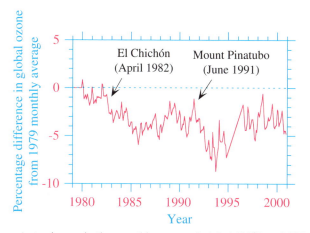

Figure 11.7. Percentage change in the monthly averaged global (90°S to 90°N) ozone column abundance between a given month and the same month in 1979. Data were obtained from the satellite-based Total Ozone Mapping Spectrometer (TOMS) and made available by NASA Goddard Space Flight Center, Greenbelt, Maryland. No data were available from December 1994 to July 1996.

Figure 11.8. Mount Pinatubo eruption, June 12, 1991. Three days later, a larger eruption, the second largest in the twentieth century, occurred. Photo by Dave Harlow, USGS.

measurements of stratospheric ozone have shown a depletion. Farman et al. (1985) first reported depletions of more than 30 percent relative to pre-1980 measurements. Since then, measurements over the South Pole have indicated depletions of up to 70 percent of the column ozone for a period of a week in early October. The largest average depletion for the month of September from 60 to 90°S since 1979 was 32.8 percent and occurred in 2000. The largest depletion for the month of October from 60 to 90°S since 1979 was 38.3 percent and occurred in 1998 [Fig. 11.9(a)]. Most ozone depletion has occurred between altitudes of 12 and 20 km. The large reduction of stratospheric ozone over the Antarctic in the Southern Hemisphere spring each year is the **Antarctic ozone hole**. The areal extent of the ozone hole is now greater than the size of North America.

Figure 11.9(b) shows the zonally and October-averaged ozone column abundance versus latitude for 1979, 1999, and 2000. The figure shows that in 1999, the October average over the South Pole was 131 DU, which compares with 286 DU in 1979. The October average was slightly higher in 2000 than in 1999, but the September average (not shown) was lower in 2000 than in 1999. October ozone levels over 45°S (the latitude of southern New Zealand, Chile, and Argentina) were 5 to 6 percent lower in

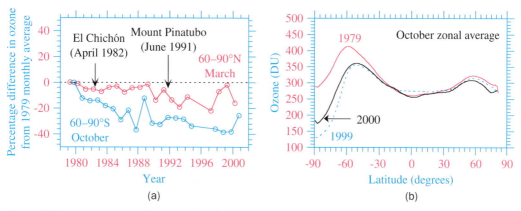

Figure 11.9. (a) Percentage difference in March-averaged 60 to 90°N and October-averaged 60 to 90°S ozone column abundances between the given year and 1979. (b) Variation with latitude of October monthly and zonally averaged column abundances of ozone in 1979, 1999, and 2000. Data were obtained from the satellite-based Total Ozone Mapping Spectrometer (TOMS) and made available by NASA Goddard Space Flight Center, Greenbelt, Maryland. No data were available from December 1994 to July 1996.

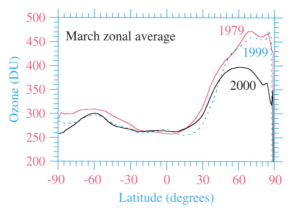

Figure 11.10. Variation with latitude of March monthly and zonally averaged column abundance of ozone in 1979, 1999, and 2000. Data were obtained from the satellite-based Total Ozone Mapping Spectrometer (TOMS) and made available by NASA Goddard Space Flight Center, Greenbelt, Maryland.

1999 and 2000 than in 1979. Ozone levels over 30°S (the latitude of Australia, South Africa, Chile, Argentina, and southern Brazil) were about 3 percent lower in 1999 and 2000 than in 1979. Temporary losses of ozone over these countries have caused concern due the effects of the resulting enhanced UV-B radiation on health (Section 11.9).

11.4.3. Arctic Stratospheric Changes

Since 1979, the stratospheric ozone layer over the North Pole has declined during the Northern Hemisphere late winter and spring (March–May). This reduction is the **Arctic ozone dent**. Figure 11.9(a) indicates that the Arctic ozone dent in March has consistently been less severe than has the corresponding springtime Antarctic ozone hole in October. Figure 11.10, however, shows that ozone levels over the Arctic during March 2000 were nearly 16 percent lower than were those during

March 1979. Ozone levels over the Arctic in March 1999 were nearly the same as those in March 1979.

11.4.4. Summary of Effects of Ozone and Air Pollution Changes on UV Radiation

The yearly global reduction and seasonal regional destruction of stratospheric ozone affect the amount of UV radiation reaching the surface. In comparison with their levels in the 1970s, levels of ground UV-B radiation in 1998 were about 7 percent higher in Northern Hemisphere midlatitudes in winter and spring, 4 percent higher in Northern Hemisphere midlatitudes in summer and fall, 6 percent higher in Southern Hemisphere midlatitudes during the entire year, 130 percent higher in the Antarctic in the Southern Hemisphere spring, and 22 percent higher in the Arctic in the Northern Hemisphere spring (Madronich et al., 1998). Localized measurements of UV at Lauder, New Zealand (45°S), for example, show that surface UV-B radiation doses during the summer of 1998–1999 were 12 percent higher than they were in the first years of the 1990s (McKenzie et al., 1999).

11.5. EFFECTS OF CHLORINE ON GLOBAL OZONE REDUCTION

Ozone reductions since the late 1970s correlate with increases in chlorine and bromine in the stratosphere. Molina and Rowland (1974) first recognized that anthropogenic chlorine compounds could destroy stratospheric ozone. Since then, scientists have strengthened the links among global ozone reduction, Antarctic ozone depletion, and the presence of chlorine- and bromine-containing compounds in the stratosphere.

11.5.1. CFCs and Related Compounds

The compounds that play the most important role in reducing stratospheric ozone are chlorofluorocarbons (CFCs). Important CFCs are identified in Table 11.2. CFCs are gases formed synthetically by replacing all hydrogen atoms in methane [$CH_4(g)$] or ethane [$C_2H_6(g)$] with chlorine and/or fluorine atoms. For example, CFC-12 [$CF_2Cl_2(g)$, dichlordifluoromethane] is formed by replacing the four hydrogen atoms in methane with two chlorine and two fluorine atoms.

11.5.1.1. Invention of CFCs

CFCs were invented on a Saturday afternoon in 1928 by Thomas Midgley and his assistants, Albert L. Henne (1901–1967) and Robert R. McNary (1903–1988), at the Thomas and Hochwalt Laboratory, 127 North Ludlow Street, Dayton, Ohio. Midgley is the same scientist who invented tetraethyl lead (Ethyl) gasoline (Section 3.6.9). Some argue that Midgley's inventions led to the two greatest environmental disasters of the twentieth century.

Midgley and his assistants developed CFC-12 effectively on the same day that a representative of General Motors' Frigidaire division asked Midgley to find a nontoxic, nonflammable substitute for an existing refrigerant, ammonia, a flammable and toxic gas. CFC-12 and subsequent CFCs were inexpensive, nontoxic, nonflammable, nonexplosive, insoluble, and chemically unreactive under tropospheric conditions; thus, they became popular. Midgley demonstrated the nontoxic and nonflammable properties of

Table 11.2. Mixing Ratios and Lifetimes of Selected Chlorocarbons, Bromocarbons, and Fluorocarbons

Chemical Formula	Trade Name	Chemical Name	Tropospheric Mixing Ratio (pptv)	Estimated Overall Atmospheric Lifetime (yrs)
CHLOROCARBONS AND CHLORINE COMPOUNDS				
Chlorofluorocarbons (CFCs)				
$CFCl_3(g)$	CFC-11	Trichlorofluoromethane	270	45
$CF_2Cl_2(g)$	CFC-12	Dichlorodifluoromethane	550	100
$CFCl_2CF_2Cl(g)$	CFC-113	1-Fluorodichloro, 2-difluorochloroethane	70	85
$CF_2ClCF_2Cl(g)$	CFC-114		15	220
$CF_2ClCF_3(g)$	CFC-115		5	550
Hydrochlorofluorocarbons (HCFCs)				
$CF_2ClH(g)$	HCFC-22	Chlorodifluoromethane	130	11.8
$CH_3CFCl_2(g)$	HCFC-141b		6	9.2
$CH_3CF_2Cl(g)$	HCFC-142b	2-Difluorochloroethane	8	18.5
Other Chlorocarbons				
$CCl_4(g)$		Carbon tetrachloride	100	35
$CH_3CCl_3(g)$		Methyl chloroform	90	4.8
$CH_3Cl(g)$		Methyl chloride	610	1.3
Other Chlorinated Compounds				
$HCl(g)$		Hydrochloric acid	10–1,000	<1
BROMOCARBONS				
Halons				
$CF_3Br(g)$	H-1301	Trifluorobromomethane	2	65
$CF_2ClBr(g)$	H-1211	Difluorochlorobromomethane	2	11
CF_2BrCF_2Br	H-2402	1-Difluorobromo, 2-difluorobromoethane	1.5	22–30
Other Bromocarbons				
$CH_3Br(g)$		Methyl bromide	12	0.7
FLUOROCARBONS AND FLUORINE COMPOUNDS				
Hydrofluorocarbons (HFCs)				
$CH_2FCF_3(g)$	HFC-134a	1-Fluoro, 2-trifluoroethane	4	13.6
Perfluorocarbons (PFCs)				
$C_2F_6(g)$		Perfluoroethane	4	10,000
Other Fluorinated Compounds				
$SF_6(g)$		Sulfur hexafluoride	3.7	3,200

Sources: Shen et al. (1995); Singh (1995); WMO (1998); Mauna Loa Data Center (2001).

his invention to the American Chemical Society in April 1930 by inhaling CFC-12, then blowing it over a candle flame, extinguishing the flame. The invention was not disclosed previously because the Frigidaire department of General Motors needed time to file patents on a family of compounds related to the invention (Bhatti, 1999).

In 1931, **CFC-12** was produced by the DuPont chemical manufacturer under the trade name **Freon**, a name chose by Midgley and his assistants. Its first use was in small ice cream cabinets. In 1934, it was used in refrigerators and whole-room coolers. Soon after, it was used in household and automotive air conditioning systems. In 1932, **CFC-11** [$CFCl_3(g)$, trichlorfluoromethane], was first produced. Its first use was in large air conditioning units. CFCs became airborne only when coolants leaked or were drained.

In 1943, Goodhue and Sullivan of the U.S. Department of Agriculture developed a method to use CFC-11 and -12 as a propellant in **spray cans**. CFCs flowed out of a spray can's nozzle, carrying with them a mist containing other ingredients. Spray cans were used to propel hair sprays, paints, deodorants, disinfectants, polishes, and insecticides.

CFC-11 and 12 have also been used as blowing agents in foam production. Foam is used in insulation, disposable cups and cartons, and fire extinguishers. CFCs are released to the air during foam-production, itself. CFCs in the air spaces of foam are usually confined and not an important source of atmospheric CFCs.

Figure 11.11 shows the reported sales of CFC-11 and -12 in 1976 and 1998. The figure shows that in 1976, almost 58 percent of CFC-11 and -12 were sold for use as propellants in spray cans. Secondary use for CFC-11 was as a blowing agent and for CFC-12 was as a refrigerant. Regulation of CFCs (Section 11.10) resulted in large decreases in the use of CFC-11 and -12 between 1976 and 1998 and increases in replacement compounds. By 1998, total sales of CFCs were about 6 percent of those in 1976.

Several other CFCs were developed, some of which are listed in Table 11.2. CFC-113, for example, was first produced in 1934 and used in air conditioning units. It has since been used primarily as a solvent in the microelectronics industry and in the dry-cleaning industry. It has also been used as a spray-can propellant and a blowing agent in foam production.

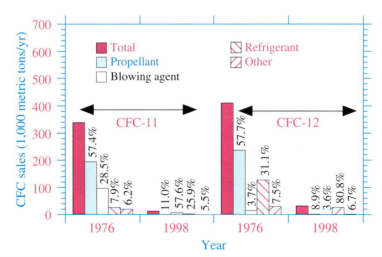

Figure 11.11. Reported sales of CFC-11 and -12 in 1976 and 1998. Percentages are of the total for the year.

Source: AFEAS (2001).

11.5.1.2. Other Chlorine Compounds

Chlorofluorocarbons are a subset of chlorocarbons, which are compounds containing carbon and chlorine. Hydrochlorofluorocarbons (HCFCs) are another subset of chlorocarbons. HCFCs are similar to CFCs except that HCFCs have at least one hydrogen atom. The hydrogen atom allows HCFCs to be broken down in the troposphere by reaction with OH(g). OH(g) does not readily break down CFCs. Because HCFCs break down more readily than do CFCs, a smaller percentage of emitted HCFCs than CFCs reaches the stratosphere. Nevertheless, because HCFCs contain chlorine and some HCFCs reach the stratosphere, HCFCs are still a danger to stratospheric ozone. HCFC-22, first produced in 1943, is the most abundant HCFC in the air today. HCFC-22 has been used as a refrigerant, spray-can propellant, and blowing agent in foam production.

Other chlorocarbons include carbon tetrachloride [$CCl_4(g)$], methyl chloroform [$CH_3CCl_3(g)$], and methyl chloride [$CH_3Cl(g)$]. Carbon tetrachloride is used as an intermediate in the production of CFCs and HCFCs and as a solvent and grain fumigant. Methyl chloroform is used as a degreasing agent, a dry-cleaning solvent and an industrial solvent. Methyl chloride is produced synthetically only in small quantities and used in the production of silicones and tetramethyl lead intermediates (Singh, 1995). Most methyl chloride in the air is produced biogenically in the oceans.

Another chlorine-containing gas in the troposphere is hydrochloric acid [$HCl(g)$]. $HCl(g)$ has larger natural than anthropogenic sources. Natural sources include evaporation of chloride from sea-spray and volcanic emissions. Although some anthropogenic emissions of $HCl(g)$ are from waste incineration, about 98 percent are from coal combustion (Saxena et al. 1993).

11.5.1.3. Bromine Compounds

Although chlorine-containing compounds are more abundant than are bromine-containing compounds, the latter compounds are more efficient, molecule for molecule, at destroying ozone. The primary source of stratospheric bromine is methyl bromide [$CH_3Br(g)$], which is produced biogenically in the oceans and emitted as a soil fumigant. Other sources of bromine are a group of synthetically produced compounds termed Halons, which are used in fire extinguishers and as fumigants. The most common Halons are H-1301 [$CF_3Br(g)$], H-1211 [$CF_2ClBr(g)$], and H-2402 [$CF_2BrCF_2Br(g)$]. Methyl bromide and Halons are bromocarbons because they contain both bromine and carbon.

11.5.1.4. Fluorine Compounds

Compounds that contain hydrogen, fluorine, and carbon but not chlorine or bromine are hydrofluorocarbons (HFCs). HFCs were produced in abundance only recently as a replacement for CFCs and HCFCs. Because the fluorine in HFCs has little effect on ozone, production of HFCs may increase in the future. Unfortunately, because they absorb thermal-IR radiation, HFCs will enhance global warming if their use increases. The most abundantly emitted HFC to date has been HFC-134a [$CH_2FCF_3(g)$]. Related to HFCs are perfluorcarbons (PFCs), such as perfluoroethane [$C_2F_6(g)$], and sulfur hexafluoride [$SF_6(g)$].

11.5.2. Lifetimes and Mixing Ratios of Chlorinated Compounds

Once emitted, CFCs take about one year to mix up to the tropopause. Because they are chemically unreactive and cannot be broken down by solar wavelengths that reach the troposphere, CFCs are not removed chemically from the troposphere. Instead, they become well mixed in the troposphere and slowly penetrate to the stratosphere. Today, the tropospheric mixing ratios of CFC-11 and CFC-12, the two most abundant CFCs, are about 270 and 550 pptv, respectively (Table 11.2 and Fig. 11.12).

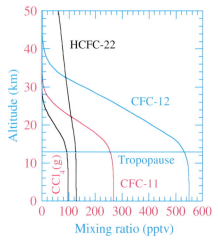

Figure 11.12. Variation of CFC-11, CFC-12, HCFC-22, and CCl_4(g) with altitude at 30°N latitude. Smoothed and scaled from Jackman et al. (1996) to present-day near-surface mixing ratios.

11.5.2.1. Lifetimes of CFCs

Because the stratosphere is one large temperature inversion, vertical transport of ozone through it is slow. About 10 Mt of chlorine in the form of CFCs reside in the troposphere, and the transfer rate of CFC-chlorine from the troposphere to the middle stratosphere is about 0.1 Mt per year. In this simplified scenario, the average time required for the transfer of a CFC molecule from the troposphere to the middle stratosphere is about 100 years.

CFCs are broken down in the stratosphere only when they are exposed to far-UV radiation (wavelengths of 0.01 to 0.25 μm), and this exposure occurs at an altitude of 12 to 20 km and higher. At such altitudes, far-UV wavelengths photolyze CFC-11 and CFC-12 by

$$CFCl_3(g) + h\nu \longrightarrow \overset{\bullet}{C}FCl_2(g) + \overset{\bullet}{C}l(g) \qquad \lambda < 250 \text{ nm}$$
$$\text{CFC-11} \qquad\qquad \text{Dichlorofluoro-} \quad \text{Atomic} \qquad\qquad\qquad\qquad (11.21)$$
$$\text{methyl radical} \quad \text{chlorine}$$

$$CF_2Cl_2(g) + h\nu \longrightarrow \overset{\bullet}{C}F_2Cl(g) + \overset{\bullet}{C}l(g) \qquad \lambda < 230 \text{ nm}$$
$$\text{CFC-12} \qquad\qquad \text{Dichlorofluoro-} \quad \text{Atomic} \qquad\qquad\qquad\qquad (11.22)$$
$$\text{methyl radical} \quad \text{chlorine}$$

At 25 km, e-folding lifetimes of CFC-11 and CFC-12 against photolysis under maximum-sunlight conditions are on the order of 23 and 251 days, respectively. Average lifetimes are on the order of two to three times these values.

In sum, the limiting factor in CFC decomposition in the stratosphere is not transported from the surface to the tropopause or photochemical breakdown in the stratosphere, but transported from the tropopause to the middle stratosphere. Table 11.2 indicates that the overall lifetimes of CFC-11 and CFC-12 between release at the surface and destruction in the middle stratosphere are about 55 and 116 years, respectively. The lifetime of CFC-12 is longer than that of CFC-11, partly because the former compound must climb to a higher altitude in the stratosphere before breaking apart than must the latter. Because of their long overall lifetimes, some CFCs emitted in the 1930s through 1950s are still present in the stratosphere. Those emitted today are likely to remain in the air until the second half of the twenty-first century.

11.5.2.2. Lifetimes of Non-CFCs

Lifetimes of non-CFC chlorinated compounds are often shorter than are those of CFCs. The lifetimes of CCl_4(g), HCFC-22(g), CH_3CCl_3(g), CH_3Cl(g), and HCl(g) between emission and chemical destruction are about 35, 12, 5, 1.3, and less than 0.1 year, respectively. Non-CFCs generally have shorter lifetimes than do CFCs because they react faster with OH(g) than do CFCs and are often more water soluble than are CFCs. The benefit of a shorter lifetime for a chlorine-containing compounds is that, if breakdown occurs in the troposphere, the chlorine released can be converted to HCl(g), which is highly soluble and can be removed readily by rainout. Because the stratosphere does not contain clouds, except for ice-containing clouds that form seasonally over the poles, HCl(g) cannot be removed from the stratosphere by rainout. Some non-CFCs, such as HCFC-22, photolyze slower than do CFCs, so once HCFC-22 reaches the middle stratosphere, its concentration builds up there to a greater extent than do concentrations of several CFCs, as seen in Fig. 11.12.

Of non-CFC chlorine compounds, $CH_3Cl(g)$, and HCl(g) have the largest natural sources. The tropospheric e-folding chemical lifetime of $CH_3Cl(g)$ against reaction by OH(g) is about 1.5 years; that of HCl(g) against reaction by OH(g) is about 15 to 30 days. HCl(g) is also soluble in water and is absorbed by clouds. Volcanos, which emit water vapor and hydrochloric acid, produce clouds and rain that remove HCl(g), preventing most of it from reaching the stratosphere (Lazrus et al., 1979; Pinto et al., 1989; Tabazadeh and Turco, 1993). The facts that the two major natural sources of chlorine, $CH_3Cl(g)$ and HCl(g), have short chemical lifetimes against destruction by OH(g) and that HCl(g) is soluble in water, whereas CFCs have long chemical lifetimes and are insoluble, support the contention that CFCs and not naturally emitted chlorine compounds are responsible for most ozone destruction in the stratosphere.

11.5.2.3. Emissions of Chlorine Compounds to the Stratosphere

Table 11.3 summarizes the relative emissions of anthropogenic and natural chlorine-containing compounds into the stratosphere in 1994. About 82 percent of chlorine entering the stratosphere originated from anthropogenic sources. Of the remainder, about 15 percent was methyl chloride, emitted almost exclusively by biogenic sources in the oceans, and 3 percent was hydrochloric acid, emitted by volcanos, evaporated from sea spray, and otherwise produced naturally. The relatively large anthropogenic versus natural

Table 11.3. Relative Emissions of Selected Chlorine Compounds into the Stratosphere

Trade Name or Chemical Name	Chemical Formula	Contribution to Stratospheric Emissions (Percent)
Anthropogenic Sources		
CFC-12	$CF_2Cl_2(g)$	28
CFC-11	$CFCl_3(g)$	23
Carbon tetrachloride	$CCl_4(g)$	12
Methyl chloroform	$CH_3CCl_3(g)$	10
CFC-113	$CFCl_2CF_2Cl(g)$	6
HCFC-22	$CF_2ClH(g)$	3
Natural Sources		
Methyl chloride	$CH_3Cl(g)$	15
Hydrochloric acid	HCl(g)	3
Total		100

Source: WMO (1995).

source of chlorine into the stratosphere supports the contention that stratospheric ozone reductions result primarily from anthropogenic chlorine, not natural chlorine.

11.5.3. Catalytic Ozone Destruction by Chlorine

Once released from their parent compounds in the stratosphere, chlorine atoms from CFCs and non-CFCs react along one of several pathways. Chlorine reacts in a **catalytic ozone destruction cycle**,

$$\dot{C}l(g) + O_3(g) \longrightarrow Cl\dot{O}(g) + O_2(g)$$

| Atomic | Ozone | Chlorine | Molecular |
| chlorine | | monoxide | oxygen |

$$\qquad\qquad\qquad\qquad\qquad\qquad\qquad\qquad (11.23)$$

$$Cl\dot{O}(g) + \cdot\dot{O}(g) \longrightarrow \dot{C}l(g) + O_2(g)$$

Chlorine	Ground-	Atomic	Molecular
monoxide	state atomic	chlorine	oxygen
	oxygen		

$$\qquad\qquad\qquad\qquad\qquad\qquad\qquad\qquad (11.24)$$

$$\cdot\dot{O}(g) + O_3(g) \longrightarrow 2O_2(g)$$

Ground-	Ozone	Molecular
state atomic		oxygen
oxygen		(net process)

$$\qquad\qquad\qquad\qquad\qquad\qquad\qquad\qquad (11.25)$$

At midlatitudes, the chain length of this cycle increases from about 10 in the lower stratosphere to about 1,000 in the middle and upper stratosphere (Lary, 1997).

The primary removal mechanisms of **active chlorine** [Cl(g)+ClO(g)] from the catalytic cycle are reactions that produce **chlorine reservoirs**, HCl(g) and **chlorine nitrate** [ClONO$_2$(g)]. Chlorine reservoirs are called such because they temporarily store active chlorine, preventing it from destroying ozone. Conversion of Cl(g) to HCl(g) occurs by

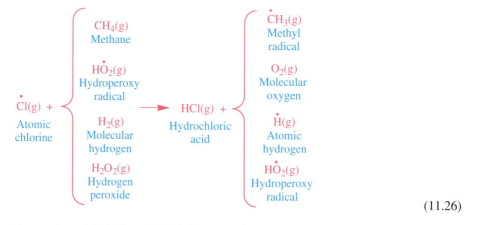

$$\qquad\qquad\qquad\qquad\qquad\qquad\qquad\qquad (11.26)$$

Conversion of ClO(g) to ClONO$_2$(g) occurs by

$$Cl\dot{O}(g) + \dot{N}O_2(g) \xrightarrow{\ M\ } ClONO_2(g)$$

| Chlorine | Nitrogen | Chlorine |
| monoxide | dioxide | nitrate |

$$\qquad\qquad\qquad\qquad\qquad\qquad\qquad\qquad (11.27)$$

At any time, about 1 percent of the non-CFC chlorine in the stratosphere is in the form of active chlorine. Most of the rest is in the form of a chlorine reservoir. Because CFCs release their chlorine by photolysis in the middle and upper stratosphere, it follows that HCl(g) mixing ratios should also peak in the middle and upper stratosphere. Indeed, observations confirm this supposition.

The HCl(g) reservoir leaks back to atomic chlorine by photolysis, reaction with OH(g), and reaction with O(g), all of which are slow processes. The e-folding lifetime of HCl(g) against photolysis, for example, is about 1.5 years at 25 km. HCl(g) also diffuses back to the troposphere, where it can be absorbed by clouds. The $ClONO_2$(g) reservoir leaks back to atomic chlorine by photolysis with an e-folding lifetime of about 4.5 hours at 25 km.

11.6. EFFECTS OF BROMINE ON GLOBAL OZONE REDUCTION

Like chlorine, bromine affects stratospheric ozone. The primary source of stratospheric bromine is **methyl bromide** [CH_3Br(g)], which is produced biogenically in the oceans and emitted as a soil fumigant. Other sources of bromine are Halons, defined in Section 11.5.1. The tropospheric mixing ratios of the most common Halons, CF_2ClBr(g) (H-1211) and CF_3Br(g) (H-1301), are both about 2 pptv, less than 1 percent of the mixing ratios of CFC-11 and -12. Nevertheless, the efficiency of ozone destruction by the bromine catalytic cycle is greater than is that by the chlorine catalytic cycle.

Methyl bromide and Halons photolyze in the stratosphere to produce atomic bromine. Photolysis of methyl bromide occurs above 20 km by

$$CH_3Br(g) + h\nu \longrightarrow \overset{\bullet}{C}H_3(g) + \overset{\bullet}{B}r(g) \qquad \lambda < 260 \text{ nm}$$

Methyl   Methyl   Atomic   (11.28)
bromide   radical   bromine

The e-folding lifetime of CH_3Br(g) against loss by this reaction is about 10 days at 25 km.

Once atomic bromine is in the stratosphere, it reacts in another **catalytic ozone destruction cycle**,

$$\overset{\bullet}{B}r(g) + O_3(g) \longrightarrow Br\overset{\bullet}{O}(g) + O_2(g)$$

Atomic   Ozone   Bromine   Molecular   (11.29)
bromine   monoxide oxygen

$$Br\overset{\bullet}{O}(g) + \bullet\overset{\bullet}{O}(g) \longrightarrow \overset{\bullet}{B}r(g) + O_2(g)$$

Bromine   Ground-   Atomic   Molecular
monoxide   state atomic   bromine   oxygen   (11.30)
              oxygen

$$\bullet\overset{\bullet}{O}(g) + O_3(g) \longrightarrow 2O_2(g)$$

Ground-   Ozone   Molecular
state atomic   oxygen   (11.31)
 oxygen   (net process)

The chain length of this cycle increases from about 100 at 20 km to about 10^4 at 40 to 50 km (Lary, 1997). The chain length of the bromine catalytic cycle is longer than is that of the chlorine catalytic cycle because Br(g) is removed more slowly from the bromine cycle by reaction with CH_4(g) and H_2(g) than Cl(g) is removed from the chlorine cycle by reaction with the same chemicals.

When atomic bromine is removed from its catalytic cycle, it forms **hydrobromic acid** [HBr(g)] by

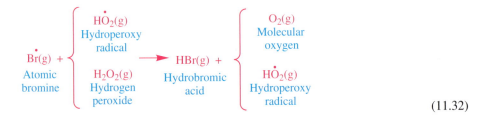

$$(11.32)$$

When BrO(g) is removed, it forms **bromine nitrate** [$BrONO_2$(g)] by

$$
\underset{\substack{\text{Bromine} \\ \text{monoxide}}}{\dot{B}\dot{r}O(g)} + \underset{\substack{\text{Nitrogen} \\ \text{dioxide}}}{\dot{N}O_2(g)} \xrightarrow{\text{M}} \underset{\substack{\text{Bromine} \\ \text{nitrate}}}{BrONO_2(g)}
$$

$$(11.33)$$

The HBr(g) reservoir leaks slowly back to atomic bromine by reaction with OH(g). The $BrONO_2$(g) reservoir quickly leaks back to atomic bromine by photolysis. The e-folding lifetime of $BrONO_2$(g) against photolysis is about 10 minutes at 25 km.

11.7. REGENERATION RATES OF STRATOSPHERIC OZONE

The presence of chlorine and bromine has decreased levels of ozone in the stratosphere. If chlorine and bromine could be removed easily from the stratosphere, the stratospheric ozone layer could regenerate quickly. The problem is that the overall lifetimes of certain chlorocarbons and bromocarbons are on the order of 50 to 100 years; thus, the natural removal rate of chlorine- and bromine-containing compounds from the stratosphere is slow.

Suppose that all ozone in the stratosphere were destroyed, all ozone-destroying compounds were removed, but all oxygen remained. How long would the ozone layer take to regenerate? An estimate can be obtained from Fig. 11.13. This figure shows results from two computer simulations of the global atmosphere in which all ozone in the present-day atmosphere was initially removed, but oxygen was not. In the first simulation, ozone regeneration was simulated in the absence of chlorine and bromine. In the second, ozone regeneration was simulated in the presence of 1989 concentrations of chlorine, but in the absence of bromine. In both simulations, the globally averaged column abundance of ozone regenerated to relatively steady values in less than a year. Regeneration during the simulation in which chlorine was initially present was about 2 to 3 percent less than that during the no-chlorine case, consistent with the estimated global reduction in ozone of 2 to 3 percent between the 1970s and 1989 due to chlorine-containing compounds.

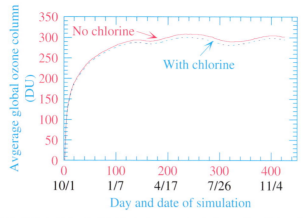

Figure 11.13. Change in ozone column abundance, averaged over the globe, during two global model simulations in which chlorine was present and absent, respectively. In both cases, ozone was removed initially from the model atmosphere on October 1, 1988. Bromine was not included in either simulation.

Source: Jacobson (1999a).

11.8. ANTARCTIC OZONE DEPLETION

Every September through November since 1980, the minimum ozone column abundance over the Antarctic decreases below its yearly average. Figure 11.14 shows that in 2000, the lowest measured column abundance over the Antarctic was about 94 DU (occurring on September 29, 2000), which was 68 percent less than 293.4 DU, the globally ($90°$S to $90°$N) and year-2000 averaged ozone column abundance. Between 1981 and 2000, the area over which ozone depletes (the **Antarctic ozone hole**) increased, as shown in the figure. The *ozone hole area* is defined as the area of the globe over which the ozone column abundance decreases below 220 DUs. The ozone hole in 2000 covered nearly 30×10^6 km^2, an area larger than the size of North America. Most Antarctic ozone depletion occurs between 14 and 18 km in altitude. Beginning in 1992, springtime ozone decreases were observed up to 24 km and down to 12 to 14 km.

Figure 11.15 shows the extent of the Antarctic ozone hole on October 1, 2000. Whereas the minimum on that date was near 94 DU within the polar vortex, an ozone maximum of 478 DU was observed just outside the vortex. The proximity of the maximum to the minimum is one factor that allows the ozone hole to replenish itself from October to December.

The Antarctic ozone hole now appears every year during the Southern Hemisphere spring. A smaller **Arctic ozone dent** (reduction in ozone to 240 to 260 DU) appears during the Northern Hemisphere late winter and spring (March–May). In 1997, when ozone reductions over the Arctic were the most severe on record, the minimum ozone value in the dent on April 1 was only 247 DU, appearing at $78.5°$N, $5.625°$W. This minimum was not low enough to qualify as a depletion (less than 220 DU). Nevertheless, the average ozone column abundance 60 to $90°$N in March 1997 was 22 percent lower than that in March 1979 [Fig. 11.9(a)], indicating that Arctic loss in 1997 was nontrivial. The ozone dent seemed to disappear in 1999, when the ozone column abundance 60 to $90°$N in

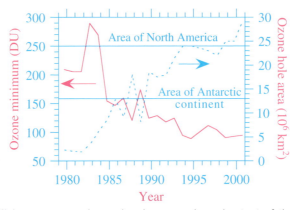

Figure 11.14. Minimum ozone column abundances and areal extent of the ozone hole over the Antarctic region from 1979 to 2000. Data from NASA Goddard Space Flight Center. For comparison, the area of the Antarctic is about 13×10^6 km^2 and the area of North America is about 24×10^6 km^2.

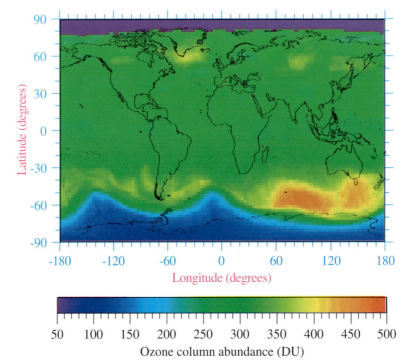

Figure 11.15. Ozone column abundance (in DU) on October 1, 2000. The figure shows an ozone hole over the Antarctic (with a minimum value of 94 DU at 86.5°S, 64.4°W) and an ozone maximum of 478 DU at 58.5°S, 83.125°E. No data were available over the North Pole. Data were obtained from the Total Ozone Mapping Spectrometer (TOMS) satellite and were made available by NASA Goddard Space Flight Center, Greenbelt, Maryland.

March was only 2.2 percent lower than it was in March 1979. The dent reappeared in March 2000, when the column abundance was 16 percent lower than that in 1979. The hole and dent are caused by set of interlinked factors. One factor linking global ozone reductions to polar ozone depletion is the presence of chlorine and bromine in the stratosphere.

Figure 11.16. Polar stratospheric clouds, photographed in spring 2000 in the Arctic. Courtesy NASA.

11.8.1. Polar Stratospheric Cloud Formation

The ozone hole over the Antarctic appears in part because the Antarctic winter (June–September) is very cold. Temperatures are low because much of the polar region is exposed to 24 hours of darkness each day during the winter, and a wind system, the **polar vortex**, circles the Antarctic. The vortex is a polar front jet-stream wind system that flows around the Antarctic continent, trapping cold air within the polar region and preventing an influx of warm air from outside this region.

Because temperatures are low in the Antarctic stratosphere, optically thin clouds, called **polar stratospheric clouds (PSCs)** form (Fig. 11.16). These clouds have few particles per unit volume of air in comparison with tropospheric clouds. Two major types of clouds form. When temperatures drop to below about 195 K, nitric acid and water vapor grow on small sulfuric acid–water aerosol particles (Toon et al. 1986). Initially, it was thought that nitric acid and water molecules deposited to the ice phase in the ratio 1:3. Such ice crystals have the composition $HNO_3 \cdot 3H_2O$ and are called **nitric acid trihydrate** (NAT) crystals. More recently, it was found that these particles contain a variety of phases. Some contain **nitric acid dihydrate** (NAD) (Worsnop et al. 1993), and others contain supercooled liquid water (liquid water present at temperatures below the freezing point of water), sulfuric acid, and nitric acid (Tabazadeh et al. 1994). Together, nitrate containing cloud particles that form at temperatures below about 195 K in the winter polar stratosphere are called **Type I polar stratospheric clouds**.

When temperatures drop below the frost point of water, which is about 187 K for typical polar stratospheric conditions, a second type of cloud forms. These clouds contain pure water ice and are **Type II polar stratospheric clouds**. Usually, about 90 percent of PSCs are Type I and 10 percent are Type II (Turco et al. 1989). Typical diameters and number concentrations of a Type I PSC are 1 μm and $\leq$1 particle cm^{-3}, respectively, although diameters vary from 0.01 to 3 μm. Typical diameters and number concentrations of a Type II PSC are 20 μm and $\leq$0.1 particle cm^{-3}, respectively, although diameters vary from 1 to 100 μm.

11.8.2. PSC Surface Reactions

Once PSC particles form in the polar winter stratosphere, chemical reactions take place on their surfaces. Such reactions are called **heterogeneous reactions** and occur after at least one gas has diffused to and adsorbed to a particle surface. **Adsorption** is a process by which a gas collides with and bonds to a surface. The primary heterogeneous reactions that occur on Type I and II PSC surfaces are

$$ClONO_2(g) + H_2O(s) \longrightarrow HOCl(g) + HNO_3(a)$$

Chlorine Water-ice Hypochlorous Adsorbed
nitrate   acid   nitric
            acid (11.34)

$$ClONO_2(g) + HCl(a) \longrightarrow Cl_2(g) + HNO_3(a)$$

Chlorine Adsorbed Molecular Adsorbed
nitrate hydrochloric chlorine nitric
    acid   acid (11.35)

$$N_2O_5(g) + H_2O(s) \longrightarrow 2HNO_3(a)$$

Dinitrogen Water-ice Adsorbed
pentoxide   nitric
         acid (11.36)

$$N_2O_5(g) + HCl(a) \longrightarrow ClNO_2(g) + HNO_3(a)$$

Dinitrogen Adsorbed Chlorine Adsorbed
pentoxide hydrochloric nitrite nitric
    acid   acid (11.37)

$$HOCl(g) + HCl(a) \longrightarrow Cl_2(g) + H_2O(s)$$

Hypochlorous Adsorbed Molecular Water-ice
acid   hydrochloric chlorine
     acid (11.38)

In these reactions, (g) denotes a gas, $H_2O(s)$ denotes a water–ice surface, HCl(a) denotes HCl adsorbed to either a Type I or II PSC, and $HNO_3(a)$ denotes HNO_3 adsorbed to a Type I or II PSC. Additional reactions exist for bromine. Laboratory studies show that HCl(g) readily coats the surfaces of Types I and II PSCs. When $ClONO_2(g)$, $N_2O_5(g)$, or HOCl(g) impinges upon the surface of a Type I or II PSC, it can react with $H_2O(s)$ or HCl(a) already on the surface. The products of these reactions are adsorbed species, some of which stay adsorbed, whereas others desorb to the vapor phase.

  In sum, heterogeneous reactions convert relatively inactive forms of chlorine in the stratosphere, such as HCl(g) and $ClONO_2(g)$, to photochemically active forms, such as $Cl_2(g)$, HOCl(g), and $ClNO_2(g)$. This conversion process is **chlorine activation**. The most important heterogeneous reaction is Reaction 11.35 (Solomon et al., 1986; McElroy et al., 1986), which generates $Cl_2(g)$. Reaction 11.36 does not activate chlorine. Its only effect is to remove nitric acid from the gas phase. When nitric acid adsorbs to a Type II PSC, which is larger than a Type I PSC, the nitric acid can sediment out along with the PSC to lower regions of the stratosphere. This removal process is **stratospheric denitrification**. Denitrification is important because it removes nitrogen that might otherwise reform Type I PSCs or tie up active chlorine as $ClONO_2(g)$.

11.8.3. Springtime Polar Chemistry

Chlorine activation occurs during the winter over the polar stratosphere. When the sun rises in early spring, Cl-containing gases created by PSC reactions photolyze by

$$Cl_2(g) + h\nu \longrightarrow 2\overset{\bullet}{C}l(g) \qquad \lambda < 450 \text{ nm}$$
Molecular   Atomic
chlorine   chlorine   (11.39)

$$HOCl(g) + h\nu \longrightarrow \overset{\bullet}{C}l(g) + \overset{\bullet}{O}H(g) \qquad \lambda < 375 \text{ nm}$$
Hypochlorous   Atomic   Hydroxyl
acid   chlorine   radical   (11.40)

$$ClNO_2(g) + h\nu \longrightarrow \overset{\bullet}{C}l(g) + \overset{\bullet}{N}O_2(g) \qquad \lambda < 370 \text{ nm}$$
Chlorine   Atomic   Nitrogen
nitrite   chlorine   dioxide   (11.41)

Once Cl has been released, it attacks ozone. The catalytic cycle that destroys ozone in the springtime polar stratosphere differs from that shown in Reactions 11.23 to 11.25, which reduces ozone on a global scale. A polar stratosphere catalytic ozone destruction cycle is

$$2 \times (\overset{\bullet}{C}l(g) + O_3(g) \longrightarrow Cl\overset{\bullet}{O}(g) + O_2(g))$$
    Atomic   Ozone   Chlorine   Molecular
    chlorine   monoxide   oxygen   (11.42)

$$Cl\overset{\bullet}{O}(g) + Cl\overset{\bullet}{O}(g) \overset{M}{\longrightarrow} Cl_2O_2(g)$$
   Chlorine   Dichlorine
   monoxide   dioxide   (11.43)

$$Cl_2O_2(g) + h\nu \longrightarrow Cl\overset{\bullet}{O}O(g) + \overset{\bullet}{C}l(g) \qquad \lambda < 360 \text{ nm}$$
Dichlorine   Chlorine   Atomic
dioxide   peroxy   chlorine   (11.44)
                  radical

$$Cl\overset{\bullet}{O}O(g) \overset{M}{\longrightarrow} \overset{\bullet}{C}l(g) + O_2(g)$$
Chlorine   Atomic   Molecular
peroxy   chlorine   oxygen   (11.45)
radical

$$\overline{2O_3(g) \longrightarrow 3O_2(g)}$$
Ozone   Molecular
             oxygen   (11.46)

This mechanism, called the **dimer mechanism** (Molina and Molina, 1986), is important in the springtime polar stratosphere because at that location and time, the ClO(g) required for Reaction 11.43 is concentrated enough for the reaction to proceed rapidly. A second cycle is

$$\overset{\bullet}{C}l(g) + O_3(g) \longrightarrow Cl\overset{\bullet}{O}(g) + O_2(g)$$

Atomic chlorine   Ozone   Chlorine monoxide   Molecular oxygen   (11.47)

$$\overset{\bullet}{B}r(g) + O_3(g) \longrightarrow Br\overset{\bullet}{O}(g) + O_2(g)$$

Atomic bromine   Ozone   Bromine monoxide   Molecular oxygen   (11.48)

$$Br\overset{\bullet}{O}(g) + Cl\overset{\bullet}{O}(g) \longrightarrow \overset{\bullet}{B}r(g) + \overset{\bullet}{C}l(g) + O_2(g)$$

Bromine monoxide   Chlorine monoxide   Atomic bromine   Atomic chlorine   Molecular oxygen   (11.49)

$$2O_3(g) \longrightarrow 3O_2(g)$$

Ozone   Molecular oxygen   (11.50)

(McElroy et al., 1986), which is important in the polar lower stratosphere. In sum, chlorine activation and springtime photochemical reactions convert chlorine from reservoir forms, such as HCl(g) and $ClONO_2$(g), to active forms, such as Cl(g) and ClO(g), as shown in Fig. 11.17. The active forms of chlorine destroy ozone in catalytic cycles.

Every November, the Antarctic warms up sufficiently for the polar vortex to break down and PSCs to melt, evaporate, and sublimate. Ozone from outside the polar region advects into the region. Ozone also regenerates chemically, and chlorine reservoirs of $ClONO_2$(g) and HCl(g) reestablish themselves. Thus, the Antarctic ozone hole is an annual, regional phenomenon that is controlled primarily by the temperature of the polar stratosphere and the presence of chlorine and bromine. The radial extent of the hole has

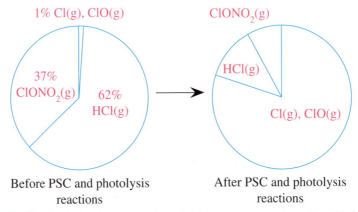

Before PSC and photolysis reactions   After PSC and photolysis reactions

Figure 11.17. Pie chart showing conversion of chlorine reservoirs to active chlorine. During chlorine activation on PSCs, HCl(g) and $ClONO_2$(g) are converted to potentially active forms of chlorine that are broken down by sunlight in springtime to form Cl(g). Cl(g) forms ClO(g), both of which react catalytically to destroy ozone.

grown and minimum ozone values have decreased steadily over the past several years. The ozone dent over the Arctic is not nearly so large or regular as is that over the Antarctic because the vortex around the Arctic is much weaker and temperatures over the Arctic do not drop so low as they do over the Antarctic. Thus, PSC formation and subsequent chemical reaction is less widespread over the Arctic than over the Antarctic.

11.9. EFFECTS OF ENHANCED UV-B RADIATION ON LIFE AND ECOSYSTEMS

In the absence of the stratospheric ozone layer, most UV-C radiation incidents at the TOA would penetrate to the surface of the Earth, destroying bacteria, protozoa, algae, fungi, plants, and animals in a short time. Fortunately, the ozone layer absorbs almost all UV-C radiation. Ozone also absorbs most UV-B radiation, but some of this radiation penetrates to the surface. UV-B radiation affects human and animal health, terrestrial ecosystems, aquatic ecosystems, biogeochemical cycles, air quality, and materials (UNEP, 1998).

11.9.1. Effects on Humans

Increases in UV-B radiation have potential to affect the skin, eyes, and immune system of humans. The layers of skin affected by UV-B are the epidermis (the outer, nonvascular, protective layer of skin that covers the dermis) and the stratum corneum (the top, horny layer of the epidermis, made mainly of peeling or dead cells). The layers of the eye affected by UV-B are the cornea, the iris, and the lens. In the immune system, the Langerhans cells, which migrate through the epidermis, are most susceptible to damage.

11.9.1.1. Effects on Skin

The severity of effects of UV-B radiation on skin depends on skin pigmentation. During the evolution of *Homo sapiens sapiens* (the variant of *Homo sapiens* to which every human belongs), humans who lived near equatorial Africa developed a dark pigment, melanin, in their skin to protect the skin against UV radiation. As humans migrated to higher latitudes, where temperatures were colder, dark pigmentation prevented what little sunlight was available from catalyzing essential chemical reactions in the skin to produce vitamin D. Thus, the skin of people who migrated poleward became lighter through natural selection (Leakey and Lewin, 1977). As populations moved across Asia, into North America, and down toward equatorial South America, the production of melanin again became an advantage. The fact that equatorial South Americans have slightly lighter skin than do equatorial Africans suggests that the former population has not been exposed to the intense equatorial UV radiation for nearly so long as the latter population (Leakey and Lewin, 1977). In recent generations, segments of light-skinned populations whose ancestors inhabited higher latitudes, such as 50 to 60°N in Northern Europe, have migrated to lower latitudes, such as 15 to 35°S in South Africa, Australia, and New Zealand. Such migration has increased the susceptibility of these populations to the many effects of UV-B radiation on skin. Enhancements of UV-B radiation due to reduced ozone further increase the susceptibility of light-skinned populations to UV-B–related skin problems.

UV-B effects on human skin include sunburn (erythema), photoaging of the skin, and skin cancer. Symptoms of **sunburn** include reddening of the skin and, in severe cases, blisters. Susceptibility to sunburn depends on skin type. People with the most sensitive skin obtain a moderate to severe sunburn in less than an hour (Longstreth et al., 1998). People who are most resistant to sunburn usually have dark-pigmented skin and become more deeply pigmented with additional exposure to UV-B.

Photoaging is the accelerated aging of the skin due to long-term exposure to sunlight, particularly UV-B radiation. Symptoms include loss of skin elasticity, wrinkles, altered pigmentation, and a decrease in collagen, a fibrous protein in connective tissue.

Skin cancer is the most common cancer among light-pigmented (skinned) humans. Three types of skin cancers occur most frequently: **basal cell carcinoma** (BCC), **squamous cell carcinoma** (SCC), and **cutaneous melanoma** (CM). Of all skin cancers, about 79, 19, and 2 percent are BCC, SCC, and CM, respectively. BCC is a tumor that develops in basal cells, which reside deep in the skin. As the tumor evolves, it protrudes through the skin, growing to a large mass that scabs over. BCC rarely spreads and can be removed by surgery or radiation treatment, so it is rarely fatal. SCC is a tumor that develops in squamous cells, which reside on the outside of the skin. SCC tumors appear as red marks. SCC can spread, but it is readily removed by surgery or radiation treatment and is rarely fatal. CM is a dark-pigmented and often malignant tumor arising from a **melanocyte**, which is a cell that produces the pigment melanin in the skin. CM tumors spread quickly and grow into dark, protruding masses that can appear anywhere on the skin. CM is fatal about one-third of the time. In some locations, such as Northern Europe, CM is at least as common as is SCC, the less common carcinoma. Susceptibility to skin cancer depends on a combination of skin pigmentation and exposure. For people with sensitive skin, it is not necessary to be exposed to UV-B over a lifetime for a person to develop skin cancer. Skin cancer rates usually increase from high latitudes (toward the poles) to lower latitudes (toward the equator).

11.9.1.2. Effects on Eyes

With respect to the eye, the **cornea**, which covers the iris and the lens, is the tissue most susceptible to UV-B damage. Little UV-B radiation penetrates past the lens to the vitreous humor or the retina, the tissues behind the lens. The most common eye problem associated with UV-B exposure is **photokeratitis** or "**snowblindness**," an inflammation or reddening of the eyeball. Other symptoms include a feeling of severe pain, tearing, avoidance of light, and twitching (Longstreth et al., 1998). These symptoms are prevalent not only among skiers, but also among people who spend time at the beach or other outdoor locations with highly reflective surfaces. From a public cost perspective, the most expensive eye-related disease associated with UV-B radiation is **cataract**, a degenerative loss in the transparency of the lens that frequently results in blindness unless the damaged lens is removed. Worldwide, cataract is the leading cause of blindness. More severe, but less widespread, eye-related diseases are squamous cell carcinoma, which affects the cornea, and **ocular melanoma**, which affects the iris and related tissues.

11.9.1.3. Effects on the Immune System

Human skin contains numerous cells to fight infection that are produced by the immune system. Enhanced UV-B radiation has been linked to suppression of these cells, reducing resistance to certain tumors and infections. Suppressed immune responses to

UV-B have been reported for herpes, tuberculosis, leprosy, trichinella, candidiasis, leish-maniasis, listeriosis, and Lyme disease (Longstreth et al., 1998).

11.9.2. Effects on Microorganisms, Animals, and Plants

Increases in UV-B radiation affect microorganisms, animals, and plants in a variety of ways. Phytoplankton, which live on the surfaces of the oceans, are susceptible to slowed growth, reproductive problems, and changes in photosynthetic energy harvesting enzymes and pigment contents (Hader et al., 1998). Because these microorganisms are near the bottom of the food chain, their deaths affect higher organisms. Algae and seagrasses, which cannot avoid exposure to the sun, are also susceptible to damage by UV-B.

Animals are susceptible to several of the same UV-B hazards as are humans. Squamous cell carcinomas have been found in cats, dogs, sheep, goats, horses, and cattle, usually on unprotected skin, such as eyelids, noses, ears, and tails (Hargis, 1981, Mendez et al., 1997; Teifke and Lohr, 1996). Cataracts and skin lesions have been found in fish (Mayer, 1992).

UV damage to crops varies with species and the crop's ability to adapt. UV-B affects the DNA of some crops and the rates of photosynthesis of others. For many crops, UV-B changes life-cycle timing, plant form, and production of plant chemicals (Caldwell et al., 1998). For others, it makes plants more susceptible to disease and attack by insects. Crop yields of some plants are affected by enhanced UV-B, but those of others are not.

11.9.3. Effects on the Global Carbon and Nitrogen Cycles

Changes in UV-B radiation affect the global carbon and nitrogen cycles. UV-B damages phytoplankton, reducing their consumption of carbon dioxide gas [CO_2(g)]. UV-B also enhances photodegradation (breakdown by light) of dead plant material, increasing release of CO_2(g) back to the air. UV-B enhances the release of carbon monoxide gas [CO(g)] from charred vegetation (Zepp et al., 1998). With respect to the nitrogen cycle, UV-B affects the rate of nitrogen fixation by cyanobacteria (Hader et al., 1998).

11.9.4. Effects on Tropospheric Ozone

Increases in UV-B radiation increase photolysis rates of UV-B absorbing gases, such as ozone, nitrogen dioxide, formaldehyde, hydrogen peroxide, acetaldehyde, and acetone. Increases in photolysis rates of nitrogen dioxide, formaldehyde, and acetaldehdye enhance rates of free-tropospheric ozone formation (Tang et al., 1998).

Whereas reductions in stratospheric ozone increase UV-B radiation reaching the free troposphere, increases in aerosol loadings in urban air can either decrease or increase UV-B radiation (Section 7.1.3.3). Reductions in UV-B in polluted air depress ozone formation; increases in UV-B have the opposite effect.

11.10. REGULATION OF CFCs

The effects of CFCs on ozone were first hypothesized by Molina and Rowland in June 1974. As early as December 1974, legislation was introduced in the U.S. Congress to study the problem further and give the U.S. EPA authority to regulate

CFCs. This bill died before any action was taken. In 1975, the Congress set up a committee that ultimately recommended that spray cans, the primary sources of CFCs, be labeled in such a way to identify whether they contained CFCs or an alternative compound. In 1976, the U.S. National Academy of Sciences released a report suggesting that CFC emissions were large enough to cause a long-term 6 to 7.5 percent decrease in stratospheric ozone, potentially increasing surface UV-B radiation by 12 to 15 percent. On the basis of this report, the U.S. Food and Drug Administration (U.S. FDA), the U.S. EPA, and the Consumer Product Safety Commission (CPSC) issued a joint decision suggesting the phase-out of CFCs from spray cans. In October 1978, the manufacture and sale of CFCs for spray cans was banned in the United States. At the time, the United States was responsible for half the global production of CFCs used in spray cans.

Although overall CFC emissions decreased as a result of the ban of CFCs in spray cans, tropospheric mixing ratios of CFCs continued to increase due to the fact that emission of CFCs from other sources still occurred. Through the late 1970s and much of the 1980s, the use of CFCs for refrigeration and foam production and as solvents increased. To limit damage due to CFCs, it was necessary to reduce their emission from all sources. In 1980, the U.S. EPA proposed limiting emission of CFCs from refrigeration to current levels at the time, but the proposed regulations were thwarted by the new presidential administration.

On the international front, the Vienna Convention for the Protection of the Ozone Layer was convened in March 1985 to discuss CFCs. The result of this convention was an agreement, signed initially by twenty countries, stating that signatory countries had an obligation to reduce CFC emissions and to study further the effect of CFCs on ozone. In September 1987, a second international agreement, the **Montreal Protocol**, was signed initially by twenty-seven countries, limiting the production of CFCs and Halons. The Montreal Protocol has since been modified by the London Amendments (1990), the Copenhagen Amendments (1992), and the Montreal Amendments (1997). The purpose of the amendments was to accelerate the phase-out of CFCs. Table 11.4 shows

Table 11.4. Percentage of 1990 CFC Production Allowed in Developed Countries under Different Agreements/Laws

Year	Montreal Protocol (1987)	London Amendments (1990)	U.S. Clean Air Act Amendments (1990)	Copenhagen Amendments (1992)	European Union Schedule (1994)
1990	100				
1991	100	100	85		
1992	100	100	80		
1993	80	80	75		50
1994	80	80	25	25	15
1995	80	50	25	25	0
1996	80	50	0	0	
1997	80	15			
1998	80	15			
1999	50	15			
2000	50	0			

Source: AFEAS (2001).

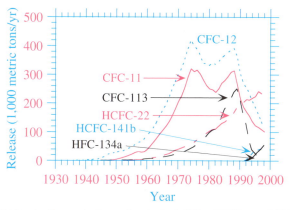

Figure 11.18. Changes in the emissions of selected CFCs, HCFCs, and HFCs, produced by reporting companies only, between 1931 and 1998.

Source: AFEAS (2001). Nonreported emissions are estimated to account for 60 percent of CFC production, 10 percent of HCFC-22 production, and 0 percent of HCFC-141b and HFC-134a production.

the phaseout schedule of CFCs resulting from the Montreal Protocol, the London Amendments, and the Copenhagen Amendments. Also shown in the table are phaseout schedules under the U.S. Clean Air Act Amendments of 1990 and the European Union Schedule of 1994. The Copenhagen Amendments of 1992 also called for the complete phase out of Halons by 1994, the complete phase out of carbon tetrachloride and methyl chloroform by 1996, and the freeze in HCFC production to specified values by 1996.

The major effect of the Montreal Protocol, later amendments, and earlier regulations to ban the use of CFCs in spray cans has been a dramatic decrease in the emission of CFCs and a corresponding increase in the emission of HCFCs and HFCs, as illustrated in Fig. 11.18. The figure shows emissions from CFCs produced by reporting companies between 1931 and 1998. Reporting companies include eleven major manufacturers who are estimated to account for 40 percent of worldwide production of CFCs, 90 percent of worldwide production of HCFC-22, and 100 percent of worldwide production of other HCFCs and all HFCs. Most remaining CFCs and HCFC-22 are produced in Russia and China.

Figure 11.18 shows that HCFC and HFC emissions increased in the 1990s. Although HCFCs break down more readily in the troposphere than do CFCs, HCFCs still contain chlorine, some of which can reach the stratosphere and damage the ozone layer. Because of their potential to destroy ozone, HCFC production rates were frozen under the Copenhagen amendments. HFCs, however, contain no chlorine and do not pose a known threat to the ozone layer. Unfortunately, they are strong absorbers of thermal-IR radiation and contribute to global warming.

Because CFC emissions have declined due to the Montreal Protocol, tropospheric mixing ratios of CFC-11 have also begun to decrease and those of CFC-12 have begun to level off, as shown in Fig. 11.19(a). More substantial decreases in carbon tetrachloride $[CCl_4(g)]$ and dramatic decreases in methyl chloroform $[CH_3CCl_3(g)]$ have been measured, as shown in Fig. 11.19(b). Figures 11.19(b) and (c) show that HCFC-22, HCFC-141b, HCFC-142b, and HFC-134a mixing ratios increased in the 1990s, as expected, because the emissions of these species increased during this period.

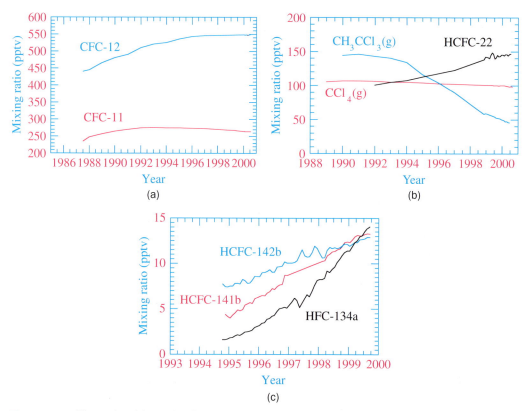

Figure 11.19. Change in mixing ratio of several (a,b) chlorinated gases at Mauna Loa Observatory, Hawaii, since 1987 and (c) fluorinated gases at Mace Head, Ireland, since 1994. Data for (a) and (b) from Mauna Loa Data Center (2001); for (c) from Prinn et al. (2000).

Because stratospheric mixing ratios of CFCs lag behind those in the troposphere, the stratospheric ozone layer is not expected to recover fully until 2050 (WMO, 1998). The annual occurrence of the Antarctic ozone hole is not expected to disappear for several decades as well.

11.11. SUMMARY

In this chapter the stratospheric ozone layer was discussed, with an emphasis on long-term global ozone reduction and seasonal Antarctic ozone depletion, both caused by enhanced levels of chlorine and bromine in the stratosphere. Damage to the ozone layer is of concern because decreases in ozone permit solar UV-B radiation to penetrate to the Earth's surface, where it harms humans, animals, microorganisms, and plants. The story of environmental damage to the ozone layer may have a positive ending. Chlorofluorocarbons, emitted since the 1930s, slowly diffused into the middle stratosphere with an overall transport time of 50 to 100 years. Most CFCs emitted in the 1930s did not reach the middle stratosphere until the 1970s, and those emitted today will not reach the middle stratosphere for another 50 to 100 years. In 1974, a relationship between CFCs and ozone loss was hypothesized. Between the 1970s and today, measurements have shown a global ozone reduction and a seasonal Antarctic depletion.

In 1978, regulations banning the use of CFCs in spray cans were implemented in the United States. Concerned that the ban on CFCs from spray cans was not enough, the international community agreed to regulate all CFC emissions through the Montreal Protocol. In 2000, reported CFC emissions were less than one-tenth those in 1976, and global ozone levels are expected to be replenished to their original levels by the year 2050 as CFCs currently in the atmosphere are slowly removed.

11.12. PROBLEMS

11.1. Explain why an ozone maximum occurs immediately outside the polar vortex during the same season that the ozone hole occurs over the Antarctic.

11.2. Briefly summarize the steps resulting in the Antarctic ozone hole and its recovery. Explain why the ozone dent is never so severe as the ozone hole.

11.3. Write down the chlorine catalytic cycles responsible for ozone reduction

   (a)   over the Antarctic
   (b)   in the global stratosphere

   Why is the Antarctic mechanism unimportant on a global scale?

11.4. What effect do you think volcanic aerosol particles that penetrate into the stratosphere have on stratospheric ozone during the daytime when chlorine reservoirs are present? How about during the nighttime?

11.5. If the entire global stratospheric ozone layer takes only a year to form from scratch chemically in the presence of oxygen, why will today's ozone layer, which has deteriorated a few percent since 1979, take up to 50 years to regenerate to its 1979 level?

11.6. Why does bromine destroy ozone more efficiently than does chlorine?

11.7. Why do CFCs not cause damage to ozone in the troposphere?

11.8. Some people argue that natural chlorine from volcanic eruptions and the ocean is responsible for global ozone reduction and Antarctic ozone depletion. What argument contradicts this contention?

11.9. Explain why the ratio of UV-B to total solar radiation reaching the top of the atmosphere is greater than that reaching the ground, as seen in Table 11.1.

11.10. Explain how UV-B levels at the ground in polluted air can be less than those in unpolluted air, if both the polluted air and clean air are situated under an ozone layer that is partially depleted by the same amount.

THE GREENHOUSE EFFECT AND GLOBAL WARMING

he two major global-scale environmental issues of international concern since the 1970s have been the threats of global stratospheric ozone reduction and global warming. As discussed in Chapter 11, the threat to the global ozone layer has been controlled to some degree because national and international regulations have required the chemical industry to use alternatives to chlorocarbons and bromocarbons, the chemicals primarily responsible for much of the ozone-layer reductions. Regulations are similarly responsible for improvements in air quality and acid deposition problems in the United States and many western European countries since the early 1970s (e.g., the U. S. Clean Air Act Amendments of 1970 motivated U.S. automobile manufacturers to develop the catalytic converter in 1975, which led to improvements in urban air quality; regulations through the Clean Air Act Amendments and the 1979 Geneva Convention on Long-Range Transboundary Air Pollution led to reductions in sulfur emissions and an amelioration of some acid deposition problem in the 1980s and 1990s). The second major issue of international concern, the threat of global warming, has been addressed at international meetings, but progress in regulating emissions responsible for the problem has been slow. In this chapter, global warming is discussed and distinguished from the natural greenhouse effect. Climate responses to enhanced gas and aerosol particle concentrations are examined. Historical and recent temperature trends, both in the lower and upper atmosphere, are described. National and international efforts to curtail global warming are also discussed, as are potential effects of global warming.

12.1. THE TEMPERATURE ON THE EARTH IN THE ABSENCE OF A GREENHOUSE EFFECT

The **natural greenhouse effect** is the warming of the Earth's lower atmosphere due to natural gases that transmit the sun's visible radiation, but absorb and reemit the Earth's thermal-IR radiation. Greenhouse gases cause the air below them to warm like a glass house causes a net warming of its interior. Because most incoming solar radiation can penetrate a glass house but a portion of outgoing thermal-IR radiation cannot, air inside a glass house warms during the day so long as mass (such as plant mass) is present to absorb solar and reemits thermal-IR radiation. The surface of the Earth, like plants, absorbs solar and reemits thermal-IR radiation. **Greenhouse gases**, like glass, are transparent to most solar radiation but absorb a portion of thermal-IR. **Global warming** is the increase in the Earth's temperature above that from the natural greenhouse effect due to the addition of anthropogenically emitted greenhouse gases and particulate black carbon to the air.

In the absence of greenhouse gases, the temperature of the Earth can be estimated, to first order, with a simple radiation balance model that considers incoming solar radiation and outgoing thermal-IR radiation from the Earth. This model can be applied to other planets as well. The difference between the temperature predicted by the energy balance model and the real temperature observed at the surface of the Earth is hypothesized to result from the greenhouse effect. The simple energy balance model is described next.

12.1.1. Incoming Solar Radiation

In Chapter 2, it was shown that the sun emits radiation with an effective photosphere temperature of about $T_p = 5,785$ K. The energy flux (joules per second per square meter or watts per square meter) emitted by the sun's photosphere can be calculated from the Stefan–Boltzmann law (Equation 2.2) as

$$F_p = \varepsilon_p \sigma_B T_p^4 \tag{12.1}$$

where ε_p is the emissivity of the photosphere. The emissivity is near unity because the sun is essentially a blackbody (a perfect absorber and emitter of radiation). Multiplying the energy flux by the spherical surface area of the photosphere, $4\pi R_p^2$, where $R_p = 6.96 \times 10^8$ m (696,000 km) is the effective **radius of the sun** (distance from the center of the sun to the top of the photosphere), gives the total energy per unit time (J s^{-1} or W) emitted by the photosphere as $4\pi R_p^2 F_p$. Energy emitted from the photosphere propagates through space on the edge of an ever-expanding concentric sphere originating from the photosphere. Because conservation of energy requires that the total energy per unit time passing through a concentric sphere any distance from the photosphere equals that originally emitted by the photosphere, it follows that the total energy per unit time passing through a sphere with a radius corresponding to the **Earth–sun distance** (R_{es}) is

$$4\pi R_{es}^2 F_s = 4\pi R_p^2 F_p \tag{12.2}$$

where F_s is the solar energy flux (J s^{-1} m^{-2} or W m^{-2}) on a sphere with a radius corresponding to the Earth–sun distance. Rearranging Equation 12.2 and combining the result with Equation 12.1 gives

$$F_s = \left(\frac{R_p}{R_{es}}\right)^2 F_p = \left(\frac{R_p}{R_{es}}\right)^2 \sigma_B T_p^4 \tag{12.3}$$

which indicates that the energy flux from the sun decreases proportionally to the square of the distance away from the sun. This can be illustrated with a simple example. If you put your hand over a lightbulb, you will feel the heat from the bulb, but as you move your hand away from the bulb, the heat that you feel decreases proportionally to the square of the distance away from the bulb.

The average Earth–sun distance is about 1.5×10^{11} m (150 million km), giving the average energy flux at the top of the Earth's atmosphere from Equation 12.3 as $F_s = 1{,}365$ W m^{-2}, which is called the **solar constant**. Figure 12.1 shows that the

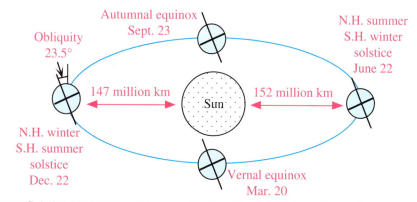

Figure 12.1. Relationship between the sun and Earth at the times of solstices and equinoxes.

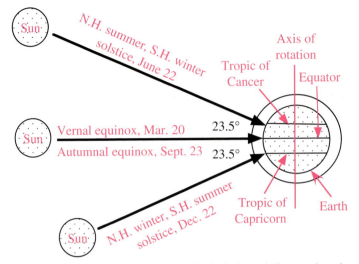

Figure 12.2. Positions of the sun relative to the Earth during solstices and equinoxes. Of the four times shown, the Earth–sun distance is greatest at the N.H. summer, S.H. winter solstice.

Earth–sun distance is 147 million km in December (Northern Hemisphere winter) and 152 million km in June (Northern Hemisphere summer) due to the fact that the Earth rotates around the sun in an elliptical pattern with the sun at one focus. If these distances are used in Equation 12.3, $F_s = 1,411$ W m^{-2} in December and $1,321$ W m^{-2} in June. Thus, a difference of 3.3 percent in Earth–sun distance between December and June corresponds to a difference of 6.6 percent in solar radiation reaching the Earth between these months. In other words, 6.6 percent more radiation falls on the Earth in December than in June.

Despite the excess radiation reaching the top of the Earth's atmosphere in December, the Northern Hemisphere (N.H.) winter still starts in December because the Southern Hemisphere (S.H.) is tilted toward the sun in December, as shown in Fig. 12.1. The figure shows that the axis of rotation of the Earth is tilted all year 23.5° from a line perpendicular to the plane of the Earth's orbit around the sun. This angle is called the **obliquity** of the Earth's axis of rotation. As a result of the Earth's obliquity, the sun's direct rays hit their farthest point south, 23.5°S latitude, on December 22 (**N.H. winter, S.H. summer solstice**) and their farthest point north, 23.5°N latitude, on June 22 (**N.H. summer, S.H. winter solstice**) (Fig. 12.1 and 12.2). The latitude at 23.5°S is called the **Tropic of Capricorn** and that at 23.5°N is the **Tropic of Cancer**. The winter and summer solstices are the shortest and longest days of the year, respectively. On March 20 (**vernal equinox**) and September 23 (**autumnal equinox**), the sun is directly over the equator and the length of day equals that of night.

Because the sun's rays are most intense over the Southern Hemisphere in December and because Northern Hemisphere days are shorter in December than in June, temperatures in the Northern Hemisphere are cooler in December than in June, even though the Earth receives more radiation in December than in June.

From the sun's point of view, the Earth is merely a circular disk (rather than a sphere) absorbing some of its radiation. Thus, the quantity of incoming radiation received by the Earth is the solar constant multiplied by the cross-sectional area of the Earth, πR_e^2 (m^2), where $R_e = 6.378 \times 10^6$ m (6378 km) is the **Earth's radius**. Not all

Table 12.1. Solar Albedos and Thermal-IR Emissivities for Several Surface Types

Surface Type	Albedo (Fraction)	Emissivity (Fraction)
Earth and atmosphere	0.3	0.90–0.98
Liquid water	0.05–0.2	0.92–0.96
Fresh snow	0.7–0.9	0.82–0.995
Old snow	0.35–0.65	0.82
Thick clouds	0.3–0.9	0.25–1.0
Thin clouds	0.2–0.7	0.1–0.9
Sea ice	0.25–0.4	0.96
Soil	0.05–0.2	0.9–0.98
Grass	0.16–0.26	0.9–0.95
Desert	0.20–0.40	0.84–0.91
Forest	0.10–0.25	0.95–0.97
Concrete	0.1–0.35	0.71–0.9

Estimates from Liou (1992), Hartmann (1994), Oke (1978), and Sellers (1965).

incoming solar radiation is absorbed by the Earth. Some is reflected by snow, deserts, and other light-colored ground surfaces. In the real atmosphere, clouds also reflect incoming solar radiation. The fraction of incident energy reflected by a surface is the **albedo** or **reflectivity** of the surface. Table 12.1 gives mean albedos in the visible spectrum for several surface types. The table shows that the albedo of the Earth and atmosphere together (**planetary albedo**) is about 30 percent. More than two-thirds of the Earth's surface is covered with water, which has an albedo of 5 to 20 percent (typical value of 8 percent), depending largely on the angle of the sun. Soils and forests also have low albedos. Much of the Earth-atmosphere reflectivity is due to clouds and snow, which have high albedos.

Taking into account the cross-sectional area of the Earth and the Earth's albedo (A_e, fraction), the total energy per unit time (W) absorbed by the Earth in the simple energy balance model is

$$E_{\mathrm{in}} = F_s(1 - A_e)(\pi R_e^2) \tag{12.4}$$

This incoming radiation must be equated with radiation emitted by the Earth to derive an equilibrium temperature.

12.1.2. Outgoing Thermal-IR Radiation

The Earth emits radiation as a sphere, not a cross section. The energy flux emitted by the Earth can be estimated with the Stefan–Boltzmann law, given in Equation 2.2. Applying this law to the Earth and multiplying through by the surface area of the Earth, $4\pi R_e^2$ (m²), gives the energy per unit time (W) emitted by the Earth as

$$E_{\mathrm{out}} = \varepsilon_e \sigma_B T_e^4 (4\pi R_e^2) \tag{12.5}$$

where ε_e is the globally averaged thermal-IR emissivity of the Earth (dimensionless) and T_e is the equilibrium temperature (K) of the Earth's surface. Table 12.1 gives thermal-IR emissivities for different surfaces. The actual globally averaged emissivity of the Earth is about 0.9 to 0.98, but for the basic calculation presented here, it is assumed to equal 1.0.

12.1.3. Equilibrium Temperature of the Earth

Equating the incoming solar radiation from Equation 12.4 with the outgoing thermal-IR radiation from Equation 12.5 and solving for the **equilibrium temperature** of the Earth's surface in the absence of an atmosphere gives

$$T_e = \left[\frac{F_s(1-A_e)}{4\varepsilon_e\sigma_B} \right]^{1/4} \tag{12.6}$$

EXAMPLE 12.1

Estimate the equilibrium temperature of the Earth given F_s = 1,365 W m^{-2}, A_e = 0.3, and ε_e = 1.0.

Solution

From Equation 12.6, the equilibrium temperature of the Earth is T_e = 254.8 K, which would be the temperature of the Earth in the absence of an atmosphere.

Example 12.1 shows that the equilibrium temperature of the Earth, 255 K, is about 18 K below the freezing temperature of water and would not support most life on Earth. Fortunately, the actual globally averaged near-surface air temperature is about 288 K. The 33 K difference between the predicted equilibrium temperature and the actual temperature of the Earth results from the fact that the Earth has an atmosphere that is transparent to most incoming solar radiation but selectively absorbs a portion of outgoing thermal-IR radiation. Some of the absorbed radiation is reemitted back to the surface, warming the surface. The resulting 33 K increase in temperature over the equilibrium temperature of the Earth is called the **natural greenhouse effect**.

Table 12.2 compares the equilibrium temperatures, calculated from Equation 12.6 with actual temperatures for several planets. The table indicates that the planet with the largest difference between its actual and equilibrium temperatures is Venus [Fig. 12.3(a)].

Table 12.2. Equilibrium and Actual Temperatures of the Surfaces of Planets and the Moon

Object	Distance from Sun (10^9 m)	Equatorial Radius (10^6 m)	Surface Pressure (Bars)	Major Atmospheric Gases (Volume Mixing Ratios)	Albedo	Equilibrium Temperature (K)	Actual Temperature (K)
Mercury	57.9	2.44	2×10^{-15}	He(0.98), H(0.02)	0.11	436	440
Venus	108	6.05	90	CO_2(0.96), N_2(0.034), H_2O(0.001)	0.65	252	730
Earth	150	6.38	1	N_2(0.77), O_2(0.21), Ar (0.0093)	0.30	255	288
Moon	150	1.74	2×10^{-14}	Ne(0.4), Ar(0.4), He(0.2)	0.12	270	274
Mars	228	3.39	0.007	CO_2(0.95), N_2(0.027), Ar(0.016)	0.15	217	218
Jupiter	778	71.4	>100	H_2(0.86), He(0.14)	0.52	102	129
Saturn	1,427	60.3	>100	H_2(0.92), He(0.74)	0.47	77	97
Uranus	2,870	26.2	>100	H_2(0.89), He(0.11)	0.51	53	58
Neptune	4,497	25.2	>100	H_2(0.89), He(0.11)	0.41	45	56
Pluto	5,900	1.5	1×10^{-5}	?	0.30	41	50

Data compiled from Lide et al. (1998), except that equilibrium temperatures were calculated from Equation 12.6.

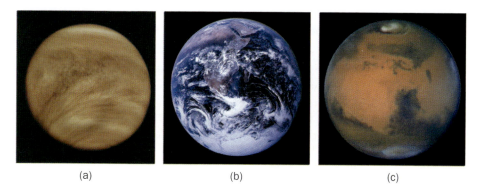

(a)   (b)   (c)

Figure 12.3. (a) Ultraviolet image of Venus's clouds as seen by the Pioneer Venus Orbiter, February 26, 1979. (b) Earth, as seen from the Apollo 17 spacecraft on December 7, 1972. (c) Sharpest image to date of Mars, taken by Hubble telescope on March 10, 1997. The North Pole dry-ice cap is rapidly sublimating during the Martian Northern Hemisphere spring. Courtesy of the National Space Science Data Center.

The surface temperature of Venus is more than 470 K warmer than is its equilibrium temperature. Because Venus is closer to the sun than is the Earth, Venus receives more incident solar radiation than does the Earth, and Venus's temperature, early in its evolution, was higher than was that of the Earth. As a result, liquid water and ice on the surface of Venus, if ever present, melted and evaporated, respectively. The resulting water vapor, exposed to intense far-UV radiation from the sun, photolyzed to atomic hydrogen [H(g)] and the hydroxyl radical [OH(g)]. Over time, atomic hydrogen escaped Venus's gravitational field to space, depleting the atmosphere of its ability to reform water vapor. Because the surface of Venus lost all liquid water from evaporation, there was no mechanism to dissolve and convert $CO_2(g)$, which was building up in the atmosphere due to volcanic outgassing, back to carbonate rock. As its mixing ratio increased, $CO_2(g)$ absorbed more thermal-IR radiation, heating the atmosphere, preventing condensation of water, preventing further removal of $CO_2(g)$. The nearly endless positive feedback cycle that occurred on Venus is called the **runaway greenhouse effect**. Today, Venus has a surface air pressure ninety times that of the Earth, and its major atmospheric constituent is $CO_2(g)$.

Table 12.2 shows that Mercury, the Moon, Mars, and Pluto have thin atmospheres and little greenhouse effect. Their atmospheres are thin because light gases have escaped their weak gravitational fields. Although Mars' surface pressure is less than one percent that of the Earth, Mars' $CO_2(g)$ partial pressure is about 20 times that of the Earth. Because $CO_2(g)$ is relatively heavy, it has not entirely escaped Mars' atmosphere, and because Mars has no oceans, $CO_2(g)$ cannot be removed by dissolution. Some $CO_2(g)$ deposits seasonally over Mars's poles as **dry ice** (solid carbon dioxide) [Figure 12.3(c)], which forms from the gas phase when the temperature cools to 194.65K. Despite its abundance of $CO_2(g)$, Mars has only a small greenhouse effect because it emits little thermal-IR due to its low surface temperature and its atmosphere contains no water vapor.

The main gases on Jupiter, Saturn, Uranus, and Neptune are molecular hydrogen and helium. These planets are so large that their gravitation prevents the escape of even light gases. Because neither hydrogen nor helium is a strong absorber of thermal-IR radiation, the greenhouse effect on these planets is small. The high surface pressure on these planets compresses hydrogen into oceans of liquid hydrogen. High pressures in the interior of Jupiter cause solid hydrogen, and possibly metallic solid hydrogen, to form.

12.1.4. The Goldilox Hypothesis

Figure 12.3 shows images of Venus, Earth, and Mars. At the surface temperature of Venus (730 K), water is in the form of vapor. At the surface temperature of Mars (218 K), water, if ever present, would be in the form of ice. At the surface temperature of the Earth (288 K), water is in the form of liquid. Whereas Venus is too hot and Mars is too cold, the Earth is ideal for supporting liquid water. The presence of liquid water and amiable temperatures have allowed life on Earth to flourish.

If Earth did not have an atmosphere that absorbed thermal-IR radiation, its equilibrium temperature (255 K) would be too cold to support liquid water or much life. Fortunately, Earth's atmosphere contains water vapor, initially outgassed from volcanos but now evaporated from the oceans as well, that absorbs thermal-IR radiation, keeping the planet warm. The atmosphere also contains carbon dioxide gas, which contributes to the Earth's natural warmth.

If the Earth were closer to or farther from the sun, its temperature would be too warm or cold, respectively, to support life. The hypothesis that the Earth is the ideal distance from the sun, in comparison with Venus or Mars, for sustaining life, is called the **Goldilox hypothesis**. Venus is too hot because of its proximity to the sun; Mars is too cold because of its distance from the sun; but the Earth is the ideal distance, so that the addition of natural greenhouse gases into its atmosphere has put the Earth's temperature in a range that allowed water to exist as a gas, liquid, and solid and life to flourish.

12.2. THE GREENHOUSE EFFECT AND GLOBAL WARMING

Greenhouse gases are relatively transparent to incoming solar radiation but opaque to certain wavelengths of IR radiation. The term *relatively transparent* is used because all greenhouse gases absorb far-UV radiation (which is a trivial fraction of incoming solar radiation). In addition, ozone strongly absorbs UV-B and UV-C radiation and weakly absorbs visible radiation. Water vapor and carbon dioxide absorb solar near-IR radiation. However, as shown in Fig. 11.5, gases affect only a fraction of total solar radiation incident at the top of the Earth's atmosphere.

12.2.1. Greenhouse Gases and Particles

The **natural greenhouse effect** is the warming of the Earth's atmosphere due to the presence of background greenhouse gases, primarily water vapor, carbon dioxide, methane, ozone, nitrous oxide, and methyl chloride. The natural greenhouse effect is responsible for about 33 K of the Earth's average near-surface air temperature of 288 K. Without the natural greenhouse effect, Earth's average near-surface temperature would be about 255 K, which is too cold to support most life. Thus, the presence of natural greenhouse gases is beneficial. Anthropogenic emissions have increased the mixing ratios of greenhouse gases and particulate black carbon (BC), causing global warming. Whereas greenhouse gases transmit solar radiation and absorb thermal-IR radiation, BC strongly absorbs solar radiation and weakly absorbs thermal-IR radiation. Thus, greenhouse gases and particulate BC both warm the air, but by different mechanisms. **Global warming** is the increase in the Earth's temperature above the natural greenhouse effect temperature as a result of the emission of anthropogenic greenhouses gases and BC.

Table 12.3 shows that the most important greenhouse gas is **water vapor**, which accounts for approximately 89 percent of the 33 K temperature increase resulting from

Table 12.3. Estimated Percentages of Natural Greenhouse Effect and Global Warming Temperature Changes Due to Greenhouse Gases and Particulate Black Carbon since the mid-1800s

Compound Name	Formula	Current Total Tropospheric Mixing Ratio (ppmv) or Loading (Tg)	Natural Percentage of Current Total Mixing Ratio or Loading	Anthropogenic Percentage of Current Total Mixing Ratio or Loading	Percentage of Natural Greenhouse Effect Temperature Change Due to Component	Percentage of Global Warming Temperature Change Due to Component
Water vapor	$H_2O(g)$	10,000	>99	<1	88.9	0
Carbon dioxide	$CO_2(g)$	370	75.7	24.3	7.5	46.6
Black carbon (BC)	$C(s)$	0.15–0.3 Tg	10	90	0.2	16.4
Methane	$CH_4(g)$	1.8	39	61	0.5	14.0
Ozone	$O_3(g)$	0.02–0.07	50–100	0–50	1.1	11.9
Nitrous oxide	$N_2O(g)$	0.314	87.6	12.4	1.5	4.2
Methyl chloride	$CH_3Cl(g)$	0.0006	100	0	0.3	0
CFC-11	$CFCl_3(g)$	0.00027	0	100	0	1.8
CFC-12	$CF_2Cl_2(g)$	0.00054	0	100	0	4.2
HCFC-22	$CF_2ClH(g)$	0.00013	0	100	0	0.6
Carbon tetrachloride	$CCl_4(g)$	0.00010	0	100	0	0.3

Source: Derived from Jacobson (2001b).

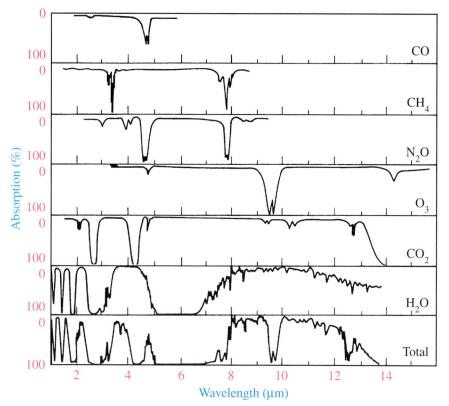

Figure 12.4. Percentage absorption of radiation by greenhouse gases at infrared wavelengths.
Source: Valley (1965).

natural greenhouse warming. Carbon dioxide is the second most important and abundant greenhouse gas, accounting for about 7.5 percent of the natural greenhouse effect. Black carbon, whose major natural source is natural forest fires, is estimated to be responsible for only 0.2 percent of the Earth's natural warming above its equilibrium temperature.

Figure 12.4 shows the percentage absorption by different greenhouse gases at IR wavelengths. Water vapor is a strong absorber at several wavelengths between 0.7 and 8 μm and above 12 μm. Carbon dioxide absorbs strongly at 2.7 and 4.3 μm and above 13 μm. The figure indicates that little thermal-IR absorption occurs between 8 and 12 μm. This wavelength region is called the **atmospheric window**. In the atmospheric window, gases are relatively transparent to the Earth's outgoing thermal-IR radiation, allowing the radiation to escape to space. Of the natural gases in the Earth's atmosphere, ozone and methyl chloride absorb radiation within the atmospheric window. Nitrous oxide and methane absorb at the edges of the window. An increase in the mixing ratios of any gas that absorbs in the window region will increase global near-surface temperatures.

12.2.2. Historical Aspects of Global Warming

The first scientist to consider the Earth as a greenhouse was **Jean Baptiste Fourier** (1768–1830), a French mathematician and physicist known for his studies of heat conductivity and diffusion. In 1827, Fourier suggested that the atmosphere behaved like the glass in a hothouse, letting through "light" rays of the sun but retaining "dark rays" from the ground.

The first to recognize that specific gases in the atmosphere selectively absorb thermal-IR radiation was **John Tyndall** (1820–1893; Fig. 12.5), the English experimental physicist who was also known for studying the interactions of light with small particles (Section 7.1.5). Near 1865, Tyndall discovered that water vapor absorbs more thermal-IR radiation than does dry air, and postulated that water vapor moderates the Earth's climate.

The first to propose the theory of global warming was Swedish physical chemist **Svante August Arrhenius** (1859–1927; Fig. 12.6). In 1896, he suggested that a doubling of $CO_2(g)$ mixing ratios due to coal combustion since the Industrial Revolution in the 18th-century United Kingdom would lead to temperature increases of 5°C (Arrhenius, 1896). This estimate

Figure 12.5. John Tyndall (1820–1893).

Figure 12.6. Svante August Arrhenius (1859–1927).

is higher than recent estimates that a doubling of $CO_2(g)$ will result in a 2°C temperature rise, but the theory is still consistent. Arrhenius also theorized that reductions in
$CO_2(g)$ caused ice ages to occur. This theory was incorrect because changes in the
Earth's orbit are responsible for the ice ages. The decrease in $CO_2(g)$ mixing ratios
during an ice age is an effect rather than a cause of temperature changes.

12.2.3. The Leading Causes of Global Warming

Since the time of Arrhenius, carbon dioxide has been considered the leading cause and
methane the second leading cause of global warming. Indeed, climate feedback studies
indicate that carbon dioxide does cause the most near-surface global warming. The
second leading cause of near-surface global warming may be particulate black carbon,
not methane (Jacobson, 2000, 2001b). Black carbon is emitted during coal, diesel fuel,
jet fuel, natural gas, kerosene, and biomass burning. Table 12.3 indicates that BC may
be responsible for 16 percent of near-surface global warming today. Carbon dioxide is
responsible for about 47 percent of such warming.

Although $CO_2(g)$ is the most abundant and important anthropogenically emitted
gas in terms of its effects on global warming, other gases are more efficient, molecule
for molecule, at absorbing thermal-IR radiation, and the enhanced emissions of such
gases is a cause for concern. A $CH_4(g)$ molecule is approximately twenty-five times
more efficient at absorbing radiation than is a $CO_2(g)$ molecule. $N_2O(g)$ and $CFCl_3(g)$
molecules are 270 and 12,500 times more efficient at absorbing, respectively, than is a
$CO_2(g)$ molecule. Thus, controlling the emission of all greenhouse gases is important
if global warming is to be remedied (Hayhoe et al., 1999; Hansen et al., 2000).

12.2.4. Trends in Greenhouse Gases

Since the mid-1800s, the tropospheric mixing ratios of $CO_2(g)$, $CH_4(g)$, and $N_2O(g)$
have increased by 30 percent, 143 percent, and 14 percent, respectively. These gases
are relatively well mixed in the lower atmosphere.

Black carbon concentrations have not been tracked so far back, but emission data
give an estimate of the change in BC atmospheric loading since prior to the Industrial
Revolution. Today, about half of BC originates from fossil fuels and half originates
from biomass burning. Since all fossil-fuel BC and 80 percent of biomass-burning BC
today are anthropogenic, 90 percent of total BC is anthropogenic. In 1850 (and presumably prior to that), biomass burning, predominantly from forest fires at the time,
was about half that today (Houghton et al., 1991). As such, before fossil-fuel combustion, total BC emissions may have been about 25 percent of those today, and all such
emissions were natural.

Since the beginning of the nineteenth century fossil fuel emissions of BC have
risen. Rates of coal combustion, one major source of BC, increased globally from
10 to 1,000 to 5,000 million metric tons per year from 1800 to 1900 to 1990 (McNeill,
2000). Coal production in the United States alone increased between 1984 and 1999
by 25 percent, and refining (for end use and resale) of No. 2 diesel fuel and jet fuel
increased by 69 and 57 percent, respectively (EIA, 2000). Because BC is a particle
component much heavier than a gas molecule, BC atmospheric lifetimes are shorter
and its spatial distributions more variable than are those of greenhouse gases.

Figure 12.7 shows changes in the mixing ratios of $CO_2(g)$, $CH_4(g)$, and $N_2O(g)$
from 1750, 1840, and 1988, respectively, to the present. The historical $CO_2(g)$ and

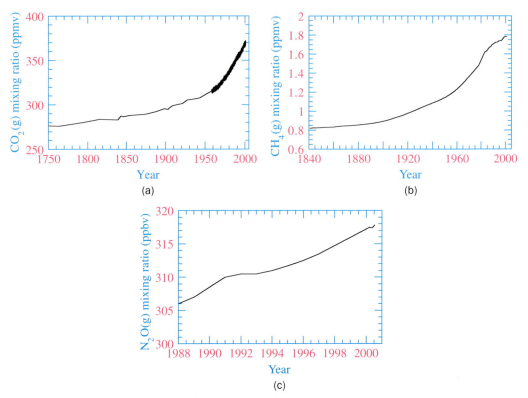

Figure 12.7. Temporal changes in (a) carbon dioxide, (b) methane, and (c) nitrous oxide mixing ratios. Data for carbon dioxide for 1744–1953 from Siple Station, Antarctica, ice core (Friedli et al., 1986), for 1958–1999 from Mauna Loa Observatory (Keeling and Whorf, 2000), and for 2000 from Mauna Loa Data Center (2001). Data for methane for 1841–1978 from Law Dome ice core, Antarctica (Ethridge et al., 1992) and for 1983–2000 from Mauna Loa Data Center (2001). Nitrous oxide data from Mauna Loa Data Center (2001).

$CH_4(g)$ data originate from ice-core measurements. The more recent data in all cases originate from ground-based ambient measurements. In 1958, **Charles David Keeling** began tracking the mixing ratio of $CO_2(g)$ at Mauna Loa Observatory, Hawaii. His record, shown in detail in Fig. 3.11 and in less detail in Fig. 12.7(a), is the longest continuous ground-based record of the gas.

The reason for the increases in tropospheric mixing ratios of $CO_2(g)$, $CH_4(g)$, and $N_2O(g)$ over time is the increase in their anthropogenic emission rates. Figure 12.8 shows the anthropogenic emission rates of $CO_2(g)$ between 1750 and 1997 (Marland et al., 2000). During this period, more than 270,000 Tg (teragrams; 1 Tg = 10^6 metric tons = 10^{12} g) of total carbon were released anthropogenically. Half of the emissions have occurred since the 1970s. The 1997 emission rate was 6,593 Tg/yr, an increase of 1.2 percent over the 1996 emission rate and the largest yearly emission rate up to that time. Liquid- and solid-fuel combustion accounted for 77.6 percent of anthropogenic $CO_2(g)$ emissions in 1997. Cement production and gas flaring (burning of natural gas emitted from coal mines and oil wells) contributed to less than 5 percent of the total $CO_2(g)$ emissions. Rates of such burning have decreased as a result of improved technologies enabling the capture of natural gas during coal and oil mining.

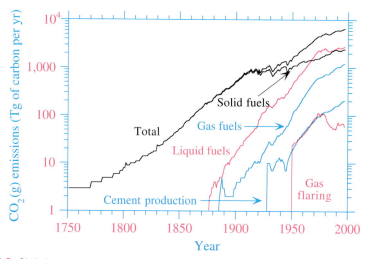

Figure 12.8. Global anthropogenic emissions of carbon dioxide from 1860 to 1997. Source of data: Marland et al. (2000). $1 \text{ Tg} = 10^6$ metric tons $= 10^{12}$ g.

Anthropogenic sources of $CO_2(g)$ not accounted for in Fig. 12.8 are biomass burning and deforestation. Biomass burning was defined in Section 5.2.1.4. **Deforestation** is the clearcutting of forests for their wood and the burning of forests to make room for farming and cattle grazing. When forests are burned, $CO_2(g)$ is emitted. Whether forests are burned or cut for their wood, their loss prevents photosynthesis from converting $CO_2(g)$ to organic material. Biomass burning rates are high in tropical rain forests of South America, Africa (Fig. 12.9), and Indonesia. Clearcutting of forests for their wood in the Pacific Northwest of the United States and in parts of Canada and in other forested regions of the world is also common. Biomass burning and deforestation result in a global emission of 1,800 to 3,700 Tg-C/yr as carbon dioxide (Houghton, 1994; Lobert et al., 1999; Andreae and Merlet, 2001). Thus, deforestation accounts for one-fifth to one-third of the total annual anthropogenic emissions of $CO_2(g)$.

Deforestation is one mechanism of **desertification**, the conversion of viable land into desert. Desertification also results from overgrazing and is a problem at the border of the Sahara Desert, which is in continuous expansion. Desertification increases the Earth's albedo and decreases the specific heat of the ground. Desert sand is more reflective than is forest or vegetation cover, which is a factor that would tend to cool the climate. On the other hand, sand has a lower specific heat than do trees that it replaces, thus, the expansion of deserts can also warm the climate. The net effect of desertification on temperatures is uncertain.

Figure 12.10 shows the increase in anthropogenic emissions of $CH_4(g)$ from different sources between 1860 and 1994. Total anthropogenic methane emissions increased from 29.3 to 371 Tg/yr during this period. The largest component of anthropogenic methane emissions in 1994 was livestock farming, which recently overtook rice farming as the leading methane emission source.

Anthropogenic emissions of $N_2O(g)$ also increased in the nineteenth and twentieth centuries. Nitrous oxide is emitted mostly by bacteria in fertilizers and sewage and during biomass burning and fossil-fuel combustion.

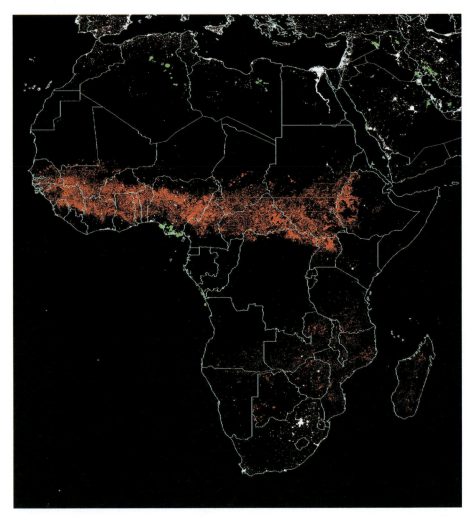

Figure 12.9. Fires in Africa, October 1994–March 1995. Fires are denoted by red; flares from oil and gas exploration/extraction by green. The more intense the color, the greater the frequency of fires. The data were collected by the U.S. Air Force Weather Agency. Image and data processing was by NOAA's National Geophysical Data Center.

In the 1930s and 1940s, synthetic long-lived chlorinated compounds, including $CFCl_3(g)$ (CFC-11), $CF_2Cl_2(g)$ (CFC-12), $CF_2ClH(g)$ (HCFC-22), and $CCl_4(g)$, were developed for industrial uses. These compounds not only contribute to stratospheric ozone destruction, but are also greenhouse gases. CFC-11, CFC-12, and HCFC-22 absorb thermal-IR radiation within the atmospheric window. $CCl_4(g)$ absorbs at the upper edge of the window. Figure 11.19 shows the changes in mixing ratios of these gases in recent years. Whereas the mixing ratios of CFC-11, CFC-12, and $CCl_4(g)$ are leveling off or decreasing, those of HCFC-22, other HCFCs, and HFCs are increasing. HFCs and other fluorine-containing compounds, such as sulfur hexafluoride [$SF_6(g)$] and perfluoroethane [$C_2F_6(g)$], are strong absorbers in the atmospheric window; thus, the buildup of these compounds is a cause for concern.

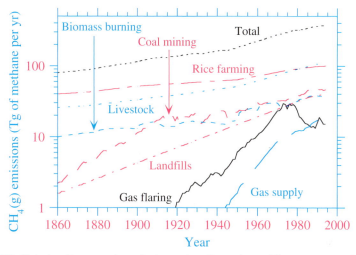

Figure 12.10. Global anthropogenic emissions of methane from different sources from 1860 to 1994. Source of data: Stern and Kaufman (1998). 1 Tg = 10^6 metric tons = 10^{12} g.

12.3. RECENT AND HISTORICAL TEMPERATURE TRENDS

To assess whether global warming is occurring, it is necessary to investigate the temperature record of the Earth. In the following subsections, recent and historic changes in global temperature are examined.

12.3.1. The Recent Temperature Record

Three types of datasets are currently used to assess global and regional air temperature changes during the twentieth century. All three show evidence of global warming.

The first type of dataset is one that uses temperature measurements taken between 2 and 10 m above the surface at land-based meteorological stations and fixed-position weather ships. Two datasets using these types of measurements include the Global Historical Climate Network dataset (GHCN, Peterson and Vose, 1997) and the dataset of Jones et al. (2000). Both datasets include measurements back to the 1850s. The number of measurement stations included in each dataset changes yearly. The GHCN dataset includes data from less than 500 stations prior to 1880, a high of 5,464 stations in 1966, and about 2,000 stations today. The Jones et al. dataset includes data from less than 250 stations prior to 1880, up to 1,800 stations in the 1950s, and less than 1,000 stations today.

Figure 12.11 shows changes in globally averaged near-surface air temperatures between 1856 and 1999, relative to the mean temperature between 1961 and 1990, from the Jones et al. dataset. The trend from the GHCN dataset is similar. Temperatures in both datasets were relatively stable between 1860 and 1910 but steadily increased from 1910 to 1940. Temperatures were stable or slightly decreased between 1940 and the mid-1970s, but they increased rapidly from the mid-1970s through 1998. The 1998 globally averaged temperature was the highest on record through that date. The six warmest years during the 143-year record were in the 1990s. Since 1856, the average global near-surface air temperature has warmed by about 0.5

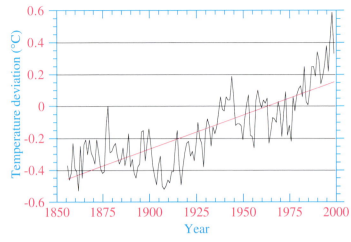

Figure 12.11. Globally averaged annual temperature variations between 1856 and 1999 relative to the 1961–1990 mean temperature. Source of data: Jones et al. (2000). Measurements were from land-based meteorological stations and fixed-position weather ships. The slope of the linear fit through the curve is 0.00429°C per year. The slope of the data from 1958 to 1999 is 0.011°C per year.

to 0.8°C. The linear fit through the data in Fig. 12.11 indicates that the rate of increase in the globally averaged near-surface temperature between 1856 and 1999 was 0.00425°C per year; between 1958 and 1999, the rate of increase in temperature was 0.011°C/yr. Additional statistics indicate that the three warmest winters in the contiguous 48 states in the United States from 1895 to 2000 were in 1999–2000, 1998–1999, and 1997–1998, respectively. The warmest nine-month period on record in the United States was January–October, 2000.

A second type of dataset is one that uses temperature measurements at different altitudes taken from balloons (radiosondes). One example is the 63-station, globally distributed radiosonde dataset compiled by Angell, (1999, 2000). Temperatures for this dataset are divided into near-surface (0 to 1.5 km), 850 to 300 mb (about 1.5 to 9 km), 300 to 100 mb (about 9 to 16 km), and 100 to 30 mb (about 16 to 24 km) temperatures. Temperatures in each pressure division are considered an average for the division. Figure 12.12 shows the vertical variation in the change in globally averaged temperature between 1958 and 1999 from the dataset. The figure shows that temperatures near the surface have increased and temperatures in the upper troposphere (300 to 100 mb) and stratosphere (100 to 30 mb) have decreased from 1958 to 1999. Both trends (near-surface warming and stratospheric cooling) are consistent with the theory of global warming.

The reason for the upper troposphere/stratospheric cooling relative to the surface warming in Fig. 12.12 is that absorption of thermal-IR radiation by greenhouse gases, particularly carbon dioxide, in the lower and midtroposphere reduces transfer of that radiation to the upper troposphere and stratosphere, where some of it would otherwise be absorbed by greenhouse gases. In other words, the addition of greenhouse gases to the air lowers the altitude of heating that occurs in the atmosphere, warming air near the surface and cooling the stratosphere. Surface warming and stratospheric cooling, observed in the radiosonde data, are simulated in models when carbon dioxide and other greenhouse gases are added to the air. The near-surface globally distributed

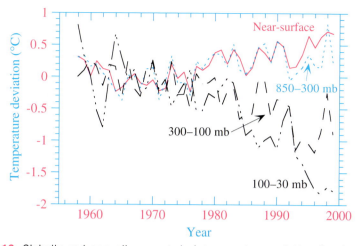

Figure 12.12. Globally and annually averaged air-temperature variations at the surface, at 850 to 300 mb, at 300 to 100 mb, and at 100 to 30 mb, derived from radiosonde records from 1958 to 1999. Source of data: Angell (1999, 2000). Deviations are relative to means derived from 1958–1977 data. Linear fits through the curves indicate that the air temperature changes between 1958 and 1999 were 0.014°C/yr at the surface, 0.01°C/yr at 850–300 mb, −0.024°C/yr at 300–100 mb, and −0.045°C/yr at 100–30 mb.

radiosonde temperature trend of 0.014°C/yr, seen in Fig. 12.12, is close to the surface-station trend of 0.011°C/yr from 1958 to 1999 (Fig. 12.11).

The third type of temperature dataset is the microwave sounding unit (MSU) satellite dataset, compiled since 1979 (Spencer and Christy, 1990). Temperatures for this dataset are derived from a comparison of the brightness or intensity of microwaves (with a wavelength of 5,000 μm) emitted by molecular oxygen in the air. Reported temperatures from this dataset are averages over a horizontal footprint of 110 km in diameter and a vertical thickness at 0 to 8 km (1,000 to 350 mb). Thus, the satellite dataset represents temperatures high above the boundary layer (e.g., on the order of 4 km), not at the surface of the Earth. Some error in the MSU measurements arises because a portion of the radiation measured by the MSU originates from the Earth's surface, not from the atmosphere (Hurrell and Trenberth, 1997). In addition, raindrops and large ice crystals can cause the MSU instrument to slightly underpredict temperatures in rain-forming clouds. The MSU dataset shows that, between 1979 and February 2001, global temperatures at 0 to 8 km increased by 0.0035°C/yr. The MSU satellite dataset also shows that stratospheric temperatures (14 to 22 km) cooled between 1979 and 1999 by −0.054°C/yr, a trend consistent with the −0.045°C/yr stratospheric cooling trend between 1958 and 1999 found in the radiosonde data and a trend consistent with the theory of global warming.

In sum, three types of global temperature measurements all show global warming. The two types that measure near-surface temperatures (surface and radiosonde) both show strong global warming. The two types that measure stratospheric temperatures (radiosonde and satellite) both show cooling there, which is consistent with the theory of global warming. The two that measure temperatures between the surface and stratosphere (radiosonde and satellite) both show global warming, but lower than at the surface. Thus, global near-surface warming and stratospheric cooling is evident in all temperature records. The underlying reason for the surface warming and simultaneous upper-atmospheric cooling is the increase in human-emitted pollutants. This physical phenomenon has been demonstrated repeatedly by computer models since the early 1970s.

12.3.2. The Historical Temperature Record

Whereas Figs. 12.11 and 12.12 indicate that recent increases in the Earth's near-surface temperature have occurred, it is important to put the temperature changes in perspective by examining past climates of the Earth. This is done next.

Historical temperatures are determined in many ways. Two techniques are ice-core analysis and oceanfloor sediment analysis. Other methods include analyses of lake levels, lake-bed sediments, tree rings, rocks, pollens in soil deposits, and pollens in deep-sea sediments.

Ice-Core Analysis

When snow accumulates over the Antarctic, it slowly compacts and recrystallizes into ice by a process called **sintering**, which is the chemical bonding of a material by atomic or molecular diffusion. During the ice-formation process, air becomes isolated in bubbles or pores within the ice. The trapped air contains roughly the composition of the air at the time the ice was formed. Among the constituents of the air trapped in the ice bubbles are carbon dioxide, methane, and molecular oxygen. When vertical ice cores are extracted, the composition of their air bubbles can be analyzed. Whereas most oxygen atoms in molecular oxygen contain 8 protons and 8 neutrons in their nucleus, giving them an atomic mass of 16, about 1 in every 1,000 such atoms contains 10 neutrons, giving it an atomic mass of 18. Such atoms are called **isotopically enriched atomic oxygen** (^{18}O), which is heavier than is standard atomic oxygen (^{16}O). Generally, the warmer the snow or air temperature, the greater the relative ratio of ^{18}O to ^{16}O in molecular oxygen trapped in snow. In fact, a relatively linear relationship has been found between the annual average snow surface temperature and the ^{18}O to ^{16}O ratio. Thus, from ice-core measurements of ^{18}O, past air temperatures can be estimated (Jouzel et al., 1987, 1993).

Ocean-Floor Sediment Analysis

A related method of obtaining temperature data is by drilling deep into the sediment of ocean floors and analyzing the composition and distribution of calcium carbonate shells. When water, which contains hydrogen and oxygen, evaporates from the ocean surface, the ratio of ^{18}O to ^{16}O in the ocean increases because ^{16}O is lighter than is ^{18}O, and water containing ^{16}O is more likely to evaporate than is water containing ^{18}O. Because calcium carbonate in shells contains oxygen from seawater, and because higher water temperatures correlate with lower ^{18}O to ^{16}O ratios in water, greater ^{18}O to ^{16}O ratios in shells correlate with higher water temperatures. Because certain shelled organism live within specified temperature ranges, the distribution of shells in a vertical core sample also gives insight into historical water temperatures.

12.3.2.1. From the Origin of the Earth to 570 Million Years Ago

Figure 12.13 shows the geologic time scale on the Earth. Since the Earth's formation 4.6 billion years ago (b.y.a.), near-surface air temperatures have gone through great swings. Between 4.5 and 4.0 b.y.a., surface temperatures increased due to

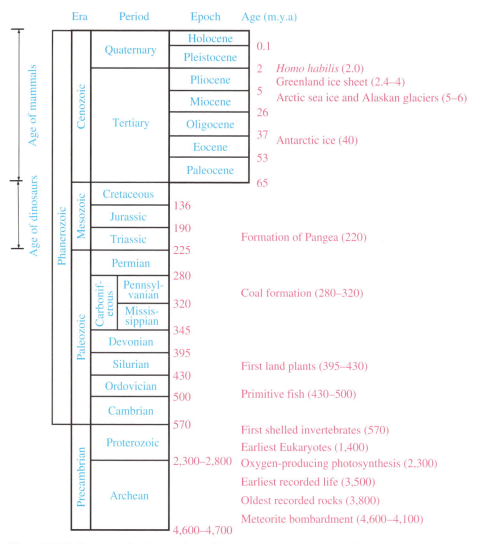

Figure 12.13. The geologic time scale and important events since the formation of the Earth. Adapted from Crowley (1983). Age is in units of million years ago (m.y.a.).

conduction of energy from the Earth's core to its surface, creating magma oceans (Section 2.3.1) that resulted in air temperatures possibly 300 to 400°C greater than those today (Crowley and North, 1991). During that time, energy released by meteorite impacts also contributed to high temperatures on the surface.

Between 4.5 and 2.5 b.y.a. (**Archean period**), the intensity of solar radiation was about 20 to 30 percent lower than it is today. Yet, even after the magma oceans solidified to form the Earth's crust 3.8 to 4 b.y.a., air temperatures were as high as 58°C, in comparison with 15°C today (Crowley and North, 1991). The low sun intensity coupled with the high air temperatures following crustal formation is referred to as the **Faint–Young Sun paradox** (Ulrich, 1975; Newman and Rood, 1977; Gilliland, 1989). The reason for the high temperatures may be an enhanced greenhouse effect. During solidification of the Earth's crust and mantle, outgassing, particularly of carbon dioxide and water vapor, occurred. Whereas much of the water vapor condensed to form the oceans, most carbon dioxide remained in the air. Figure 2.12 shows that the Earth's atmosphere prior to the oxygen age (2.3 b.y.a.) may have contained 10–80 percent carbon dioxide by volume. In terms of absolute quantities, partial pressures of $CO_2(g)$ during the Archean period may have been 0.01 to 0.1 atm, 30 to 300 times their current values (Garrels and Perry, 1974; Pollack and Yung, 1980). The high partial pressures of $CO_2(g)$ enhanced the greenhouse effect.

Over time, chemical weathering converted $CO_2(g)$ to carbonate-containing rocks. Carbonate formation may have been enhanced by cyanobacteria and other autotrophic bacteria, which use carbon dioxide as an energy source. Dead cyanobacteria pile up to form laminated, bounded structures of trapped carbonaceous material called **stromatolites**. Stromatolites are usually found in shallow marine waters in warm regions, and some are still being formed.

During the **Proterozoic period** (2.5 to 0.57 b.y.a.), two phases of continental glaciation may have occurred (2.5 and 0.9–0.6 b.y.a., respectively) (Frakes, 1979). The glaciations may have been due to low $CO_2(g)$ partial pressure coupled with the relatively low solar intensity. Between the two glaciations, the Earth was ice-free. During the second glaciation, ice formed on continents down to low latitudes, as evidenced by rock analysis (Williams, 1975). Because there is no corresponding evidence of sea ice appearing at low latitudes, it is unlikely that the entire Earth was covered with ice (Crowley and North, 1991), although many scientists have speculated that it was.

12.3.2.2. From 570 to 100 Million Years Ago

Between 570 and 100 m.y.a., which encompasses the **Paleozoic era** (570 to 225 m.y.a) and much of the **Mesozoic era** (226 to 65 m.y.a.), the Earth's climate was relatively mild, but with interruptions by two important glaciations.

From 570 to 460 m.y.a., global $CO_2(g)$ mixing ratios were thirteen to fourteen times their current value, sea levels were at an all-time high, and the Northern Hemisphere was covered with water above 30°N latitude (due to the configuration of the continents, which were concentrated in the Southern Hemisphere). The high $CO_2(g)$ mixing ratios and sea levels suggest that temperatures were high during this period.

Near 440 m.y.a., the first glaciation of the Paleozoic era occurred over parts of the Earth, but $CO_2(g)$ partial pressures were still 10 times their current value. The reason for this paradox is still uncertain.

During the **Denovian period** (395 to 345 m.y.a.), the land-plant coverage of the Earth increased, decreasing continental albedo by up to 10 to 15 percent (Posey and Clapp, 1964), warming air temperatures and causing glaciers to melt. Plant uptake of $CO_2(g)$ by photosynthesis offset some of the warming.

The second period of glaciation during the Paleozoic era began in the **Carboniferous period** (345 to 280 m.y.a.). This glaciation lasted 100 million years and decreased sea levels.

Evidence suggests warm temperatures and no glaciation during the **Permian** (280 to 225 m.y.a.), **Triassic** (225 to 190 m.y.a.), or **Jurassic** (190 to 136 m.y.a.) periods. Around 220 m.y.a., during the Triassic period, the continents finally merged together to form one supercontinent, **Pangea**. Since then, the continents have slowly refractured and drifted apart, giving us our current continental distribution. The Pangean climate was drier and warmer than is our present climate (Crowley and North, 1991).

12.3.2.3. From 100 to 3 Million Years Ago

Between 100 and 50 m.y.a., during the Cretaceous (136 to 65 m.y.a) and early Tertiary (65 to 2 m.y.a) periods, temperatures were again mild and the Earth was without glaciers. In the mid-Cretaceous period (120 to 90 m.y.a.), temperatures increased above those today. This may have been the last period of an ice-free globe. Sea levels were high, covering about 20 percent of continental areas (Barron et al., 1980).

At the end of the Cretaceous period (65 m.y.a.), a mass extinction of 75 percent of all plant and animal species, including the dinosaurs, occurred. This extinction is called the Cretaceous–Tertiary (K-T) extinction, and may have been due to an asteroid 10 km wide hitting the Earth (Alvarez et al., 1980). Evidence for the asteroid theory includes a worldwide layer of the element, iridium (Ir) in clays at depths below the Earth's surface corresponding to the K-T transition. Although iridium has been measured in a Hawaiian volcanic eruption plume (Zoller et al., 1983), Hawaiian volcanic plumes are not known to penetrate to the stratosphere, which would be necessary for the iridium to be distributed globally (Crowley and North, 1991). Currently, the most accepted explanation for the iridium layer at the K-T boundary is an extraterrestrial source, such as an asteroid. The asteroid impact is thought to have created a large dust cloud that blocked the sun for a period of weeks to months, lowering the surface temperature by tens of degrees (Toon et al., 1982). The reduction in surface temperature may have been responsible for the mass extinction. Because some extinctions occurred before the K-T transition, the asteroid theory is still open to debate.

Temperatures continued to cool from the Cretaceous period into the Paleocene epoch (65 to 53 m.y.a.). Temperatures abruptly increased from the late Paleocene to the early Eocene epoch (53 to 37 m.y.a.). From 55 to 53 m.y.a., temperatures were higher than during any other time in the Cenozoic era, although they were lower than during the Cretaceous period in the late Mesozoic era. At the onset of the early Eocene warming, the carbon content of deep-sea bulk sediments decreased, indicating an increase in $CO_2(g)$ (Berner et al., 1983; Shackleton, 1985). During the onset of warming, the Norwegian–Greenland sea opened (Talwani and Eldholm, 1977). This and other tectonic activity may have resulted in enhanced volcanism, increasing $CO_2(g)$ levels.

Following the early Eocene temperature maximum, a gradual period (50 to 3 m.y.a.) of global cooling occurred. During this period, continents drifted toward higher latitudes, cooling mean temperatures over land. A sharp cooling event occurred near 40 to 38 m.y.a., in the late Eocene epoch, resulting in one of the largest extinctions during the Cenozoic era.

Cenozoic era ice may have first appeared over the Antarctic near 40 m.y.a., forming the base of the present-day Antarctic Ice Sheets. An ice sheet is a broad, thick sheet of glacier ice that covers an extensive land area for a long time. An ice sheet can also form on top of sea ice if the sea ice is adjacent to land. A glacier is a large mass of land ice formed by the compaction and recrystallization of snow. Glaciers flow slowly downslope or outward in all directions under their own weight. Sea ice is ice formed by the freezing of sea water (Fig. 12.14). Today, the Antarctic is covered by two ice sheets, the East and West Sheets, which are separated by the Transantarctic Mountains. The West Sheet, which lies over the Ross Sea and over land (Fig. 12.15), is an order of magnitude smaller than is the East Sheet, which lies exclusively over land.

(a)   (b)

Figure 12.14. (a) Spring (1950) melt of sea ice in the Beaufort Sea near Tigvariak Island, Alaska North Slope. (b) Winter (1950) sea ice near the same location. Photos by Rear Admiral Harley D. Nygren, NOAA Corps (ret.), available from NOAA Central Library.

(a)   (b)

Figure 12.15. (a) Edge of Ross Sea Ice Sheet, December 1996. (b) Icebergs grounded on Pennel Bank, Ross Sea, Antarctica, January 1999. Photos by Michael Van Woert, NOAA NESDIS, ORA. available from NOAA Central Library.

During the **Oligocene epoch** (37 to 26 m.y.a.), temperatures continued their downward trend that started during the Eocene epoch. Around 31 to 29 m.y.a., Antarctic ice cover expanded. From 26 to 15 m.y.a., during the early **Miocene epoch** (26 to 5 m.y.a.) temperatures temporarily stabilized. From 15 to 10 m.y.a., temperatures dropped, increasing Antarctic ice coverage and decreasing sea levels. The cause of this temperature decrease may have been a decrease in $CO_2(g)$ due to changes in ocean water vertical mixing (Crowley and North, 1991).

From 5 to 4 m.y.a., in the early **Pliocene epoch** (5 to 2 m.y.a.), temperatures rebounded slightly, reducing the sizes of the Antarctic ice sheets. Although much of the Northern Hemisphere may have been warm during the early Pliocene, seasonal **sea ice over the Arctic Ocean** and **glaciers over Alaska** may have first formed 5 to 6 m.y.a. and persisted thereafter (Clark et al., 1980; Zubakov and Borzenkova, 1988). From 2.6 to 2.4 m.y.a., a sharp decrease in temperatures caused trees to disappear in Northern Greenland and tundra to expand in Siberia. The **Greenland ice sheet** may have formed during this cooling period, although evidence for its formation between 3 to 4 m.y.a. has also been suggested (Leg 105 Shipboard Scientific Party, 1986).

12.3.2.4. From 3 Million Years Ago to 20,000 Years Ago

The past 3 million years have been characterized by advances and retreats of Northern Hemisphere ice sheets. During the late Pliocene epoch and early **Pleistocene epoch** (2 m.y.a. to 10,000 years ago, y.a.), fluctuations in the advances and retreats of

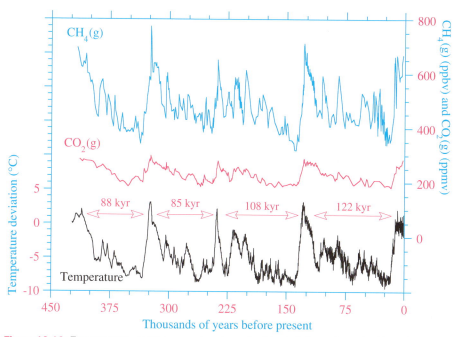

Figure 12.16. Temperature, methane, and carbon dioxide variations from the Vostok ice core over the past 420,000 to 450,000 years. Sources: Jouzel et al. (1987, 1993, 1996); Petit et al. (1999). Temperature variations are relative to a modern surface–air temperature over the ice of $-55°C$.

ice sheets occurred with a period of about 40,000 years (Shackleton and Opdyke, 1976). During the past 700,000 years, such advances and retreats have occurred with periods of about 100,000 years. The reasons for the 100,000-year cycle can be elucidated from Fig. 12.16.

Figure 12.16 shows the temperature trend over the past 450,000 years, a period covering nearly one-quarter of the Pleistocene epoch. The figure also shows carbon dioxide and methane trends over the past 420,000 years. Data for the table were obtained from the Vostok, Antarctica, ice core. To date, drilling of the Vostok ice core has extended down to a depth of more than 3.6 km. Data from the core enabled the reconstruction of 450,000 years of atmospheric history.

Figure 12.16 shows that the Earth has gone through four glacial and interglacial periods during the past 450,000 years. Within these major periods are minor periodic oscillations. The major and minor temperature oscillation can be explained, for the most part, in terms of three cycles related to the Earth's orbit, called **Milankovitch cycles**. Milankovitch cycles are named after Serbian astronomer **Milutin Milankovitch** (Milankovitch, 1930, 1941). They are caused by gravitational attraction between the planets of the solar system and the Earth. Milankovitch was the first to calculate the effects of these cycles on incident solar radiation reaching the Earth. The cycles are discussed briefly next.

Changes in the Eccentricity of the Earth's Orbit

Figure 12.17 shows that the Earth travels around the sun in an elliptical pattern, with the sun at one focus. In the figure, *a* and *b* are the lengths of the major and minor semiaxes, respectively. The distance between the center of the ellipse (point C) and

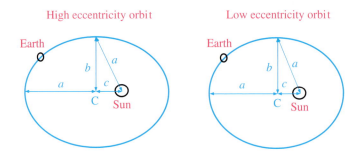

Figure 12.17. Earth's orbit around the sun during periods of high and low orbital eccentricity. The eccentricities are exaggerated in comparison with real eccentricities of the Earth's orbit.

either focus is *c,* which is related to *a* and *b* by the Pythagorean relation, $a^2 = b^2 + c^2$. The **eccentricity** (*e*) of the ellipse is

$$e = \frac{c}{a} \tag{12.7}$$

The eccentricity varies between 0 and 1. A circle has an eccentricity of 0. Earth's eccentricity is currently low, 0.017 (Example 12.2), indicating that the earth's orbit is nearly circular but noncircular enough to create a 3.3 percent difference in distance and a 6.6 percent difference in incoming solar radiation between June and December.

EXAMPLE 12.2

Calculate the current eccentricity of the Earth.

Solution

Figure 12.1 shows that the Earth-sun distance during the winter solstice (when the Earth is closest to the sun) is currently a − c = 147 million km and that during the summer solstice (when the Earth is furthest from the sun) is a + c = 152 million km. Solving these two equations gives c = 2.5 million km and a = 149.5 million km. Substituting these numbers into Equation 12.7 gives the current eccentricity of the Earth's orbit around the sun as *e* ≈ 0.017.

The eccentricity of the Earth varies sinusoidally with a period of roughly 100,000 years. The minimum and maximum eccentricities during each 100,000-year period vary as well. The minimum eccentricity is usually greater than 0.01 and the maximum is usually less than 0.05. Whereas today the Earth is in an orbit of relatively low eccentricity, in 50,000 years it will be in an orbit of high eccentricity. During orbits of high eccentricity, the Earth–sun distance can be about 10 percent greater in June than in December, and the incident solar radiation can be about 21 percent less in June than in December. Today, the Earth–sun distance is about 3 percent greater in June than in December, and incident solar radiation is about 7 percent less in June than in December.

Because the yearly averaged distance of the Earth from the sun is less in a period of low eccentricity than in a period of high eccentricity, yearly averaged temperatures are higher in periods of low eccentricity than in periods of high eccentricity. This can be seen in Fig. 12.16, which shows that natural interglacial temperature maxima occurred 122,000, 230,000, 315,000, and 403,000 years ago, all times of low eccentricity.

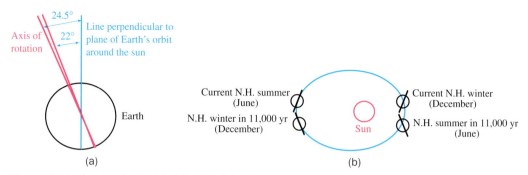

Figure 12.18. Changes in the (a) obliquity of the Earth's axis of rotation and (b) precession of the Earth's axis relative to a point in space.

Changes in the Obliquity of the Earth's Axis

The **obliquity** of the Earth's axis of rotation is the angle of the axis relative to a line perpendicular to the plane of the Earth's orbit around the sun. Figure 12.18(a) shows the variation in the Earth's obliquity. Every 41,000 years, the Earth goes through a complete cycle where the obliquity changes from 22° to 24.5° and back to 22°. Currently, the obliquity is 23.5°, in the middle of a cycle. When the obliquity is low (22°), more sunlight hits the equator, increasing south–north temperature contrasts. When the obliquity is high (24.5°), more sunlight reaches higher latitudes, reducing temperature contrasts. Thus, the obliquity affects the seasons and temperatures at each latitude on Earth. Superimposed on the large temperature variations seen in Fig. 12.16 are smaller variations, some resulting from changes in obliquity.

Precession of the Earth's Axis of Rotation

Precession is the angular motion (wobble) of the axis of rotation of the Earth about an axis fixed in space [Fig. 12.18(b)]. It is caused by the gravitational attraction between the Earth and other bodies in the solar system. Currently, the Northern Hemisphere is further from the sun in summer than in winter. In 11,000 years, angular motion of the Earth's axis of rotation will cause the Northern Hemisphere to be closer to the sun in summer than in winter. The complete cycle of the precession of the Earth's axis is 22,000 years. Precession of the Earth's axis does not change the yearly or globally averaged incident solar radiation at the top of the Earth's atmosphere. Instead, it changes the quantity of incident radiation at each latitude during a season. In 11,000 years, for example, Northern Hemisphere summers will be warmer and Southern Hemisphere summers will be cooler than they are today. Because seasonal changes in temperature result in yearly averaged changes in temperature at a given latitude, some of the cyclical changes in temperatures seen in the Antarctic data shown in Fig. 12.16 are due to changes in precession.

Effects of the Milankovitch Cycles

The Milankovitch cycles appear to be responsible for the cyclical changes in the Earth's temperature, seen in Figure 12.16, and the corresponding advances and retreats of glaciers during the Pleistocene epoch. As such, the Milankovitch cycles must also be responsible for changes in $CO_2(g)$, and $CH_4(g)$, which are correlated with changes in temperature. Increases in temperature decrease the solubility of $CO_2(g)$ in seawater, increasing the atmospheric loading of $CO_2(g)$. Changes in temperature also change vertical mixing rates of ocean water, nutrient uptake rates by phytoplankton, and rates of erosion of continental shelves (which affect biomass loadings), thereby affecting mixing ratios of $CO_2(g)$ (e.g., Crowley and North, 1999). Changes in microbiological activity resulting from changes in temperature may explain the correlation between temperatures and $CH_4(g)$.

Figure 12.16 shows that a temperature minimum occurred about 150,000 y.a. Near that time, glaciers extended down to Wisconsin in the United States, and possibly further south in Europe (Kukla, 1977). Temperatures increased about 130,000 y.a., causing deglaciation. Over the Antarctic, temperatures rose 2 to 3°C above what they are today (Fig. 12.16). As the eccentricity of the Earth's orbit increased, temperatures cooled again, causing a renewed period of glaciation. During this period (the last glacial period), two major stages of glaciation occurred, the first starting 115,000 y.a. and the second starting 75,000 y.a. The second stage continued until about 6,000 years ago.

12.3.2.5. From 20,000 to 9,000 Years Ago

The last glacial maximum (last ice age) occurred 22,000 to 14,000 y.a. (Fig. 12.16). Depending on whether glaciation over eastern North America, western Europe, or the Alps is considered, this maximum is called the Wisconsin, Weichselian, or Würm. During the maximum, an ice sheet called the Laurentide Ice Sheet covered North America, and another called the Fennoscandian Ice Sheet covered much of Northern Europe. These ice sheets were about 3,500 to 4,000 m thick and drew up enough ocean water to decrease the sea level by about 120 m (CLIMAP, 1981; Fairbanks, 1989). The decrease in sea level was sufficient to expose land connecting Siberia to Alaska, creating the Bering land bridge. This land bridge allowed humans to migrate from Asia to North America and, ultimately, to Central and South America. The Laurentide sheet extended from the Rocky Mountains in the west to the Atlantic Ocean in the east, but only as far south as the Missouri and Ohio Valleys.

Temperatures during the last ice age were about 4°C less than they are today over the Northern Hemisphere and 8°C less than they are today over the Antarctic (Fig. 12.19). During the last ice age, Antarctic ice coverage expanded as did Arctic sea ice coverage. In the tropics, precipitation decreased, resulting in lower inland lake and river levels. Globally, near-surface winds may have been 20 to 50 percent higher than those today. $CO_2(g)$ mixing ratios were about 200 ppmv, as seen in Fig. 12.16, almost half their current value. $CH_4(g)$ mixing ratios were about 0.35 ppmv, 20 percent of their current value.

Figure 12.19 shows temperature change estimates in the Northern Hemisphere and from the Vostok ice core in the Antarctic during the last 20,000 years. The ice core data indicate that temperatures over the Antarctic increased between 17,000 and 11,000 y.a., with a hiatus between 13,500 and 12,000 y.a. The increases in temperatures were caused by Milankovitch cycle variations and were responsible for the melting of ice over the Antarctic starting 16,000 to 17,000 y.a. (Labeyrie et al., 1986; Jones and Keigwin, 1988).

In the Northern Hemisphere, temperature increases and deglaciation started around the same time as they did over the Antarctic, near 17,000 y.a. At first, Northern Hemisphere deglaciation was slow. From 13,000 to 12,000 y.a., an abrupt increase in temperatures hastened deglaciation. Around 12,000 y.a., temperatures dropped slightly, then plunged 10,900 y.a. This strong cooling, which lasted until 10,100 y.a., is called the Younger Dryas period. Dryas is the name of an Arctic flower. The Younger Dryas cooling period followed a shorter Older Dryas cooling period, which occurred

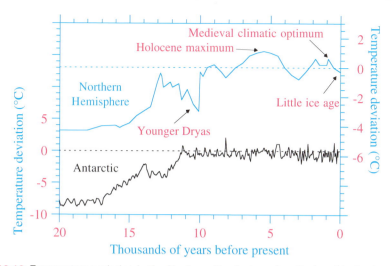

Figure 12.19. Temperature variation in the Northern Hemisphere (top line) and in the Antarctic (bottom line) during the last 20,000 years. Antarctic data were from the Vostok ice core (Jouzel et al., 1987, 1993, 1996; Petit et al., 1999). The deviations for the ice core data are relative to a modern surface-air temperature over the ice of $-55°C$. The Northern Hemisphere figure was modified from Ahrens (1994). The deviations for these data are relative to current surface-air temperatures.

12,100–11,950 y.a. The Younger Dryas cooling may have been due to changes in atmospheric circulation resulting from large releases of meltwater from the glaciers. During glacial retreat, bursts of meltwater flooded the Columbia Plateau in eastern Washington, the largest flood in known history. Up to forty similar bursts may have occurred during this period (Waitt, 1985). Some of the bursts may have poured water into the North Atlantic Ocean.

Following the Younger Dryas, temperatures increased again. Most of the Northern Hemisphere ice sheets disappeared by 9,000 y.a., although remnants of the Laurentide sheet remained until 6,000 y.a. In sum, the bulk of the two major ice sheets over the Northern Hemisphere disappeared in a period of 5,000 years, from 14,000 to 9,000 y.a.

12.3.2.6. The Holocene Epoch

The period from 10,000 y.a. to the present is the **Holocene epoch**. Temperatures during the first 5,500 years of the Holocene were warmer than were those during the second 4,500 years. Humans developed agriculture about 8,000 y.a. Temperatures in the Northern Hemisphere 6,000 to 5,000 y.a. were about 1°C warmer than they are today, as seen in Fig. 12.19. This warm period is referred to as the **mid-Holocene maximum**. Because plants flourished, the period is also known as the **climatic optimum**. Corresponding temperature changes over the Antarctic are less obvious.

After the mid-Holocene, temperatures cooled. Between 1100 and 1300 A.D., temperatures in the Northern Hemisphere, particularly in Europe, warmed again, becoming warmer than at any time since the mid-Holocene. During this warm period, called the **medieval climatic optimum**, Arctic ice retreated, Iceland and Greenland were colonized by the Vikings, Alpine passes between Germany and Italy became free of ice, and grapes were harvested for wine production in England (Le Roy Ladurie, 1971; Crowley and North, 1991).

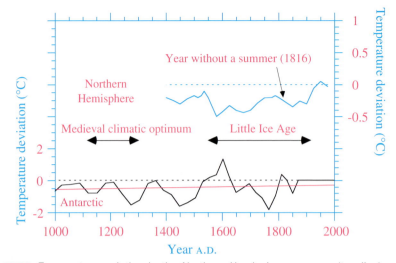

Figure 12.20. Temperature variation in the Northern Hemisphere summer (top line) and in the Antarctic (bottom line) during the last 1,000 years. Antarctic data were from the Vostok ice core (Jouzel et al., 1987, 1993, 1996; Petit et al., 1999). The deviations for the ice core data are relative to a modern surface-air temperature over the ice of $-55°C$. The slope of the linear fit through the curve (solid flat line) is $0.000519°C$ per year. The Northern Hemisphere figure was modified from Bradley and Jones (1993). The deviations for these data are relative to 1961–1990 surface-air temperatures.

Between 1450 and 1890 A.D., temperatures in North America, particularly in Europe, cooled, and mild glaciation reappeared. This period is called the **Little Ice Age** (Matthes, 1939). The coldest temperatures during this period were during the mid-to late 1600s, early 1800s, and late 1800s (Crowley and North, 1991). During the Little Ice Age, glaciers advanced in the European Alps, Sierra Nevada Mountains, Rocky Mountains, Himalayas, southern Andes Mountains, and mountains in eastern Africa, New Guinea, and New Zealand. The area of sea ice around Iceland increased (Bergthorsson, 1969). Although temperatures decreased over much of the world, temperatures over the Antarctic warmed during part of the Little Ice Age, as seen in Fig. 12.20. Temperature decreases during the Little Ice Age were smaller in magnitude than those during the Younger Dryas period.

From April 5–12, 1815, during the Little Ice Age, the most deadly volcano in human history, **Tambora**, Indonesia, erupted, killing an estimated 92,000 islanders directly and through disease and famine. The volcano emitted an estimated 1.7 million tons of ash and aerosol particles that traveled globally, cooling temperatures in North America and Europe during the following 2 years (Stommel and Stommel, 1981; Stothers, 1984). In 1816, frost in the spring and summer killed crops in the United States, Canada, and Europe, sparking famine in some areas of Europe. In the northeast United States and Canada, 1816 was known as the **year without a summer**. Colorful sunsets caused by the volcanic particles were captured in paintings by J. M. W. Turner. In the dark summer of 1816, George Gordon (Lord) Byron (1788–1824) met and became friends with Percy Bysshe Shelley (1792–1822) in Geneva, Switzerland. Because of the depressing nature of the weather, Lord Byron and Shelley's wife, Mary Wollstonecraft Shelley (1797–1851) entered into a competition to write the most depressing work. Lord Byron wrote the poem *Darkness*, and Mary Shelley started the novel, *Frankenstein: The Modern Prometheus*, which she finished in 1818.

Table 12.4. Rates of Temperature Change during Different Historical Periods

Period	Years ago	Temperature Change (K) per 100 Years
From 1958 to 1999 (radiosonde records, Fig. 12.12)	42–2	+1.4
From 1958 to 1999 (land and ship measurements, Fig. 12.11)	42–2	+1.1
From 1856 to 1999 (land and ship measurements, Fig. 12.11)	144–2	+0.43
During the last 1,000 years (Vostok core, Fig. 12.20)	1,000–0	+0.052
Deglaciation after Younger Dryas (Vostok core, Fig. 12.19)	12,632–11, 191	+0.3
Last years of deglaciation after Younger Dryas (Vostok core, Fig. 12.19)	11,237–11,191	+2.2
Deglaciation leading to last interglacial period (Vostok core, Fig. 12.16)	138,000–128,000	+0.13

12.3.2.7. Comparison of Temperatures Today with Historical Temperatures

Temperatures today are not unprecedented in the Earth's history, but the rate of temperature increase is unusual in comparison with historical rates of temperature increase. Table 12.4 shows that the rate of temperature increase from 1856 to 1999 was about 0.43°C per 100 years, and that from 1958 to 1999 was about 1.1 to 1.4°C per 100 years. For comparison, the rate of temperature increase during the last millennium was about 0.052°C per 100 years, that during the deglaciation after the Younger Dryas period was 0.3°C per 100 years, and that during the deglaciation leading to the last interglacial period was 0.13°C per 100 years. In all cases, rates of temperature increase were less than are those today. Yet, a careful examination of the Vostok core data indicates that during deglaciations, temperature increases over short time periods (as opposed to averaged over the entire deglaciation period) could be quite rapid. For example, during the last years of deglaciation after the Younger Dryas, the rate of temperature increase was 2.2°C per 100 years. The difference today, is that the Earth is in an interglacial, not a deglaciation period. **The current rapid rate of increase in temperature is unusually high for an interglacial period.**

12.4. FEEDBACKS AND OTHER FACTORS THAT MAY AFFECT GLOBAL TEMPERATURES

Whereas recent near-surface air temperatures appear to have increased at unprecedented rates in comparison with the known climate record and levels of $CO_2(g)$, $BC(s)$, $CH_4(g)$, and $N_2O(g)$ have increased since the mid-1800s, questions are still asked as to whether any factor aside from pollutants could cause global warming. In addition, the full effect of continued emission of greenhouse gases and particles on climate is still uncertain. These topics are discussed next.

12.4.1. Alternative Arguments Used to Explain Global Warming

A common alternative explanation for global warming is that increased temperatures are due to natural Milankovitch cycle variations. One argument is that because the current eccentricity of the Earth's orbit (0.017) is in a declining stage, temperatures should naturally increase over the next several hundred years, and such increases might explain global warming. Because the eccentricity has also been declining for the last 1,000 years, and the rates of temperature increase today are much higher than were those during the last 1,000 years (Table 12.4), this argument does not explain global warming.

Another proposed explanation for global warming has been that it is due to the natural variation in solar output. When sunspots appear, the intensity of the sun's output increases. A **sunspot** is a large magnetic solar storm that consists of a dark, cool central core, called an **umbra**, surrounded by a ring of dark fibrils, called a **penumbra**. As a result of the magnetic activity associated with sunspots, regions near the umbra are hot, resulting in more net energy emitted by the sun when sunspots are present than when they are absent. Sunspot number and size peak every 11 years, but because the sun's magnetic field reverses itself every 11 years, a complete sunspot cycle actually takes 22 years. The difference in solar intensity at the top of the Earth's atmosphere between times of sunspot maxima and minima is about 1.4 W m^{-2}, or 0.1 percent of the solar constant. Although sunspots have been linked to changes in climate during the course of a sunspot cycle, the fact that sunspot intensity varies relatively consistently from cycle to cycle makes it difficult to link sunspots to multidecade increases in temperatures.

A third proposed explanation has been that warming is due to natural **internal variability** of the ocean atmosphere system. Internal variability results in sinusoidal warming and cooling of the Earth's climate. Whereas internal variability explains part of the warming and cooling cycles between the 1850s and the present (seen in Fig. 12.11), changes due to internal variability are on the order of $+/-0.3°C$ (Stott et al., 2000), smaller than global observed temperature changes of $+0.6$ to $+0.9°C$ since 1856.

12.4.2. Feedbacks of Gases to Climate

Whereas the simple radiative energy balance model introduced at the beginning of this chapter described the equilibrium temperature of the Earth in the absence of natural greenhouse gases, temperature changes in the real atmosphere depend not only on greenhouse gases, but also on their feedbacks to clouds, winds, and the oceans. Feedbacks will affect the magnitude of future global warming.

If an increase in emissions initially forces a warming of the climate, the climate may respond either positively, enhancing the warming, or negatively, canceling the warming. A **positive feedback mechanism** is a climate response mechanism that causes temperatures to change further in the same direction as that of the initial temperature perturbation. A **negative feedback mechanism** is a mechanism that causes temperatures to change in the opposite direction from that of the initial temperature perturbation. Positive feedbacks can lead to a runaway greenhouse effect, such as that which occurred on Venus. Negative feedbacks tend to mitigate potential effects of global warming. Next, some positive and negative feedback mechanisms resulting from an initial increase in temperatures by greenhouse gases are listed.

- **Water vapor – temperature rise positive feedback**
  If air temperatures initially increase, water evaporates from the oceans, increasing atmospheric water vapor (a greenhouse gas), increasing temperatures further.
- **Snow-albedo positive feedback**
  If air temperatures initially increase, sea ice and glaciers melt, decreasing the Earth-atmosphere albedo, increasing incident solar radiation to the surface, increasing temperatures further.

- **Water vapor – high cloud positive feedback**
  If air temperatures initially increase, water evaporates from the oceans, increasing the cover of high clouds (made primarily of ice, which is relatively transparent to solar radiation but absorbent of thermal-IR radiation), increasing the absorption and reemission of the Earth's thermal-IR radiation, increasing temperatures further.
- **Solubility – carbon dioxide positive feedback**
  If air temperatures initially increase, the solubility of carbon dioxide in ocean water decreases, increasing the transfer of carbon dioxide from water to air, increasing temperatures further.
- **Saturation vapor pressure – water vapor positive feedback**
  If air temperatures initially increase, the saturation vapor pressure of water increases, increasing the quantity of uncondensed water vapor in the air, increasing temperatures further.
- **Bacteria – carbon dioxide positive feedback**
  If air temperatures initially increase, the rate at which soil bacteria decompose dead organic matter into carbon dioxide and methane increases, increasing atmospheric carbon dioxide levels, increasing temperatures further.
- **Permafrost – methane positive feedback**
  If air temperatures initially increase, Arctic permafrost melts, enhancing release of methane stored under the permafrost, increasing temperatures further.
- **Water vapor – low cloud negative feedback**
  If air temperatures initially increase, water evaporates from the oceans, increasing the cover of low clouds (made of liquid water, which reflects solar radiation), increasing the effective Earth-atmosphere albedo, decreasing incident solar radiation, decreasing temperatures.
- **Plant – carbon dioxide negative feedback**
  If air temperatures initially increase, plants and trees flourish and photosynthesize more, decreasing the quantity of carbon dioxide in the air, decreasing temperatures.

The only practical way to elucidate the relative importance of the different feedback mechanisms is through computer modeling of the climate. To date, climate models accounting for feedbacks have found that increases in the atmospheric greenhouse gases increase global near-surface temperatures and cool stratospheric temperatures, consistent with theory and observations described in Section 12.3.1.

12.4.3. Effects of Aerosol Particles on Climate

Although all greenhouse gases warm near-surface air, some aerosol particle components warm and others cool the air. Particle components that warm the air include black carbon (BC), iron, aluminum, polycyclic aromatic compounds, and nitrated aromatic compounds. These components warm the air primarily by absorbing solar radiation, although they also absorb thermal-IR radiation. Most other particle components, including water, sulfate, nitrate, and most organic compounds, cool near-surface air by backscattering incident solar radiation to space more than they absorb thermal-IR radiation from the Earth. A major unresolved issue is the extent to which warming from absorbing particle components offsets cooling from other components. If only instantaneous radiative effects are considered and time-dependent effects are ignored, cooling caused by all reflective components appears to exceed the warming caused by all absorbing components in the global average, but not by much. Instantaneous radiative effects, however, tell only part of the story. A true estimate of the effect of aerosol particles on climate requires the consideration of time-dependent effects as well. Some time-dependent effects are summarized next.

The "Self-Feedback Effect"

When particles are emitted into to the air, they change the air temperature, relative humidity, and surface area available for gases to condense upon, all of which affect the composition, liquid water content, size, and optical properties of both the new and existing particles. This process is called the *self-feedback effect* (Jacobson, 2002). For example, when BC warms the air, it decreases the relative humidity, decreasing the liquid water content and reflectivity of particles containing sulfate and nitrate, warming the air further. Reduced aerosol particle liquid-water also decreases the liquid-phase chemical conversion of sulfur dioxide to sulfate and the dissolution of ammonia, nitric acid, and hydrochloric acid into particles, further reducing particle size and reflectivity. In addition, when BC is emitted in one location, it increases the surface area available for sulfuric acid to condense upon, increasing the formation of sulfate upwind and decreasing it downwind.

The "Photochemistry Effect"

Aerosol particles alter photolysis coefficients of gases, affecting their concentrations and those of other gases (through chemical reactions). Because many gases absorb solar and/or thermal-IR radiation, changing the concentration of such gases affects temperatures. The process by which aerosol particles change photolysis coefficients, thereby affecting temperatures, is the *photochemistry effect* (Jacobson, 2002).

The "Smudge-Pot Effect"

During day and night all aerosol particles trap the Earth's thermal-IR radiation, warming the air (Bergstrom and Viskanta, 1973; Zdunkowski et al., 1976). This warming is well known to citrus growers who, at night, used to burn crude oil in smudge pots to fill the air with smoke and trap thermal-IR radiation, preventing crops from freezing. The warming of air relative to a surface below increases the stability of air, slowing surface winds (and increasing them aloft), reducing the wind speed dependent emission rates of sea spray, soil dust, road dust, pollens, spores and some gas-phase particle precursors. The reduction in concentration of these particles affects daytime solar reflectivity and day-and nighttime thermal-IR heating. Changes in stability and winds due to thermal-IR absorption by aerosols also affect energy and pollutant transport. The effect of thermal-IR absorption by particles on emissions of other particles and gases and on local energy and pollutant transport is referred to as the *smudge-pot effect* (Jacobson, 2002).

The "Daytime Stability Effect"

If airborne particles absorb solar radiation, the air warms. Whether the particles absorb or only scatter, they prevent solar radiation from reaching the surface, cooling the surface and increasing the air's stability (Bergstrom and Viskanta, 1973; Venkatram and Viskanta, 1977; Ackerman, 1977). Like with the *smudge-pot effect*, enhanced daytime stability slows surface winds, reducing emissions of wind-driven particles and gases and affecting local pollutant and energy transport. This effect is called the *daytime stability effect* (Jacobson, 2002).

The "Particle Effect Through Surface Albedo"

During the day, airborne BC reduces sunlight to and cools the ground, increasing the lifetime of existing snow. Conversely, because BC warms the air, snow passing through a BC layer is more likely to melt. At night, airborne BC also enhances downward thermal-IR, melting snow on the ground. Because the albedo of new snow exceeds that of sea ice, which exceeds those of soil or water, the melting of snow or sea ice increases sunlight to the surface. The effect of aerosol particles on temperatures through their change in snow and sea-ice cover is the *particle effect through surface albedo.*

The "Particle Effect Through Large-Scale Meteorology"

Aerosol particles affect local temperatures, which affect local air pressures, winds, relative humidities, and clouds. Changes in local meteorology slightly shift the locations and magnitudes of semipermanent and thermal pressure systems and jet streams. The effect of local particles on large-scale temperatures is the *particle effect through large-scale meteorology.*

The "Indirect Effect"

When pollution particles are emitted, more cloud-condensation nuclei (CCN) are available for water vapor to condense on. If all else is the same, the addition of more CCN produces more small cloud drops and fewer large cloud drops. For the same total volume of water, a large number of small drops has a greater cross-sectional area, summed among all drops, than does a small number of large drops. The greater cross-sectional area of small drops in comparison with large drops means that adding CCN to the air increases the reflectivity of sunlight, cooling the ground during the day (Twomey, 1977). An increase in the number of small drops and a decrease in the number of large drops due to the addition of pollution particles also reduces rates of drizzle in low clouds, thereby increasing the liquid water content and fractional cloudiness of such clouds, further cooling the surface during the day (Albrecht, 1989). These two effects of aerosol particles on clouds and, therefore, temperatures, are referred to as *indirect effects.*

The "Semidirect Effect"

Solar absorption by a low cloud increases stability below the cloud, reducing vertical mixing of moisture, thinning the cloud (Nicholls, 1984). Decreases in relative humidity correlate with decreases in low cloud cover (Bretherton et al., 1995; Klein, 1997). Similarly, absorbing particles warm the air, decreasing its relative humidity and increasing its stability, reducing low-cloud cover (Hansen et al., 1997; Ackerman et al., 2000). Reduced cloud cover increases sunlight reaching the surface, warming the surface in a process called the *semidirect effect* (Hansen et al., 1997).

The "BC-Low-Cloud-Positive Feedback Loop"

When BC reduces low-cloud cover by increasing stability and decreasing relative humidity, enhanced sunlight through the air is absorbed by BC, further heating the air and reducing cloud cover in a positive feedback loop, called the *BC-low-cloud positive feedback loop* (Jacobson, 2002). Whereas CO_2 also warms the air by absorbing thermal- and solar-near-IR radiation, reducing low cloud cover and enhancing surface solar radiation in some cases, it absorbs solar radiation much less effectively than does BC, so it partakes less in this positive feedback loop than does BC.

12.5 POSSIBLE CONSEQUENCES OF GLOBAL WARMING

$CO_2(g)$ mixing ratios are expected to increase from about 355 ppmv in 1990 to 540 to 970 ppmv in 2100. During that same period, global near-surface temperatures are estimated to increase by 1.4 to 5.8°C (IPCC, 2001). The possible consequences of global warming are discussed next.

12.5.1. Rise in Sea Level

Increases in temperatures affect sea levels in at least two ways. First, higher temperatures enhance the melting of ice sheets and glaciers, adding water to the oceans. Second, because liquid water density decreases with increasing temperature, higher temperatures cause water to expand and sea levels to rise. Historical changes in global temperature have been correlated with changes in sea levels. When temperatures peaked 120 to 90 m.y.a., during the mid-Cretaceous period, the Earth's polar caps melted, sea levels rose to unprecedented levels, and 20 percent of continental land flooded. Today, snow and ice cover 3.3 percent of the Earth's total surface area. The total ice volume is about 25 million km^3. If this ice melts, the sea level will rise 65 m above its current level. During the twentieth century, the sea level rose by about 10 to 25 cm. By the year 2100, the sea level is expected to rise by another 10 to 90 cm (IPCC, 2001).

Although the melting of ice sheets, glaciers, and sea ice and the corresponding rise in sea level are of concern, a large increase in sea level is unlikely to occur during the next 500 years. The largest sources of sea level rise would be the melting of the East and West Antarctic Ice Sheets, the Greenland Ice Sheet, sea ice over the Arctic, the large valley and piedmont glaciers of southeast Alaska, and the glaciers of central Asia. The West Antarctic Ice Sheet, based over water, is an order of magnitude smaller than is the East Sheet, based over land. Thus, the West Sheet is less stable than is the East Sheet (Stuiver et al., 1981). If melted, the East and West Sheets would raise the sea level 55 to 60 m (Denton et al., 1971), with the West Sheet responsible for about 5 m of this rise (Mercer, 1978). If extreme global warming occurs, the West Sheet, as a result of its relative instability, is more likely to collapse than is the East Sheet. A collapse of the West Sheet would probably take about 500 years (Bentley, 1984). Currently, the East Sheet may be increasing in size because of an increased water vapor supply to the sheet resulting from higher global temperatures (Bentley, 1984). Extended global warming could reverse this trend and ultimately cause a collapse of the sheet, increasing sea levels by 50 to 55 m. Such a process, though, is likely to take thousands of years (Crowley and North, 1991).

The main effect of sea level rise, even in small quantities, is the flooding of low-lying coastal areas and the elimination of a few flat islands that lie just above sea level. Bangladesh, the most densely populated country in the world, is particularly at risk. A 1 m rise in sea level would displace about 17 million people from their homes. New Orleans, Louisiana, which already lies below sea level, would similarly face a danger of flooding. Tuvulu (Fig. 12.21) is a chain of nine coral atolls in the South Pacific Ocean, about halfway between Hawaii and Australia. Tuvulu has a total land area of 26 km^2, about 0.1 times the size of Washington DC, a coastline that stretches for 24 km and a population of about 10,000. An increase in sea level of 2 m could eliminate the country.

Figure 12.21. Island of Fualopa, part of the Funafuti Atoll, one of nine coral atolls comprising the country of Tuvulu. Fualopa is set aside for conservation and uninhabited, but it and all other islands making up Tuvulu would be substantially eliminated if sea levels rose just a couple of meters. Photo courtesy of Peter Bennetts.

12.5.2. Changes in Regional Climate and Agriculture

Global warming is likely to cause regional and temporal variations in temperature. The number of extremely hot days is likely to increase and the number of extremely cold days is likely to decrease. Droughts will increase in some areas and floods in others. Precipitation intensity, averaged over the globe, and the number of extreme rainfall events are expected to increase (IPCC, 1995). Changes in regional climates are likely to shift the location of viable agriculture. Crops may flourish in areas that were once too cold or too dry, but they may also die in regions that become too hot or too wet. It is difficult to determine whether global warming will cause a net long-term increase or decrease in food supply, but it is fairly certain that locations of crop viability will shift (Wuebbles, 1995).

Because plants grow faster when temperatures, carbon dioxide levels, or water vapor levels mildly increase (plant–carbon dioxide negative feedback), it is expected that in areas where only mild changes in temperature and moisture occur, agriculture will flourish. In areas where extreme variations occur, agriculture will die out. Of particular concern are subtropical desert regions of Africa, where temperatures are already hot. In these regions, agriculture is subject to the whims of the climate, and millions of people depend on the local food supply. Small changes in climate could trigger famine, as has occurred in the past.

12.5.3. Changes in Ecosystems

Rapid, continuous increases in temperature could lead to the extinction of many species that are accustomed to narrow climate conditions and are unable to migrate faster than the rate of climate change. Although enhanced $CO_2(g)$ levels invigorate forests, sharp increases in temperature could lead to forest dieback in tropical regions, affecting the rates of $CO_2(g)$ removal by photosynthesis and emission by respiration.

12.5.4. Effects on Human Health

If global temperatures increase, people living in locations where temperatures are already hot are likely to experience more stress and heat-related health problems than are people living in milder climates. People currently living in cold climates are likely to experience less stress. Heat-related health problems, such as heat rash and heat stroke, generally affect the elderly and those suffering from illnesses more than they affect the general population. Increases in precipitation as a result of global warming could increase the populations of mosquitoes and other insects that carry diseases.

$CO_2(g)$, $CH_4(g)$, and $N_2O(g)$ cause no direct harmful human health problems at ambient mixing ratios (Section 3.6). Nevertheless, increases in the mixing ratios of these gases will affect human health indirectly through the effects of these gases on climate change and the effect of the resulting climate change on health. Particulate BC, another agent of global warming, will affect human health directly. BC is emitted primarily in submicron particles. Epidemiological studies have shown that long-term exposure to particles ≤ 2.5 μm in diameter above background levels causes increased mortality, increased disease, and decreased lung function in adults and children (Özkatnak and Thurston, 1987; U.S. EPA, 1996; Pope and Dockery, 1999).

12.5.5. Effects on Stratospheric Ozone

Figure 12.12 shows that whereas global warming has caused a warming of near-surface air since at least 1958, it has caused a cooling of the stratosphere. Cooling of the stratosphere affects the ozone layer in at least three ways.

First, cooling affects the rates of chemical reactions that produce and destroy ozone. Whereas many reactions proceed more slowly when temperatures decrease, the reaction $O(g) + O_2(g) + M \longrightarrow O_3(g) + M$ proceeds more rapidly when temperatures decrease. Thus, when gas chemistry alone is considered, a cooling of the stratosphere slightly increases global stratospheric ozone.

Second, cooling decreases the saturation vapor pressure (SVP) of water, allowing sulfuric acid–water aerosol particles in the background stratosphere to grow larger. The increase in size of these aerosol particles increases the rates at which heterogeneous reactions occur on their surfaces. Because such reactions produce chlorine gases that photolyze to products that destroy ozone, a decrease in stratospheric temperatures reduces global stratospheric ozone when only this effect is considered.

Third, a cooling of the stratosphere increases the occurrence, size, and lifetime of Polar Stratospheric Clouds (PSCs). Type I PSCs form at below 195 K and Type II PSCs form at below 187 K. Stratospheric cooling decreases temperatures below these critical levels during winter more frequently and for a longer period than when no cooling occurs, increasing Type I and Type II PSC lifetime, and size, enhancing Antarctic and Arctic ozone destruction during Southern and Northern Hemisphere springtime, respectively.

In sum, stratospheric cooling resulting from near-surface global warming has opposing effects on ozone in the global stratosphere, but it causes a net destruction of ozone over the Antarctic and Arctic.

12.6. REGULATORY CONTROL OF GLOBAL WARMING

Global warming is an international problem because all nations emit greenhouse gases and BC and are affected by changes in temperatures induced by such pollutants. Figure 12.22 shows that the countries emitting the most greenhouse gases in 1997 were the United States, China, and Russia. All countries in Central and South America and Africa combined emitted only 8.8 percent of the world total.

Figure 12.23 shows carbon emissions per capita by country in 1997. The figure shows that per capita emission were highest in oil-producing countries, particularly Qatar, the United Arab Emirates, and Kuwait. Yet, the populations of these countries are relatively small, so the total emission from them was also small. The United States stands out as having a high per capita emission rate as well as a large total emission rate.

Global warming is a divisive political issue because many industries depend on the combustion of fossil fuels for their viability, and switching to alternative sources of energy is often either uneconomical or otherwise undesirable. Many newly industrialized nations, as well, find that increasing the use of fossil fuels is the most economical means of expanding their economies.

Unlike with urban air pollution, acid rain, or global ozone problems, national and international governments have done relatively little to control global warming. In the United States, for example, no federal regulation to date has directly confronted the threat of global warming. The Regulation of vehicle emissions (e.g., Table 8.1) through the Clean Air Act Amendments has improved vehicle mileage, reducing carbon dioxide indirectly. United States federal tax code incentives since the late

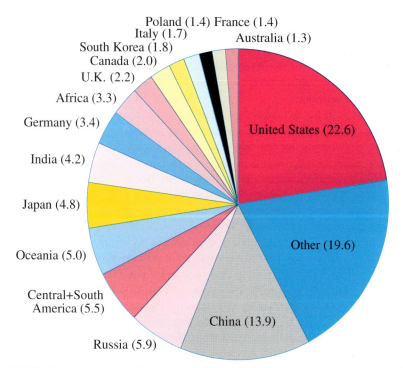

Figure 12.22. Percentage of world carbon dioxide emissions by country or continent, 1997. Source: Marland et al. (2000).

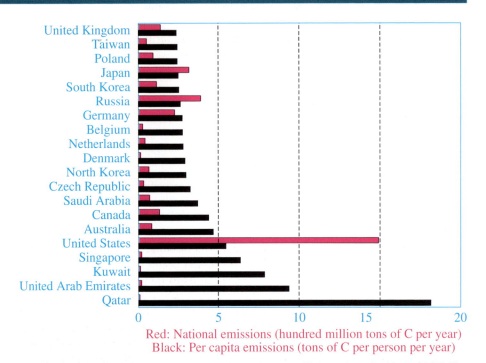

Red: National emissions (hundred million tons of C per year)
Black: Per capita emissions (tons of C per person per year)

Figure 12.23. Per capita and national emissions of carbon (C) in carbon dioxide in 1997. The figure includes only those countries emitting more than 10 million tons of C per year. Source: Marland et al. (2000).

1970s for the development of renewable energy and for improvements in energy efficiency have also addressed the issue indirectly. Nevertheless, simultaneous tax incentives have existed for the development of fossil fuel energy sources. In the 1980s, tax incentives for alternative energy sources were severely reduced, whereas those for fossil fuels were enhanced.

The 1990 Clean Air Act Amendments required the U.S. EPA to publish a list identifying the potential effect of CFCs and their alternatives on global warming, so the problem of global warming has been recognized increasingly. Whereas all countries tax fuels to some extent, Denmark, the Netherlands, Finland, Norway, and Sweden specifically implemented carbon taxes in the mid-1990s. Denmark taxed all carbon dioxide emission sources, except gasoline, natural gas, and biofuel. The Netherlands taxed all energy sources used as fuel. Finland taxed all fossil fuels. Norway taxed mineral oil, gasoline, gas burned in marine oil fields, coal, and coke. Sweden taxed oil, kerosene, natural gas, coal, coke, and other sources.

On May 9, 1992, an international agreement addressing global climate change, hashed out in Rio de Janeiro, Brazil, was adopted at the United Nations. The agreement, the **United Nations Framework Convention on Climate Change**, called on signatory nations to develop current and projected emission inventories for greenhouse gases, devise policies (to be implemented at a later meeting) for reducing greenhouse gas emissions, and promote technologies for reducing emissions. By 1994, 184 nations had signed the agreement and most had ratified it. In 1995, the nations involved in the convention met in Berlin, Germany, to discuss details of the proposed policies and target dates for implementing them.

Table 12.5. Percentage Change in Emissions Allowed for Industrialized Countries under the Kyoto Protocol

Countries	Percentage Change in Emissions
Switzerland, central Europe, and European Union	−8
United States	−7
Canada, Hungary, Japan, Poland	−6
Russia, New Zealand, Ukraine	0
Norway	+1
Australia	+8
Iceland	+10

In December 1997, the nations met again for an 11-day conference in Kyoto, Japan, to finalize the policies proposed at the Berlin meeting. The conference resulted in the Kyoto Protocol, an international agreement designed to control the emission of greenhouse gases. The Kyoto Protocol called for industrialized countries to reduce greenhouse gas emissions by 2008–2012. Such gases included carbon dioxide, methane, nitrous oxide, hydrofluorocarbons (HFCs), perfluorocarbons (PFCs), and sulfur hexafluoride (SF_6). Reductions in $CO_2(g)$, $CH_4(g)$, and $N_2O(g)$ emissions would be relative to 1990 emissions. Reductions in the emission of others gases would be relative to either 1990 or 1995 emissions. Table 12.5 lists the allocation of emission reductions mandated for industrialized countries. Some countries were allowed to increase emissions. The net change in emissions, weighted over all industrialized countries, was 5.2 percent. Countries would be allowed to meet the reductions in one of many ways. One mechanism would be to finance emission-reduction projects in developing countries, which were not subject to emission limits. Tree-planting, protecting forests, improving energy efficiency, using alternative energy sources, reforming energy and transportation sectors, creating technologies with zero emissions, and reducing emissions at their sources with existing technologies were other mechanisms.

By the end July 2001, 178 nations had signed the Kyoto Protocol. Although the United States, the largest emitter of $CO_2(g)$ in the world, had signed (but not ratified) the Protocol in 1997, the United States pulled out of the Protocol in 2001. One rationale for the pullout was that controlling $CO_2(g)$ emissions would damage the U.S. economy. The same argument was made by the automobile industry prior to the passage of the Clean Air Act Amendments of 1970, which required 90 percent reductions of three pollutants from automobiles by 1976: CO, NO_x, and hydrocarbons. Despite opposition to them, the Amendments motivated U.S. automobile manufacturers to invent the catalytic converter by 1975. The invention not only reduced pollutant emissions as mandated, but also produced profitable patents. Neither the U.S. economy nor the automobile industry suffered as a result of the 1970 regulations. From 1970–2000, for example, the number of vehicles in the United States doubled, whereas the population increased by only one-third. The U.S. Gross Domestic Product in fixed dollars also doubled and the unemployment rate decreased from 4.9 to 4.0 percent during this period. Stringent air pollution regulations under the Clean Air Act Amendments had no overall detrimental effect on the U.S. economy.

In the present case, a long-term method exists for the United States and other countries to reduce $CO_2(g)$ emissions toward satisfying the Kyoto Protocol. The method is to shift a portion of electric generation from coal and natural gas to wind power. Since the 1980s, wind turbine sizes and efficiencies have improved so that the direct cost of energy from the largest turbines is now 3 to 4 cents per kilowatt-hour (¢/kWh) when the turbines are placed in locations of sufficiently fast winds (Jacobson and Masters, 2001). This cost compares with that from new pulverized coal and natural gas power plants of 3.3 to 4 ¢/kWh. Given that estimated health, environmental, and global-warming costs of coal are another 2 to 4.3 ¢/kWh and of natural gas are another 1.2 to 2.2 ¢/kwh, the direct plus health/environmental cost of wind energy is less than is that of coal or natural gas energy. The United States could displace 10% of coal energy at no net cost if the federal government spent 3 to 4% of one year's budget on 36,000 to 40,000 large wind turbines and sold the electricity to consumers over 20 years. U.S. Kyoto Protocol greenhouse gas reductions could be met as of 1999 by replacing 59% of coal energy with energy from 214,000 to 236,000 large turbines. At 6 per square kilometer, the turbines could be spread over 194 $\times$ 194 km^2 of farmland or ocean. Whereas replacing coal with wind would displace coal workers, it would provide new jobs for other industries. More important, it would reduce the health and environmental problems associated with coal mining and combustion (black lung disease, effects of strip mining, particulate health problems and mortality, acid deposition, urban air pollution, visibility degradation, and global warming).

Whereas, the Kyoto Protocol is a necessary first step to combat global warming, the protocol must be stronger to slow global warming effectively. The protocol would be more effective if more countries were required to reduce emissions and if the level of emission reductions required of each country were increased. In addition, the protocol does not call for reducing the emissions of black carbon. Controlling BC is possibly the most effective method of slowing global warming (Jacobson, 2002). The benefit of controlling BC is illustrated in Figure 12.24. The figure compares the time-dependent effects on global temperatures of eliminating emissions of fossil-fuel black carbon plus organic matter (ff. BC+OM), anthropogenic $CO_2(g)$, and anthropogenic $CH_4(g)$. Removal of OM was included with removal of BC because all fossil-fuel sources that emit BC also emit OM, and it would be difficult to remove BC without removing OM simultaneously. The figure shows that removing f.f. BC+OM could cool climate by greater than 0.3 K after three years. This rapid decrease in temperatures arises from the fact that eliminating f.f. BC+OM emissions reduces the atmospheric concentration of f.f. BC+OM quickly (because particles have short atmospheric lifetimes), and temperatures respond quickly to changes in concentration. The atmospheric lifetime of $CO_2(g)$ ranges from 50 to 200 years. Thus, the elimination of $CO_2(g)$ emissions affects temperatures over a long period. Figure 12.24 implies that any emission reduction of f.f. BC+OM will slow global warming more than will any emission reduction of $CO_2(g)$ or $CH_4(g)$ for a specific period. When all f.f. BC+OM and anthropogenic $CO_2(g)$ and $CH_4(g)$ emissions are eliminated together, that period is 25 to 100 years.

The final cooling values in Fig. 12.24 imply that the net global warming from $CO_2(g)$, $CH_4(g)$, and f.f. BC+OM is about +1.5 K. Other greenhouse gases, if scalable to $CO_2(g)$, bring the total warming to +1.9 K. About 1.2 K of this warming is estimated to be offset by cooling and feedbacks due to other anthropogenic particles, such as sulfate, nitrate, secondary organics, fly ash, soil dust, and certain biomass-burning particles, giving the net warming due to greenhouse gases, and f.f. BC+OM as +0.7 K. Natural variations in climate may cause either warming or cooling, but the net effect between 1860–2000 may be near zero (Stott et al., 2000). Observed warming from 1856–1999, as shown in Figure 12.11 is +0.75 ± 0.15K. Thus, observed net global warming can be roughly reconciled by considering greenhouse gases, f.f. BC+OM,

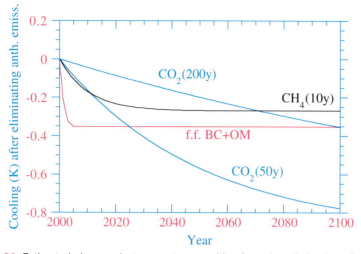

Figure 12.24. Estimated changes in temperature resulting from the elimination of all anthropogenic emissions of each CO_2(g) CH_4(g) and f.f BC+OM during the next 100 years (Jacobson, 2002). The numbers in parentheses represent estimates of the atmospheric lifetimes of the gases. In the case of CO_2(g), the lifetime is 50 to 200 years. The net cooling due to CO_2(g) after 1000 years under both assumptions is about 0.9 K. Because of BC+OM's short lifetime (about a week to a month), the elimination of its emissions cools climate almost immediately, but climate feedbacks allow its removal to affect temperatures over a longer period.

and other particulates, implying that non-BC+OM cooling masks much of greenhouse gas and f.f. BC+OM warming. The figure also implies that eliminating all f.f. BC+OM with no other change might reduce more than 40 percent of net global warming to date. Reducing CO_2(g) emissions by a third would have the same effect, but over 50 to 200 years.

Despite its drawbacks, the Kyoto Protocol, is an important first attempt to control a damaging environmental problem. Failure of countries responsible for most emissions to adopt this or a stronger treaty will undoubtedly exacerbate the problem of global warming.

12.7. SUMMARY

In this chapter, the greenhouse effect, global warming, historical temperature trends, possible consequences of global warming, and regulatory control of global warming were discussed. Greenhouse gases selectively absorb thermal-IR radiation but are transparent to visible radiation. Without the presence of natural greenhouse gases, particularly water vapor and carbon dioxide, the Earth would be too cold to support higher life forms. Global warming is the increase in the Earth's temperature above that caused by natural greenhouse gases. Greenhouse gases that contribute the most to global warming are carbon dioxide, methane, and nitrous oxide. The second-most important component of near-surface global warming, after carbon dioxide and before methane, may be particulate black carbon (BC), which absorbs solar as well as thermal IR radiation. Anthropogenic emissions and ambient levels of greenhouse gases and BC have increased since the mid-1800s. Air temperatures have also increased.

Whereas temperatures over the Earth's history have frequently been higher than they are today, the rate of temperature increase today is relatively unusual, considering the fact that the Earth is not in a period of deglaciation. The consequences of increasing temperatures over the next 100 years are expected to be a small rise in sea level, shifts in agriculture, damage to ecosystems, and heat-related health problems. Recently, the international community attempted, through the Kyoto Protocol to control global warming by reducing emissions of greenhouse gases (but not BC). Because the reductions mandated are small, this effort may result in only minor modifications to the increases in temperature expected from global warming.

12.8. PROBLEMS

12.1. Calculate the effective temperature of the Earth's surface in the absence of a greenhouse effect assuming $A_e = 0.4$ and $\varepsilon_e = 1.0$. By what factor does the result change if the Earth–sun distance is doubled?

12.2. What is the relative ratio of energy received at the top of Mars's and Venus's atmosphere compared with that at the top of the Earth's atmosphere?

12.3. Explain, in terms of atmospheric components and distance from the sun, why Mars's actual temperatures are colder and Venus's actual temperatures are much warmer than are those of the Earth.

12.4. What would the equilibrium temperature of the Earth be if it were in Mars's orbit? What about if it were in Venus's orbit? Assume no other characteristics of the Earth changed.

12.5. Discuss at least two ways that deforestation can affect global warming.

12.6. Explain how CFCs, HCFCs, and HFCs might contribute to global warming.

12.7. Explain why greenhouse gases may cause an increase in near-surface temperatures but a decrease in stratospheric temperatures.

12.8. Explain why the Earth's temperature during the Archean period might have been much warmer than today's temperature, although the solar output was lower than today's solar output.

12.9. Why are scientists more concerned about the collapse of the West than the East Antarctic Ice Sheet?

12.10. Explain the common theory as to how the dinosaurs became extinct.

12.11. Explain the fundamental reason for the four glacial periods that have occurred during the past 450,000 years. Did carbon dioxide cause these events, or did its mixing ratio change in response to them? Why?

12.12. What is unusual about the rate of change in the Earth's near-surface temperature since the late 1950s compared with historic rates of temperature change?

12.13. How would a change in the Earth's obliquity to zero (no tilt) affect seasons? How would it affect the relative amount of sunlight over the South Pole during the Southern Hemisphere winter in comparison with today?

12.14. If the Earth's temperature initially decreases and only the snow-albedo feedback is considered, what will happen to the temperature subsequently?

12.15. If the Earth's temperature initially increases and only the plant–carbon dioxide feedback is considered, what will happen to the temperature subsequently?

12.16. How will absorbing aerosol particles in the boundary layer affect air pollution buildup and cloud formation if only the effects of the aerosol particles on atmospheric stability are considered?

12.9. ESSAY QUESTIONS

12.17. Identify at least six activities that you do or products that you consume that result in the release of one or more greenhouse gases, and identify ways that you can reduce emissions.

12.18. Identify two arguments against global warming and two arguments for global warming.

12.19. Discuss three possible effects of global warming. Would any of these effects affect your life or lifestyle?

12.20. If you think the global warming problem is an important issue, what specific steps do you think your national and local governments should take to address the issue? If you do not think it is an important issue, what steps do you think scientists should take to understand the issue better?

APPENDIX
CONVERSIONS AND CONSTANTS

1 bar	$= 10^3$ mb	$= 0.986923$ atm	$= 10^5$ N m^{-2}
	$= 10^5$ J m^{-3}	$= 10^5$ Pa	$= 10^5$ kg m^{-1} s^{-2}
	$= 10^6$ dyn cm^{-2}	$= 10^6$ g cm^{-1} s^{-2}	$= 750.06$ torr
	$= 750.06$ mm Hg		
1 atm	$= 1.01325$ bar	$= 760$ torr	$= 760$ mm Hg

A.1.2. ENERGY CONVERSIONS

1 J	$= 1$ N m	$= 10^7$ erg	$= 1$ W s
	$= 10^4$ mb cm^3	$= 10^7$ dyn cm	$= 0.239$ cal
	$= 1$ kg m^2 s^{-2}	$= 10^7$ g cm^2 s^{-2}	$= 10^{-5}$ bar m^3
	$= 6.25 \times 10^{18}$ eV	$= 1$ C V	

A.1.3. SPEED CONVERSIONS

1 m s^{-1}	$= 100$ cm s^{-1}	$= 3.6$ km h^{-1}	$= 1.9459$ knots
	$= 2.2378$ mi hr^{-1}		

A.1.4. CONSTANTS

A	$=$ Avogadro's number	$= 6.02213 \times 10^{23}$ molecules mole^{-1}
c	$=$ speed of light	$= 2.99792 \times 10^8$ m s^{-1}

F_s = solar constant   = 1,365 W m^{-2}

k_B = Boltzmann's constant (R^*/A)   = 1.3807 $\times$ 10^{-23} J K^{-1}

     = 1.3807 $\times$ 10^{-23} kg m^2 s^{-2} K^{-1} molecule^{-1} = 1.3625 $\times$ 10^{-22} cm^3 atm K^{-1}

     = 1.3807 $\times$ 10^{-16} g cm^2 s^{-2} K^{-1} molecule^{-1} = 3.299 $\times$ 10^{-24} cal K^{-1}

     = 1.3807 $\times$ 10^{-19} cm^3 mb K^{-1} molecule^{-1} = 1.3625 $\times$ 10^{-25} L atm K^{-1} molecule^{-1}

     = 1.3625 $\times$ 10^{-22} cm^3 atm K^{-1} molecule^{-1} = 1.3807 $\times$ 10^{-25} m^3 mb K^{-1} molecule^{-1}

g = gravity   = 9.81 m s^{-2}

m_d = molecular weight of dry air   = 28.966 g mole^{-1}

$\overline{M}$ = mass of an air molecule (m_d/A)   = 4.8096 $\times$ 10^{-26} kg

R^* = universal gas constant   = 8.3145 J mole^{-1} K^{-1}

     = 8.3145 kg m^2 s^{-2} mole^{-1} K^{-1} = 0.083145 m^3 mb mole^{-1} K^{-1}

     = 8.3145 $\times$ 10^7 g cm^2 s^{-2} mole^{-1} K^{-1} = 0.08206 L atm mole^{-1} K^{-1}

     = 8.3145 $\times$ 10^4 cm^3 mb mole^{-1} K^{-1} = 8.3145 $\times$ 10^7 erg mole^{-1} K^{-1}

     = 82.06 cm^3 atm mole^{-1} K^{-1} = 1.987 cal mole^{-1} K^{-1}

R' = gas constant for dry air (R^*/m_d) = 287.04 J kg^{-1} K^{-1}

     = 0.28704 J g^{-1} K^{-1} = 2.8704 m^3 mb kg^{-1} K^{-1}

     = 2870.4 cm^3 mb g^{-1} K^{-1} = 287.04 m^2 s^{-2} K^{-1}

     = 2.8704 $\times$ 10^6 cm^2 s^{-2} K^{-1}

R_e = radius of the Earth   = 6.378 $\times$ 10^6 m

R_p = radius of the sun   = 6.96 $\times$ 10^8 m

R_{es} = mean Earth–sun distance   = 1.5 $\times$ 10^{11} m

T_p = temperature of the sun's photosphere = 5,785 K

σ_B = Stefan–Boltzmann constant   = 5.67051 $\times$ 10^{-8} W m^{-2} K^{-4}

REFERENCES

ACEA (1999) *ACEA Programme on Emissions of Fine Particles from Passenger Cars.* ACEA, Brussels.

Ackerman A. S., Toon O. B., Stevens D. E., Heymsfield A. J., Ramanathan V., and Welton E. J. (2000) Reduction of tropical cloudiness by soot. *Science* **288,** 1042–47.

Ackerman T. P. (1977) A model of the effect of aerosols on urban climates with particular applications to the Los Angeles Basin. *J. Atmos. Sci.* **34,** 531–47.

Ahrens C. D. (1994) *Meteorology Today,* West Publishing Co., St. Paul, Minn.

Albrecht B. A. (1989) Aerosols, cloud microphysics, and fractional cloudiness. *Science* **245,** 1227–30.

Allen P. and Wagner K. (1992) 1987 California Air Resources Board emissions inventory, magnetic tapes ARA806, ARA807.

Alvarez L. W., Alvarez W., Asaro F., and Michel H. V. (1980) Extraterrestrial cause for the Cretaceous–Tertiary extinction. *Science* **208,** 1095–1108.

Alternate fluorocarbons environmental acceptability study (AFEAS) (2001) web site at http://www.afeas.org.

American Wind Energy Association (AWEA) (1999) Global wind energy market report, http://www.awea.org.

Anderson I., Lundquist G. R., and Molhave L. (1975) Indoor air pollution due to chipboard used as a construction material. *Atmos. Environ.* **9,** 1121–27.

Andreae M. O., Charlson R. J., Bruynseels F., Storms H., Van Grieken R., and Maenhaut W. (1986) Internal mixture of sea salt, silicates, and excess sulfate in marine aerosols. *Science* **232,** 1620–23.

Andreae M. O., Andreae T. W., Annegarn H., Beer J., Cachier H., le Canut P., Elbert W., Maenhaut W., Salma I., Wienhold F. G., and Zenker T. (1998) Airborne studies of aerosol emissions from savanna fires in southern Africa: 2. Aerosol chemical composition. *J. Geophys. Res.* **103,** 32, 119–28.

Andreae M. O. and Merlet P. (2001) Emission of trace gases and aerosols from biomass burning. *Glob. Biogeochem. Cyc.* **15,** 955–66.

Andrews E. and Larson S. M. (1993) Effect of surfactant layers on the size changes of aerosol-particles as a function of relative humidity. *Environ. Sci. Technol.* **27,** 857–65.

Angell J. K. (1999) Comparison of surface and tropospheric temperature trends estimated from a 63-station radiosonde network, 1958–1998. *Geophys. Res. Lett.* **26,** 2761–64.

Angell J. K. (2000) Global, hemispheric, and zonal temperature deviations derived from radiosonde records, In *Trends Online: A compendium of data on global change.* Carbon Dioxide Information Analysis Center, Oak Ridge National Laboratory, U.S. Department of Energy, Oak Ridge, Tenn.

Arashidani K., Yoshikawa M., Kawamoto T., Matsuno K., Kayam F., and Kodama Y. (1996) Indoor pollution from heating. *Indust. Health* **34,** 205–15.

Arrhenius S. (1896) On the influence of carbonic acid in the air upon the temperature of the ground. *Philos. Mag.* **41,** 237.

Ayars G. H. (1997) Biological agents and indoor air pollution. In Bardana E. J. and Montanaro A., eds., *Indoor Air Pollution and Health.* Marcel Dekker, New York, pp. 11–60.

Barron E. J., Sloan J. L., and Harrison C. G. A. (1980) Potential significance of land–sea distribution and surface albedo variations as a climatic forcing factor: 180 m.y. to the present. *Palaeogeog. Palaeoclim. Palaeoecol.* **30,** 17–40.

Bentley C. R. (1984) Some aspects of the cryosphere and its rose in climatic change. In *Climate Processes and Climate Sensitivity,* Hansen J. E. and Takahashi T., eds. *Geophys. Mono.* **29,** AGU, Washington, DC, pp. 207–20.

Bergstrom R. and Viskanta R. (1973) Modelling of the effects of gaseous and particulate pollutants in the urban atmosphere, Part I, Thermal structure. *J. Appl. Meteorol.* **12,** 901–12.

Bergthorsson P. (1969) An estimate of drift ice and temperature in 1000 years. *Jökull* **19,** 94–101.

Berner R. A., Lasaga A. C., and Garrels R. M. (1983) The carbonate–silicate geochemical cycle and its effects on atmospheric carbon dioxide over the last 100 million years. *Am. J. Sci.* **283,** 641–83.

Berner A., Sidla S., Galambos Z., Kruisz C., Hitzenberger R., ten Brink H. M., and Kos G. P. A. (1996) Modal character of atmospheric black carbon size distributions. *J. Geophys. Res.* **101,** 19,559–65.

Bhatti M. S. (1999) A historical look at chlorofluorocarbon refrigerants. *ASHRAE Transactions* **105,** 1186–1208.

Blatchford R. (1899) *Dismal England,* p. 15.

Bornstein R. and Lin Q. (2000) Urban heat islands and summertime convective thunderstorms in Atlanta: Three case studies. *Atmos. Environ.* **34,** 507–16.

Bradley R. S. and Jones P. D. (1993) Little Ice Age summer temperature variations: their nature and relevance to recent global warming trends. *The Holocene* **3,** 367–76.

Bretherton C. S., Klinker E., Betts A. K., and Coakley J. (1995) Comparison of ceilometer, satellite, and synoptic measurements of boundary layer cloudiness and the ECMWF diagnostic cloud parameterization scheme during ASTEX. *J. Atmos. Sci.,* **52,** 2736–51.

Brimblecombe P. (1987) *The Big Smoke.* Methuen, London.

Brimblecombe P. (1999) Air pollution and health history. In *Air Pollution and Health,* Holgate S. T., Samet J. M., Koren H. S., and Maynard R. L., eds. Academic Press, San Diego, pp. 5–18.

Brock W. H. (1992) *The Norton History of Chemistry.* W. W. Norton & Company, Inc., New York.

Brook G. A., Folkoff M. E., and Box E. O. (1983) A world model of soil carbon dioxide. *Earth Surf. Proc. Landforms,* **8,** 79–88.

Brooks B. O., Utter G. M., DeBroy J. A., and Schimke R. D. (1991) Indoor air pollution: An edifice complex. *Clin. Toxicol.* **29,** 315–74.

Brown J. C. (1913) *A History of Chemistry.* P. Blakiston's Son & Co., Philadelphia.

Burr M. L., St.-Leger A. S., and Yarnell J. W. G. (1981) Wheezing, dampness, and coal fires. *Commun. Med.* **3,** 205–09.

Caldwell M. M., Bjorn L. O., Bornman J. F., Flint S. D., Kulandaivelu G., Teramura A. H., and Tevini M. (1998) Effects of increased solar ultraviolet radiation on terrestrial ecosystems. *J. Photochem. Photobiol B: Biol.* **46,** 40–52.

Campbell F. W. and Maffel L. (1974) Contrast and spatial frequency. *Sci. Am.,* **231,** 106–14.

Carter W. P. L. (1991) Development of ozone reactivity scales for volatile organic compounds, EPA-600/3-91-050. U.S. Environmental Protection Agency, Research Triangle Park, NC.

Cass G. R. (1979) On the relationship between sulfate air quality and visibility with examples in Los Angeles. *Atmos. Environ.* **13,** 1069–84.

Cattermole P. and Moore P. (1985) *The Story of the Earth.* Cambridge University Press, New York.

Chang E., Nolan K., Said M., Chico T., Chan S., and Pang E. (1991) 1987 emissions inventory for the South Coast Air Basin: Average annual day. South Coast Air Quality Management District (SCAQMD), Los Angeles.

Chang S. G., Brodzinsky R., Gundel L. A., and Novakov T. (1982) Chemical and catalytic properties of elemental carbon. In *Particulate Carbon: Atmospheric Life Cycle,* Wolff G. T. and Klimsch R. L., eds. Plenum Press, New York, pp. 158–81.

Chapman S. (1930) A theory of upper-atmospheric ozone. *Mem. Royal Met. Soc.* **3,** 104–25.

Choi I. S. (1983) Delayed neurological squelae in carbon monoxide intoxication. *Arch. Neurol.* **40,** 433–35.

Clapp B. W. (1994) *An Environmental History of Britain.* Longman, London.

Clark D. L., Whitman R. R., Morgan K. A., and Mackey S. D. (1980) Stratigraphy and glacial–marine sediments of the Amerasian Basin, central Arctic Ocean. *Geol. Soc. Am. Spec. Pap.* **181.**

CLIMAP Project Members (1984) The last interglacial ocean. *Quart. Res.* **21,** 123–224.

Cohen B. S. (1998) Deposition of charged particles on lung airways. *Health Phys.* **74,** 554–60.

Cohen A. J. and Nikula K. (1999) The health effects of diesel exhaust: Laboratory and epidemiologic studies. In *Air Pollution and Health,* Holgate S. T., Samet J. M., Koren H. S., and Maynard R. L., eds. Academic Press, San Diego, pp. 707–45.

Cooke W. F. and Wilson J. J. N. (1996) A global black carbon aerosol model. *J. Geophys. Res.* **101,** 19,395–409.

Crowley T. J. (1983) The geologic record of climatic change. *Rev. Geophys. Space Phys.* **21,** 828–77.

Crowley T. J. and North G. R. (1991) *Paleoclimatology.* Oxford University Press, New York.

Davidson C. I., Tang W., Finger S., Etyemezian V., Striegel M. F., and Sherwood S. I. (2000) Soiling patterns on a tall limestone building: Changes over 60 years. *Environ. Sci. Technol.* **34,** 560–65.

Dentener F. J., Carmichael G. R., Zhang Y., Lelieveld J., and Crutzen P. J. (1996) Role of mineral aerosol as a reactive surface in the global troposphere. *J. Geophys. Res.* **101,** 22,869–889.

Denton G. H., Armstrong R. L., and Stuiver M. (1971) The late Cenozoic glacial history of Antarctica. In *The Late Cenozoic Glacial Ages,* Turekian K. K., Ed. (1971) Yale University Press, New Naven, Conn., pp. 267–306.

Derwent R. G. and Jenkin M. E. (1991) Hydrocarbons and the long-range transport of ozone and PAN across Europe. *Atmos. Environ.* **25A,** 1661–78.

Dickerson R. R., Doddridge B. G., Kelley P., and Rhoads K. P. (1995) Large-scale pollution of the atmosphere over the remote Atlantic Ocean: Evidence from Bermuda. *J. Geophys. Res.* **100,** 8945–52.

Dickerson R. R., Kondragunta S., Stenchikov G., Civerolo K. L., Doddridge B. G., and Holben B. N. (1997) The impact of aerosols on solar UV radiation and photochemical smog. *Science* **278,** 827–30.

Didyk B. M., Simoneit B. R. T., Pezoa L. A., Riveros M. L., and Flores A. A. (2000) Urban aerosol particles of Santiago, Chile: Organic content and molecular characterization. *Atmos. Environ.* **34,** 1167–79.

Dockery D. W., Pope III C. A., Xu X., Spengler J. D., Ware J. H., Fay M. E., Ferris, Jr. B. G., and Speizer F. E. (1993) An association between air pollution and mortality in six U.S. cities. *N. Engl. J. Med.* **329,** 1753–59.

Duce R. A. (1969) On the source of gaseous chlorine in the marine atmosphere. *J. Geophys. Res.* **70,** 1775–9.

Egloff G. (December, 1940) *Motor Fuel Economy of Europe.* American Petroleum Institute, Washington, DC.

Energy Information Administration (EIA) (2000) http://www.eia.doe.gov, U.S. Energy Information Administration, Washington, DC.

Environment Canada (2000a) Government of Canada actions on clean air. http://www.ec.gc.ca/press/000519h_f_e.htm.

Environment Canada (2000b) Acid rain update. http://www.ec.gc.ca/press/000519d_f_e.htm.

Eriksson E. (1960) The yearly circulation of chloride and sulfur in nature: Meteorological, geochemical and pedological implications. Part I. *Tellus* **12,** 63–109.

Etherington D. J., Pheby D. F., and Bray F. I. (1996) An ecological study of cancer incidence and radon levels in south west England. *Euro. J. Canc.* **32,** 1189–97.

Ethridge D. M., Pearman G. I., and Fraser P. J. (1992) Changes in tropospheric methane between 1841 and 1978 from a high accumulation rate Antarctic ice core. *Tellus* **44B,** 282–94.

Fairbanks R. G. (1989) A 17,000-year glacio-eustatic sea level record: Influence of glacial melting rates on Younger Dryas event and deep-ocean circulation. *Nature* **342,** 637–42.

Fang M., Zheng M., Wang F., To K. L., Jaafar A. B., and Tong S. L. (1999) The solvent-extractable organic compounds in the Indonesia biomass burning aerosols – characterization studies. *Atmos. Environ.* **33,** 783–95.

Farman J. C., Gardiner B. G., and Shanklin J. D. (1985) Large losses of total ozone in Antarctica reveal seasonal ClO_x/NO_x interaction. *Nature* **315,** 207–10.

Federal Trade Commission (1936) Docket No. 2825, Cushing Refining & Gasoline Co, June 19, 1936, Dept. of Justice files, 60-57-107, National Archives, Washington, DC.

Ferek R. J., Reid J. S., Hobbs P. V., Blake D. R., and Liousse C. (1998) Emission factors of hydrocarbons, halocarbons, trace gases, and particles from biomass burning in Brazil. *J. Geophys. Res.* **103,** 32, 107–18.

Finlayson-Pitts B. J. and Pitts, Jr. J. N. (1986) *Atmospheric Chemistry: Fundamentals and Experimental Techniques.* John Wiley & Sons, Inc., New York.

Finlayson-Pitts B. J. and Pitts, Jr. J. N. (1999) *Chemistry of the Upper and Lower Atmosphere.* Academic Press, San Diego.

Fitzner C. A., Schroeder J. C., Olson R. F., and Tatreau P. M. (1989) Measurement of ozone levels by ship along the eastern shore of lake Michigan. *J. Air Pollut. Control Assoc.* **39,** 727–8.

Fleming E. L., Chandra S., Schoeberl M. R., and Barnett J. J. (1988) Monthly mean global climatology of temperature, wind, geopotential height, and pressure for 1–120 km. Tech. Memo. 100697, NASA, 85 pp.

Frakes L. A. (1979) *Climates Throughout Geologic Time.* Elsevier, Amsterdam.

Framton M. W., Morrow P. E., Cox C., Gibb F. R., Speers D. M., and Utell M. J. (1991) Effects of nitrogen dioxide exposure on pulmonary function and airway reactivity in normal humans. *Am. Rev. Respir. Disord.* **143,** 522–7.

Friedli H., Lötscher H., Oeschger H., Siegenthaler U., and Stauffer B. (1986) Ice core record of 13C/12C ratio of atmospheric CO_2 in the past two centuries. *Nature* **324,** 237–8.

Garrels R. M. and Perry E. A. (1974) Cycling of carbon, sulphur, and oxygen through geologic time, In The Sea v. 5, E. D. Goldberg, Ed., John Wiley and Sons, New York, pp. 303–36.

Gerritson S. L. (1993) The status of the modeling of ozone formation and geographic movement in the Midwest. In *Cost Effective Control of Urban Smog,* Kosobud R. F., Testa W. A., and Hanson D. A., eds. Federal Reserve Bank of Chicago.

Gharib S. and Cass G. R. (1984) Ammonia emissions in the South Coast Air Basin 1982, Open File Report 84-2, Environmental Quality Lab. California Institute of Technology, Pasadena, California.

Ghio A. J. and Samet J. M. (1999) Metals and air pollution particles. In *Air Pollution and Health,* Holgate S. T., Samet J. M., Koren H. S., and Maynard R. L., eds. Academic Press, San Diego, pp. 635–51.

Gillette D. A., Patterson Jr. E. M., Prospero J. M., and Jackson M. L. (1993) Soil aerosols. In *Aerosol Effects on Climate,* Jennings S. G., ed. University of Arizona Press, Tucson, pp. 73–109.

Gilliland R. L. (1989) Solar evolution. *Glob. Plan. Change* **1**, 35–56.

Glass N. R., Glass G. E., and Rennie P. J. (1979) Effects of acid precipitation. *Environ. Sci. Technol.* **13**, 1350–55.

Gold D. R. (1992) Indoor air pollution. *Clin. Chest Med.* **13**, 215–229.

Goldstein I. F., Andrews L. R., and Hartel D. (1988) Assessment of human exposure to nitrogen dioxide, carbon monoxide, and respirable particles in New York inner-city residences. *Atmos. Environ.* **22**, 2127–39.

Goody R. (1995) *Principles of Atmospheric Physics and Chemistry.* Oxford University Press, New York.

Graedel T. E. and Weschler C. J. (1981) Chemistry within aqueous atmospheric aerosols and raindrops. *Rev. Geophys. Space Phys.* **19**, 505–39.

Graves C. K. (1980) Rain of troubles. *Science* **80**, 74–9.

Green H. and Lane W. (1969) *Particle Clouds.* Van Nostrand, NJ.

Greenberg R. R., Zoller W. H., and Gordon G. E. (1978) Composition and size distributions of articles released in refuse incineration. *Environ. Sci. Technol.* **12**, 566–73.

Griffing G. W. (1980) Relations between the prevailing visibility, nephelometer scattering coefficient and sunphotometer turbidity coefficient. *Atmos. Environ.* **14**, 577–84.

Groblicki P. J., Wolff G. T., and Countess R. J. (1981) Visibility-reducing species in the Denver "brown cloud." I. Relationships between extinction and chemical composition. *Atmos. Environ.* **15**, 2473–84.

Hader D.-P., Kumar H. D., Smith R. C., and Worrest R. C. (1998) Effects on aquatic ecosystems. *J. Photochem. Photobiol. B: Biology* **46**, 53–68.

Hale G. M. and Querry M. R. (1973) Optical constants of water in the 200-nm to 200-μm wavelength region. *Appl. Opt.* **12**, 555–63.

Hall J. V., Winer A. M., Kleinman M. T., Lurmann F. W., Brajer V., and Colome S. D. (1992) Valuing the health benefits of clean air. *Science* **255**, 812–17.

Hampson S. E., Andres J. A., Lee M. E., Foster L. S., Glasgow R. E., and Lichtenstein E. (1998) Lay understanding of synergistic risk: The case of radon and cigarette smoking. *Risk Anal.* **18**, 343–50.

Hansen J., Sato M., and Ruedy R. (1997) Radiative forcing and climate response. *J. Geophys. Res.* **102**, 6831–64.

Hansen J., Sato M., Ruedy R., Lacis A., and Oinas V. (2000) Global warming in the twenty-first century: An alternative scenario. *Proc. Natl. Acad. Sci.* **97**, 9875–80.

Hargis A. M. (1981) A review of solar induced lesions in domestic animals. *Compend. Contin. Educ.* **3**, 287–300.

Harlap S. and Davies A. M. (1974) Infant admissions to hospital and maternal smoking. *Lancet* **1**, 529–32.

Hartmann D. L. (1994) *Global Physical Climatology.* Academic Press, Inc., San Diego.

Hayami H. and Carmichael G. R. (1997) Analysis of aerosol composition at Cheju Island, Korea, using a two-bin gas–aerosol equilibrium model. *Atmos. Environ.* **31**, 3429–39.

Hayhoe K., Jain A., Pitcher H., MacCracken C., Gibbs M., Wuebbles D., Harvey R., and Kruger D. (1999) Cost of multigreenhouse gas reduction targets for the U.S.A. *Science* **286**, 905.

Heinsohn R. J. and Kabel R. L. (1999) *Sources and Control of Air Pollution.* Prentice Hall, Upper Saddle River, New Jersey.

Henry W. M. and Knapp K. T. (1980) Compound forms of fossil fuel fly ash emissions. *Environ. Sci. Technol.* **14**, 450–6.

Henshaw D. L., Eatough J. P., and Richardson R. B. (1990) Radon as a causative factor in induction of myeloid leukaemia and other cancers. *Lancet* **335**, 1008–12.

Hering S. V. and Friedlander S. K. (1982) Origins of aerosol sulfur size distributions in the Los Angeles Basin. *Atmos. Environ.* **16**, 2647–56.

Hines A. L., Ghosh T. K., Loyalka S. K., and Warder R. C., eds. (1993) *Indoor Air-Quality and Control.* Prentice-Hall, Englewood Cliffs, New Jersey.

Hitzenberger R. and Puxbaum H. (1993) Comparisons of the measured and calculated specific absorption coefficients for urban aerosol samples in Vienna. *Aerosol Sci. Technol.* **18**, 323–45.

Hitchcock D. R., Spiller L. L., and Wilson W. E. (1980) Sulfuric acid aerosols and HCl release in coastal atmospheres: Evidence of rapid formation of sulfuric acid particulates. *Atmos. Environ.* **14**, 165–82.

Hong S., Candelone J.-P., Patterson C. C., and Boutron C. F. (1996) History of ancient copper smelting pollution during Roman and Medieval times recorded in Greenland ice. *Science* **272**, 246–8.

Houghton R. A. (1991) Biomass burning from the perspective of the global carbon cycle. In *Global Biomass Burning: Atmospheric, Climatic, and Biospheric Implications,* Levine J. S., ed. MIT Press, Cambridge, Mass., pp. 321–5.

Houghton R. A. (1994) Effects of land-use change, surface temperature and CO_2 concentrations on terrestrial stores of carbon. In *Biotic Feedbacks in the Global Climate System: Will the Warming Speed the Warming?* Woodwell G. M. and Mackenzie F. T., eds. Oxford University Press, Oxford.

Hughes J. D. (1994) *Pan's Travail: Environmental Problems of the Ancient Greeks and Romans.* The Johns Hopkins University Press, Baltimore.

Hurrell J. W. and Trenberth K. E. (1997) Spurious trends in satellite MSU temperatures from merging different satellite records. *Nature* **386**, 164–7.

Intergovernmental Panel on Climate Change (IPCC) (2001) *Third Assessment Report. Climate Change 2001: The Scientific Basis.* Houghton, J. T. et al., eds. Cambridge University Press, New York, 881 pp.

Islam M. S. and Ulmer W. T. (1979) Threshold concentrations of SO_2 for patients with oversensitivity of the bronchial system. *Wissenschaft and Umwelt* **1**, 41–7.

Jackman C. H., Fleming E. L., Chandra S., Considine D. B., and Rosenfield J. E. (1996) Past, present, and future modeled ozone trends with comparisons to observed trends. *J. Geophys. Res.* **101**, 28,753–67.

Jacob D. J. (1985) Comment on "The photochemistry of a remote stratiform cloud," by W. L. Chameides. *J. Geophys. Res.* **90**, 5864.

Jacob D. J., Logan J. A., and Murti P. (1999) Effect of rising Asian emissions on surface ozone in the United States. *Geophys. Res. Lett.* **26**, 22,175–78.

Jacobson M. Z. (1997a) Development and application of a new air pollution modeling system. Part II: Aerosol module structure and design. *Atmos. Environ.* **31**, 131–44.

Jacobson M. Z. (1997b) Development and application of a new air pollution modeling system. Part III: Aerosol-phase simulations. *Atmos. Environ.* **31**, 587–608.

Jacobson M. Z. (1998) Studying the effects of aerosols on vertical photolysis rate coefficient and temperature profiles over an urban airshed. *J. Geophys. Res.* **103**, 10,593–604.

Jacobson M. Z. (1999a) *Fundamentals of Atmospheric Modeling.* Cambridge University Press, New York 656 pp.

Jacobson M. Z. (1999b) Effects of soil moisture on temperatures, winds, and pollutant concentrations in Los Angeles. *J. Appl. Meteorol.* **38**, 607–16.

Jacobson M. Z. (1999c) Isolating nitrated and aromatic aerosols and nitrated aromatic gases as sources of ultraviolet light absorption. *J. Geophys. Res.* **104**, 3527–42.

Jacobson M. Z. (1999d) Studying the effects of calcium and magnesium on size-distributed nitrate and ammonium with EQUISOLV II. *Atmos. Environ.* **33**, 3635–49.

Jacobson M. Z. (2000) A physically-based treatment of elemental carbon optic: Implications for global direct forcing of aerosols. *Geophys. Res. Lett.* **27**, 217–20.

Jacobson M. Z. and Masters G. M. (2001) Exploiting wind versus coal. *Science* **293**, 1438.

Jacobson M. Z. (2001a) Global direct radiative forcing due to multicomponent anthropogenic and natural aerosols. *J. Geophys. Res.* **106**, 1551–68.

Jacobson M. Z. (2001b) Strong radiative heating due to the mixing state of black carbon in atmospheric aerosols. *Nature* **409,** 695–7.

Jacobson M. Z. (2002) Control of fossil fuel particulate black carbon plus organic matter, possibly the most effective method of slowing global warming. *J. Geophys. Res.* In press. http://www.stanford.edu/group/efmh/fossil/fossil.html.

Janerich D. T., Thompson W. D., and Varela L. R. (1990) Lung cancer and exposure to tobacco smoke in the household. *N. Engl. J. Med.* **323,** 632–6.

John W., Wall S. M., Ondo J. L., and Winklmayr W. (1989) Acidic aerosol size distributions during SCAQS. Final Report for the California Air Resources Board under Contract No. A6-112-32.

John W. Wall S. M., Ondo J. L., and Winklmayr W. (1990) Modes in the size distributions of atmospheric inorganic aerosol. *Atmos. Environ.* **24A,** 2349–59.

Jones A. P. (1999) Indoor air quality and health. *Atmos. Environ.* **33,** 4535–64.

Jones P. D. and Keigwin L. D. (1988) Evidence from Fram Strait (78°N) for early deglaciation. *Nature* **336,** 56–9.

Jones P. D., Parker D. E., Osborn T. J., and Briffa K. R. (2000) Global and hemispheric temperature anomalies – land and marine instrumental records. In *Trends Online: A Compendium of Data on Global Change.* Carbon Dioxide Information Analysis Center, Oak Ridge National Laboratory, U.S. Department of Energy, Oak Ridge, Tenn.

Jorquera H., Palma W., and Tapia J. (2000) An intervention analysis of air quality data at Santiago, Chile. *Atmos. Environ.* **34,** 4073–84.

Jouzel J., Lorius J. C., Petit J. R., Genthon C., Barkov N. I., Kotlyakov V. M., and Petrov V. M. (1987) Vostok ice core: A continuous isotope temperature record over the last climatic cycle (160,000 years). *Nature* **329,** 403–8.

Jouzel J., Barkov N. I., Barnola J. M., Bender M., Chappellaz J., Genthon C., Kotlyakov V. M., Lipenkov V., Lorius C., Petit J. R., Raynaud D., Raisbeck G., Ritz C., Sowers T., Stievenard M., Yiou F., and Yiou P. (1993) Extending the Vostok ice-core record of palaeoclimate to the penultimate glacial period. *Nature* **364,** 407–12.

Jouzel J., Waelbroeck C., Malaize B., Bender M., Petit J. R., Stievenard M., Barkov N. I., Barnola J. M., King T., Kotlyakov V. M., Lipenkov V., Lorius C., Raynaud D., Ritz C., and Sowers T. (1996) Climatic interpretation of the recently extended Vostok ice records. *Climate Dyn.* **12,** 513–21.

Junge C. E. (1961) Vertical profiles of condensation nuclei in the stratosphere. *J. Meteorol.* **18,** 501–9.

Kallos G., Kotroni V., Lagouvardos K., and Papadopoulos A. (1998) On the long-range transport of air pollutants from Europe to Africa. *Geophys. Res. Lett.* **25,** 619–22.

Katrlnak K. A., Rez P., and Buseck P. R. (1992) Structural variations in individual carbonaceous particles from an urban aerosol. *Environ. Sci. Technol.* **26,** 1967–76.

Katrlnak K. A., Rez P., Perkes P. R., and Buseck P. R. (1993) Fractal geometry of carbonaceous aggregates from an urban aerosol. *Environ. Sci. Technol.* **27,** 539–47.

Keeling C. D. and Whorf T. P. (2000) Atmospheric CO_2 concentrations (ppmv) derived from in situ air samples collected at Mauna Loa Observatory, Hawaii. http://cdiac.esd.ornl.gov/ftp/maunaloa-co2/maunaloa.co2.

Koschmieder H. (1924) Theorie der horizontalen Sichtweite. *Beitr. Phys. Freien Atm.* **12,** 33–53, 171–81.

Klein S. A. (1997) Synoptic variability of low-cloud properties and meteorological parameters in the subtropical trade wind boundary layer. *J. Clim.* **10,** 2018–39.

Kovarik B. (1998) Henry Ford, Charles Kettering and the "Fuel of the Future." http://www.runet.edu/~wkovarik /papers/fuel.html.

Kovarik B. (1999) Charles F. Kettering and the 1921 discovery of tetraethyl lead in the context of technological alternatives. http://www.runet.edu/~wkovarik/papers/kettering.html.

Krekov G. M. (1993) Models of atmospheric aerosols. In *Aerosol Effects on Climate,* Jennings S. G., ed., University of Arizona Press, Tucson, pp. 9–72.

Kukla G. J. (1977) Pleistocene land–sea correlations. I. Europe. *Earth Sci. Rev.* **13,** 307–74.

Labeyrie L. D., *et al.* (1986) Melting history of Antarctica during the past 60,000 years. *Nature* **322,** 701–6.

Lagarde F., Pershagen G., Akerblom G., Axelson O., Baverstam U., Damber L., Enflo A., Svartengren M., and Swedjemark G. A. (1997) Residential radon and lung cancer in Sweden: risk analysis accounting for random error in the exposure assessment. *Health Phys.* **72,** 269–76.

Lalas D. P., Asimakopoulos D. N., Deligiorgi D. G., and Helmis C. G. (1983) Sea-breeze circulation and photochemical pollution in Athens, Greece. *Atmos. Environ.* **17,** 1621–32.

Larson S., Cass G., Hussey K., and Luce F. (1984) Visibility model verification by image processing techniques. Final report to the California Air Resources Board under Agreement A2-077-32.

Lary D. J. (1997) Catalytic destruction of stratospheric ozone. *J. Geophys. Res.* **102,** 21, 515–26.

Lazrus A. L., Cadle R. D., Gandrud B. W., Greenberg J. P., Huebert B. J., and Rose W. I. (1979) Sulfur and halogen chemistry of the stratosphere and of volcanic eruption plumes. *J. Geophys. Res.* **84,** 7869.

Leaderer B. P., Stolwijk J. A. J., Zagraniski R. T., and Qing-Shan M. (1984) A field study of indoor air contaminant levels associated with unvented combustion sources. *Proc. 77th Ann. Meet. Air Pollut. Cont. Assoc.* San Francisco, CA.

Leaderer B. P., Koutrakis P., Briggs S. L. K., and Rizzuto J. (1990) Measurement of toxic and related air pollutants. In *Proc. EPA/Air Waste Manag. Assoc. Internat. Symp.* (VIP-17), U.S. EPA, Washington, DC, pp. 567.

Leaderer B. P., Stowe M., Li R., Sullivan J., Koutrakis P., Wolfson M., and Wilson W. (1993) Residential levels of particle and vapor phase acid associated with combustion sources. *Proc. Sixth Internat. Conf. Indoor Air Qual. Clim.* Helsinki, Finland, pp. 147–52.

Leakey R. E. and Lewin R. (1977) *Origins,* MacDonald and Janes, London.

Lee H. D. P. (trans.) (1951) *Meteorologica* by Aristotle, T. E. Page, ed., Harvard University Press, Cambridge.

Lee R. J., Van Orden D. R., Corn M., and Crump K. S. (1992) Exposure to airborne asbestos in buildings. *Regulat. Toxicol. Pharmacol.* **16,** 93–107.

Leg 105 Shipboard Scientific Party (1986) High-latitude palaeoceanography. *Nature* **230,** 17–8.

Le Roy Ladurie E. (1971) *Times of Feast, Times of Famine: A History of Climate since the Year 1000.* Doubleday, New York.

Levy H., Kasibhatla P., Moxim W. J., Klonecki A. A., Hirsch A. I., Oltmans S. J., and Chameides W. L. (1997) The global impact of human activity on tropospheric ozone. *Geophys. Res. Lett.* **24,** 791–4.

Li Y., Powers T. E., and Roth H. D. (1994) Random effects linear regression meta-analysis models with application to nitrogen dioxide health effects studies. *J. Air Waste Manag. Assn.* **44,** 261–70.

Lide D. R., ed. (1998) *CRC Handbook of Chemistry and Physics.* CRC Press, Inc., Boca Raton, FL.

Likens G. E. (1976) Acid precipitation. *Chem. Eng. News* **54,** 29–44.

Liou K. N. (1992) *Radiation and Cloud Processes in the Atmosphere.* Oxford University Press, New York.

Liousse C., Penner J. E., Chuang C., Walton J. J., Eddleman H., and Cachier H. (1996) A global three-dimensional model study of carbonaceous aerosols. *J. Geophys. Res.* **101,** 19,411–32.

Liu S. C., Trainer M., Fehsenfeld F. C., Parrish D. D., Williams E. J., Fahey D. W., Hubler G., and Murphy P. C. (1987) Ozone production in the rural troposphere and the implications for regional and global ozone distributions. *J. Geophys. Res.* **92,** 4191–207.

Lobert J. M., Keene W. C., Logan J. A., and Yevich R. (1999) Global chlorine emissions from biomass burning: Reactive chlorine emissions inventory. *J. Geophys. Res.* **104,** 8373–89.

Longstreth J., de Gruigj F. R., Kripke M. L., Abseck S., Arnold F., Slaper H. I., Velders G., Takizawa Y., and van der Leun J. C. (1998) Health risks. *J. Photochem. Photobiol. B: Biology* **46,** 20–39.

Lu R. and Turco R. P. (1995) Air pollution transport in a coastal environment – II. Three-dimensional simulations over Los Angeles Basin. *Atmos. Environ.* **29,** 1499–1518.

Lurmann F. W., Main H. H., Knapp K. T., Stockburger L., Rasmussen R. A., and Fung K. (1992) Analysis of the ambient VOC data collected in the Southern California Air Quality Study. Final Report to the California Air Resources Board under Contract A832–130.

Lyman G. H. (1997) Radon. In *Indoor Air Pollution and Health,* Bardana E. J. and Montanaro A., eds. Dekker, New York, pp. 83–103.

Lyons W. A. and Olsson L. E. (1973) Detailed mesometeorological studies of air pollution dispersion in the Chicago lake breeze. *Mon. Weath. Rev.* **101,** 387–403.

MacNee W. and Donaldson K. (1999) Particulate air pollution: Injurious and protective mechanisms in the lungs. In *Air Pollution and Health.* Holgate S. T., Samet J. M., Koren H. S., and Maynard R. L., eds. Academic Press, San Diego, pp. 653–72.

Madronich S., McKenzie R. L., Bjorn L. O., and Caldwell M. M. (1998) Changes in biologically active ultraviolet radiation reaching the Earth's surface. *J. Photochem. Photobiol. B: Biology* **46,** 5–19.

Marcinowski F., Lucas R. M., and Yeager W. M. (1994) National and regional distributions of airborne radon concentrations in U.S. homes. *Health Phys.* **66,** 699–706.

Maricq M. M., Chase R. E., Podsiadlik D. H., and Vogt R. (1999) Vehicle exhaust particle size distributions: A comparison of tailpipe and dilution tunnel measurements. SAE Technical Paper 1999-01-1461, Warrendale, Penn.

Marland G., Boden T. A., and Andres R. J. (2000) Global, regional, and national CO_2 emissions. In *Trends Online: A Compendium of Data on Global Change.* Carbon Dioxide Information Analysis Center, Oak Ridge National Laboratory, U.S. Department of Energy, Oak Ridge, Tenn.

Maroni M. B., Seifert X. X., and Lindvall T., eds. (1995) *Indoor Air Quality – A Comprehensive Reference Book.* Elsevier, Amsterdam.

Marsh A. R. W. (1978) Sulphur and nitrogen contributions to the acidity of rain. *Atmos. Environ.* **12,** 401–6.

Martins J. V., Artaxo P., Liousse C., Reid J. S., Hobbs P. V., and Kaufman Y. J. (1998) Effects of black carbon content, particle size, and mixing on light absorption by aerosols from biomass burning in Brazil. *J. Geophys. Res.* **103,** 32,041–50.

Masters G. M. (1998) *Introduction to Environmental Engineering and Science,* Second Edition. Prentice-Hall, Inc., Upper Saddle River, New Jersey.

Matthes F. E. (1939) Report of committee on glaciers, April 1939. *Trans. Am. Geophys. Union* **20,** 518–23.

Mauna Loa Data Center (2001) Data for atmospheric trace gases, http://mloserv.mlo.hawaii.gov/.

Mayer S. J. (1992) Stratospheric ozone depletion and animal health. *Vet. Rec.* **131,** 120–2.

McElroy M. B., Salawitch R. J., Wofsy S. C., and Logan J. A. (1986) Reduction of Antarctic ozone due to synergistic interactions of chlorine and bromine. *Nature* **321,** 759–62.

McBride S. J., Ferro A. R., Ott W. R., Switzer P., and Hildemann L. M. (1999) Investigations of the proximity effect for pollutants in the indoor environment. *J. Expos. Anal. Env. Epidemiol.* **9,** 602–21.

McElroy J. L. and Smith T. B. (1992) Creation and fate of ozone layers aloft in Southern California. *Atmos. Environ.* **26,** 1917–29.

McKenzie R., Connor B., and Bodeker G. (1999) Increased summertime UV radiation in New Zealand in response to ozone loss. *Science* **285,** 1709–11.

McMurry P. H. and Zhang X. Q. (1989) Size distributions of ambient organic and elemental carbon. *Aerosol Sci. Technol.* **10,** 430–7.

McNeill J. R. (2000) *Something New under the Sun,* W. W. Norton & Company Ltd., New York.

Mendez A., Perez J., Ruiz-Villamor E., Martin M. P., and Mozos E. (1997) Clinicopathological study of an outbreak of squamous cell carcinoma in sheep. *Vet. Rec.* **141,** 597–600.

Mercer J. H. (1978) West Antarctic ice sheet and CO_2 greenhouse effect: A threat of disaster. *Nature* **271,** 321–5.

Middleton W. E. K. (1952) *Vision Through the Atmosphere.* University of Toronto Press, Toronto, Canada.

Midgley Jr., T. (August 1925) Tetraethyl lead poison hazards. *Indust. Eng. Chem.* **17** (8), 827.

Midgley Jr., T. (April 7, 1925) "Radium Derivative $5,000,000 an Ounce/Ethyl Gasoline Defended," New York Times, p. 23.

Mie G. (1908) Optics of turbid media. *Ann. Phys.,* 25, 377–445.

Milankovitch M. (1930) Mathematische klimalehre und astronomische theorie der Klimaschwankungen. In *Handbuch der Klimatologie,* I. Köppen W. and Geiger R., eds. Gebruder Borntraeger, Berlin.

Milankovitch M. (1941) *Canon of Insolation and the Ice-Age Problem,* Königlich Serbische Akademie, Belgrade, 484 pp. (English translation by the Israel Program for Scientific Translation, published by the U. D. Department of Commerce and the National Science Foundation.)

Miller S. L. and Orgel L. E. (1974) *The Origins of Life on Earth.* Prentice-Hall, Englewood Cliffs, New Jersey.

Molina L. T. and Molina M. J. (1986) Production of Cl_2O_2 by the self reaction of the ClO radical. *J. Phys. Chem.* **91,** 433–6.

Molina M. J. and Roland F. S. (1974) Stratospheric sink for chlorofluoromethanes: Chlorine atom catalysed destruction of ozone. *Nature* **249,** 810–2.

Moody J. L., Davenport J. C., Merrill J. T., Oltmans S. J., Parrish D. D., Holloway J. S., Levy II H., Forbes G. L., Trainer M., and Buhr M. (1996) Meteorological mechanisms for transporting ozone over the western North Atlantic Ocean: A case study for August 24–29, 1993. *J. Geophys. Res.* **101,** 29,213–27.

Munger J. W., Jacob D. J., Waldman J. M., and Hoffmann M. R. (1983) Fogwater chemistry in an urban atmosphere. *J. Geophys. Res.* **88,** 5109–21.

Murphy D. M., Anderson J. R., Quinn P. K., McInnes L. M., Brechtel F. J., Kreidenweis S. M., Middlebrook A. M., Posfai M., Thomson D. S., and Buseck P. R. (1998) Influence of sea-salt on aerosol radiative properties in the Southern Ocean marine boundary layer. *Nature* **395,** 62–5.

Nagda N. L., Rector H. E., and Koontz M. D. (1987) *Guidelines for Monitoring Indoor Air Quality.* Hemisphere, Washington, DC.

National Atmospheric Deposition Program (2000) (NRSP-3)/National Trends Network, NADP Program Office, Illinois State Water Survey, 2204 Griffith Dr., Champaign, IL.

National Centers for Environmental Prediction (NCEP) (2000) 2.5 degree global final analyses, distributed by the Data Support Section, National Center for Atmospheric Research, Boulder, Colorado.

National Institute for Occupational Safety and Health (NIOSH) (2000) http://www.cdc.gov/niosh/.

National Mining Association (NMA) (2000) http://www.nma.org/.

Nazaroff W. W. and Nero, Jr. A. V. (1988) *Radon and Its Decay Products in Indoor Air.* Wiley-Interscience, New York.

Nero A. V., Schwehr M. B., Nazaroff W. W., and Revzan K. L. (1986) Distribution of airborne radon-222 concentrations in U.S. homes. *Science* **234,** 992–7.

Nicholls S. (1984) The dynamics of stratocumulus: Aircraft obervations and comparisons with a mixed layer model. *Quart. J. Roy. Meteor. Soc.* **110,** 783–820.

Newman M. J. and Rood R. T. (1977) Implications of solar evolution for the Earth's early atmosphere. *Science* **194,** 1413–4.

Nordo J. (1976) Long range transport of air pollutants in Europe and acid precipitation in Norway. *Water Air Soil Pollut.* **6,** 199–217.

Offenberg J. H. and Baker J. E. (2000) Aerosol size distributions of elemental and organic carbon in urban and over-water atmospheres. *Atmos. Environ.* **34,** 1509–17.

Oke T. R. (1978) *Boundary Layer Climates.* Methuen, London.

Oke T. R. (1988) The urban energy balance. *Prog. Phys. Geog.* **12,** 471–508.

Oke T. R., Spronken-Smith R. A., Jauregui E., and Grimmond C. S. B. (1999) The energy balance of central Mexico City during the dry season. *Atmos. Environ.* **33**, 3919–30.

Özkatnak H. and Thurston G. D. (1987) Association between 1980 U.S. mortality rates and alternative measures of airborne particle concentrations. *Risk Anal.* **7**, 449–61.

Pandis S. N., Harley R.A., Cass G. R., and Seinfeld J. H. (1992) Secondary organic aerosol formation and transport. *Atmos. Environ.* **26A**, 2269–82.

Paulson S. E. and Seinfeld J. H. (1992) Development and evaluation of a photooxidation mechanism for isoprene. *J. Geophys. Res.* **97**, 20,703–15.

Peterson T. C. and Vose R. S. (1997) An overview of the Global Historical Climatology Network temperature data base. *Bull. Amer. Meteorol. Soc.* **78**, 2837–49.

Petit J. R., Jouzel J., Raynaud D., Barkov N. I., Barnola J.-M., Basile I., Bender M., Chappellaz J., Davis M., Delayque G., Delmotte M., Kotlyakov V. M., Legrand M., Lipenkov V. Y., Lorius C., Pepin L., Ritz C., Saltzman E., and Stievenard M. (1999) Climate and atmospheric history of the past 420,000 years from the Vostok ice core. *Antarctica Nature* **399**, 429–39.

Pilotto L. S., Douglas R. M., Attewell R. G., and Wilson S. R. (1997) Respiratory effects associated with indoor nitrogen dioxide exposure in children. *Int. J. Epidemiol.* **26**, 788–96.

Pinto J. P., Turco R. P., and Toon O. B. (1989) Self-limiting physical and chemical effects in volcanic eruption clouds. *J. Geophys. Res.,* **94**, 11,165.

Platts-Mills T. A. E. and Carter M. C. (1997) Asthma and indoor exposure to allergens. *N. Engl. J. Med.* **336**, 1382–84.

Pollack J. B. and Yung Y. L. (1980) Origin and evolution of planetary atmospheres. *Ann. Rev. Earth Plan. Sci.,* **8**, 425–88.

Polpong P. and Bovornkitti S. (1997) Indoor radon. *J. Med. Assoc.Thailand* **81**, 47–57.

Pooley F. D. and Mille M. (1999) Composition of air pollution particles. In *Air Pollution and Health,* Holgate S. T., Samet J. M., Koren H. S., and Maynard R. L., eds., Academic Press, San Diego, pp. 619–34.

Pope C. A. (2000) Review: Epidemiological basis for particulate air pollution health standards. *Aerosol. Sci. Technol.* **32**, 4–14.

Pope C. A., Bates D. V., and Raizenne M. E. (1995) Health-effects of particulate air-pollution – time for reassessment. *Environ. Health Perspect.* **103**, 472–80.

Pope C. A. III and Dockery D. W. (1999) Epidemiology of particle effects. In *Air Pollution and Health,* Holgate S. T., Samet J. M., Koren H. S., and Maynard R. L., eds., Academic Press, San Diego, pp. 673–705.

Posey J. W. and Clapp P. F. (1964) Global distribution of normal surface albedo. *Geofis. Int.,* **4**, 33–48.

Pósfai M., Anderson J. R., Buseck P. R., and Sievering H. (1999) Soot and sulfate aerosol-particles in the remote marine troposphere. *J. Geophys. Res.* **104**, 21,685–93.

Poulos G. S. and Pielke R. A. (1994) Numerical analysis of Los Angeles Basin pollution transport to the Grand Canyon under stably stratified southwest flow conditions. *Atmos. Environ.* **28**, 3329–57.

Prados A. I., Dickerson R. R., Doddridge B. G., Milne P. A., Moody J. L., and Merrill J. T. (1999) Transport of ozone and pollutants from North America to the North Atlantic Ocean during the 1996 Atmosphere/Ocean Chemistry Experiment (AEROCE) intensive. *J. Geophys. Res.* **104**, 26,219–33.

Prinn R. G. et al., (2000) A history of chemically and radiatevely important gases in air deduced from ALE/GAGE/AGAGE. *J. Geophys. Res.* **105**, 17,751–92.

Prospero J. M. and Savoie D. L. (1989) Effect of continental sources on nitrate concentrations over the Pacific Ocean. *Nature* **339**, 687–9.

Pruppacher H. R. and Klett J. D. (1997) *Microphysics of Clouds and Precipitation,* 2nd rev. and enl. ed. Kluwer Academic Publishers, Dordrecht.

Qin Y. H., Zhang X. M., Jin H. Z., Liu Y. Q., Fan D. L., and Fan Z. J. (1993) Effects of indoor air pollution on respiratory illness of school children. *Proc. Sixth Int. Conf. Indoor Air Qual. Clim.* Helsinki, Finland, 477–82.

Rando R. J., Simlote P., Salvaggio J. E., and Lehrer S. B. (1997) Environmental tobacco smoke measurement and health effects of involuntary smoking. In *Indoor Air Pollution and Health,* Bardana E. J. and Montanaro A., eds., Marcel Dekker, New York, pp. 61–82.

Rayleigh, Lord (1871) On the light from the sky, its polarization and colour. *Phil. Mag.* **41,** 107–20.

Reid J. S. and Hobbs P. V. (1998) Physical and optical properties of young smoke from individual biomass fires in Brazil. *J. Geophys. Res.* **103,** 32,013–30.

Reid J. S., Hobbs P. V., Ferek R. J., Blake D. R., Martins J. V., Dunlap M. R., and Liousse C. (1998) Physical, chemical, and optical properties of regional hazes dominated by smoke in Brazil. *J. Geophys. Res.* **103,** 32,059–80.

Robinson J. P., Thomas J., and Behar J. V. (1991) Time spent in activities, locations, and microenvironments: A California–National comparison. Report under Contract No. 69-01-7324, Delivery Order 12, Exposure Assessment Research Division, National Exposure Research Center, U.S. Environmental Protection Agency, Las Vegas, NV.

Rodes C. E., Kamens R. M., and Wiener R. W. (1991) The significance and characteristics of the personal activity cloud on exposure assessment measurements for indoor contaminants. *Indoor Air* **2,** 123–45.

Rosenberg N. and Birdzell, Jr. L. E. (1986) *How the West Grew Rich,* Basic Books, Inc., New York.

Rushton L. and Cameron K. (1999) Selected organic chemicals. In *Air Pollution and Health,* Holgate S. T., Samet J. M., Koren H. S., and Maynard R. L., eds., Academic Press, San Diego, pp. 813–38.

Ryan P. B., Lee M. W., North B., and McMichael A. J. (1992) Risk factors for tumours of the brain and meninges: Results from the Adelaide adult brain tumour study. *Int. J. Canc.* **51,** 20–7.

Saffman P. G. and Turner J. S. (1956) On the collision of drops in turbulent clouds. *J. Fluid Mech.* **1,** 16–30.

Samet J. M., Marbury M. C., and Spengler J. D. (1987) Health effects and sources of indoor air pollution: Part 1. *Am. Rev. Resp. Dis.* **136,** 1486–1508.

Saxena P., Mueller P. K., and Hildemann L. M. (1993) Sources and chemistry of chloride in the troposphere: A review. In *Managing Hazardous Air Pollutants: State of the Art.* W. Chow and K. K Connor, eds., Lewis Publishers, Boca Raton, FL, pp. 173–90.

Schichtel B. A., Husar R. B., Falke S. R., and Wilson W. E. (2001) Haze trends over the United States, 1980–1995. *Atmos. Environ.* **35,** 5205–10.

Schlitt H. and Knöppel H. (1989) Carbonyl compounds in mainstream and sidestream tobacco smoke. In Bieva C. J., Courtois Y., and Govaerts M., eds., *Present and Future of Indoor Air Quality,* Excerpta Medica, Amsterdam, pp. 197–206.

Schroeder W. H., Dobson M., Kane D. M., and Johnson N. D. (1987) Toxic trace elements associated with airborne particulate matter: A review. *J. Air Pollut. Control Assoc.* **37,** 1267–85.

Schwarzberg M. N. (1993) Carbon dioxide level as migraine threshold factor: Hypothesis and possible solutions. *Med. Hypoth.* **41,** 35–6.

Sellers W. D. (1965) *Physical Climatology.* University of Chicago Press, Chicago, 272 pp.

Shackleton N. J. (1985) Oceanic carbon isotope constraints on oxygen and carbon dioxide in the Cenozoic atmosphere. In *The Carbon Cycle and Atmospheric CO_2: Natural Variations Archean to Present,* Sundquist E. T. and Broecker W. S., eds., Geophys. Mono. 32, AGU, Washington, DC, pp. 412–8.

Shackleton N. J. and Opdyke N. D. (1976) Oxygen isotope and paleomagnetic stratigraphy of Pacific core V28-239 late Pliocene to latest Pleistocene. In *Investigation of Late Quaternary Paleoceanography and Paleoclimatology,* Cline R. M. and Hays J. D., eds., *Geol. Soc. Am. Mem.,* **145,** Geol. Soc. Am., Boulder, CO, pp. 449–64.

Shen T.-L., Wooldridge P. J., and Molina M. J. (1995) Stratospheric pollution and ozone depletion. In *Composition, Chemistry, and Climate of the Atmosphere,* Singh H. B., ed., Van Nostrand Reinhold, New York.

Sheridan P. J., Brock C. A., and Wilson J. C. (1994) Aerosol-particles in the upper troposphere and lower stratosphere: Elemental composition and morphology of individual particles in northern midlatitudes. *Geophys. Res. Lett.* **21,** 2587–90.

Singh H. B. (1995) Halogens in the atmospheric environment. In *Composition, Chemistry, and Climate of the Atmosphere,* Singh H. B., ed., Van Nostrand Reinhold, New York.

Smith K. R. (1993) Fuel combustion, air pollution exposure, and health: The situation in the developing countries. Annual Review of Energy and Environment, 1993. Annual Reviews, Inc., Palo Alto, CA **18,** 529–66.

Sokolik I., Andronova A., and Johnson C. (1993) Complex refractive index of atmospheric dust aerosols. *Atmos. Environ.* **27A,** 2495–502.

Sollinger S., Levsen K., and Wünsch G. (1994) Indoor pollution by organic emissions from textile floor coverings: Climate test chamber studies under static conditions. *Atmos. Environ.* **28,** 2369–78.

Solomon P. A. and Thuillier R. H. (1995) SJVAQS/SUSPEX/SARMAP 1990 Air Quality Field Measurement Project Volume II: Field measurement characterization. PG&E Cost reduction projects report 009.2-94.1.

Solomon S., Garcia R. R., Rowland F. S., and Wuebbles D. J. (1986) On the depletion of Antarctic ozone. *Nature* **321,** pp. 755–7.

Somerville S. M., Rona R. J., and Chinn S. (1988) Assive smoking and resiratory conditions in primary school children. *J. Epidemiol. Comm. Health* **42,** 105–10.

Song C. H. and Carmichael G. R. (1999) The aging process of naturally emitted aerosol (sea-salt and mineral aerosol) during long range transport. *Atmos. Environ.* **33,** 2203–18.

South Coast Air Quality Management District (2000) Web site, http://www.aqmd.gov/.

Spencer R. W. and Christy J. R. (1990) Precise monitoring of global temperature trends from satellites. *Science* **247,** 1558–62.

Spengler J. D. (1993) Nitrogen dioxide and respiratory illnesses in infants. *Am. Rev. Resp. Dis.* **148,** 1258–65.

Spengler J. D. and Sexton K. (1983) Indoor air pollution: A public health perspective. *Science* **221,** 9–17.

Spengler J. D., Dockery D. W., Turner W. A., Wolfson J. M., and Ferris B. G. (1981) Long-term measurements of respirable particles, sulphates, and particulates inside and outside homes. *Atmos. Environ.* **15,** 23–30.

Steiner D., Burtchnew H., and Grass H. (1992) Structure and disposition of particles from a spark ignition engine. *Atmos. Environ.* **26,** 997–1003.

Stephens E. R., Scott W. E., Hanst P. L., and Doerr R. C. (1956) Recent developments in the study of the organic chemistry of the atmosphere. *J. Air Pollut. Contr. Assoc.* **6,** 159–65.

Stern D. I. and Kaufman R. K. (1998) Annual estimates of global anthropogenic methane emissions: 1860–1994. Trends Online: A compendium of data on global change. Carbon Dioxide Information Analysis Center, Oak Ridge National Laboratory, U.S. Department of Energy, Oak Ridge, TN.

Stockholm Environment Institute (SEI) (1998) Regional Air Pollution in Africa, http://www.sei.se

Stommel H. and Stommel E. (1981) The year without a summer. *Sci. Am.* **240,** 176–86.

Stothers R. B. (1984) The great Tambora eruption in 1815 and its aftermath. *Science* **224,** 1191–8.

Stott P. A., Tett S. F. B., Jones G. S., Allen M. R., Mitchell J. F. B., and Jenkins G. J. (2000) External control of 20th century temperature by natural and anthropogenic forcings. *Science* **290,** 2133–6.

Stradling D. (1999) *Smokestacks and Progressives.* The Johns Hopkins University Press, Baltimore.

Strawa A. W., Drdla K., Ferry G. V., Verma S., Pueschel R. F., Yasuda M., Salawitch R. J., Gao R. S., Howard S. D., Bui P. T., Loewenstein M., Elkins J. W., Perkins K. K., and Cohen R. (1999) Carbonaceous aerosol (soot) measured in the lower stratosphere during POLARIS and its role in stratospheric photochemistry. *J. Geophys. Res.* **104,** 26,753–66.

Streets D. G., Gupta S., Waldhoff S. T., Wang M. Q., Bond T. C., and Yiyun B. (2001) Black carbon emissions in China. *Atmos. Environ.* **35,** 4281–426.

Stuiver M., Denton G. H., Hughes T. J., and Fastook J. L. (1981) History of marine ice sheet in West Antarctica during the last glaciation: A working hypothesis. In *The Last Great Ice Sheets,* Denton G. H. and Hughes T. J., eds. Wiley-Interscience, New York, pp. 319–436.

Stull R. B. (1998) *An Introduction to Boundary Layer Meteorology.* Kluwer Academic Publishers, Dordrecht.

Tabazadeh A. and Turco R. P. (1993) Stratospheric chlorine injection by volcanic eruptions: HCl scavenging and implications for ozone. *Science* **260,** 1082–6.

Tabazadeh A., Turco R. P., Drdla K., and Jacobson M. Z. (1994) A study of Type I polar stratospheric cloud formation. *Geophys. Res. Lett.* **21,** 1619–22.

Tabazadeh A., Jacobson M. Z., Singh H. B., Toon O. B., Lin J. S., Chatfield B., Thakur A. N., Talbot R. W., and Dibb J. E. (1998) Nitric acid scavenging by mineral and biomass burning aerosols. *Geophys. Res. Lett.* **25,** 4185–8.

Talwani M. and Eldholm O. (1977) Evolution of the Norwegian–Greenland Sea. *Geol. Soc. Am. Bull.* **88,** 969–99.

Tang I. N., Wong W. T., and Munkelwitz H. R. (1981) The relative importance of atmospheric sulfates and nitrates in visibility reduction. *Atmos. Environ.* **15,** 2463–71.

Tang X., Madronich S., Wallington T., and Calamari D. (1998) Changes in tropospheric composition and air quality. *J. Photochem. Photobiol. B.: Biology* **46,** 83–95.

Tegen I., Lacis A. A., and Fung I. (1996) The influence on climate forcing of mineral aerosols from disturbed soils. *Nature* **380,** 419–22.

Teifke J. P. and Lohr C. V. (1996) Immunohistochemical detection of p53 over-expression in paraffin wax-embedded squamous cell carcinomas of cattle, horses, cats, and dogs. *J. Compar. Pathol.* **114,** 205–10.

Tomlinson G. G. II (1983) Air pollutants and forest decline. *Environ. Sci. Technol.* **17,** 246–56.

Toon O. B., Pollack J. B., Ackerman T. P., Turco R. P., McKay C. P., and Liu M. S. (1982) Evolution of an impact-generated dust cloud and its effects on the atmosphere. *Geol. Soc. Am. Spec. Pap.* **190,** 187–200.

Toon O. B., Hamill P., Turco R. P., and Pinto J. (1986) Condensation of HNO_3 and HCl in the winter polar stratospheres. *Geophys. Res. Lett. Nov. Supp.* **13,** 1284–7.

Turco R. P. (1997) *Earth Under Siege.* Oxford University Press, Oxford.

Turco R. P., Toon O. B., and Hamill P. (1989) Heterogeneous physiochemistry of the polar ozone hole. *J. Geophys. Res.* **94,** 16,493–510.

Twomey S. A. (1977) The effect of cloud scattering on the absorption of solar radiation by atmospheric dust. *J. Atmos. Sci.* **29,** 1156–9.

Ulrich R. K. (1975) Solar neutrinos and variations in the solar luminosity. *Science* **190,** 619–24.

United Nations Environmental Program (UNEP) (1998) Environmental effects of ozone depletion: 1998 assessment. *J. Photochem. Photobiol. B: Biology* **46,** 1–4.

U.S. Environmental Protection Agency (U.S. EPA) (1978) Air quality criteria for ozone and other photochemical oxidants. Report No. EPA-600/8-78-004.

U.S. Environmental Protection Agency (U.S. EPA) (1996) Air Quality Criteria for Particulate Matter, Research Triangle Park, NC, National Center of Environmental Assessment-RTP Office, EPA/600/P-95/001aF-cf. 3 volumes. Available from NTIS, Springfield, VA, PB96-168224.

U.S. Environmental Protection Agency (U.S. EPA) (1997) National Air Quality and Emissions Trend Report, Office of Air Quality Planning and Standards, U.S. Environmental Protection Agency.

U.S. Environmental Protection Agency (U.S. EPA) (1998) 1997 National Air Quality Status and Trends, U.S. Environmental Protection Agency, Office of Air and Radiation.

U.S. Geological Survey (U.S. GS)/University of Nebraska, Lincoln/European Commission's Joint Research Center (1999) 1-km resolution global landcover characteristics data, derived from Advanced Very High Resolution Radiometer (AVHRR) data from the period April 1992 to March 1993.

U.S. Public Health Service (U.S. PHS) (Sept. 1925) The use of tetraethyl lead gasoline in its relation to public health. Public Health Bulletin No. 163, Treasury Dept., Washington, DC.

U.S. Public Health Service (U.S. PHS) (March 30, 1959) Public health aspects of increasing tetraethyl lead content in motor fuel. Report on the Advisory Committee on Tetraethyl Lead to the Surgeon General, PHS publication No. 712, Washington, DC.

Valley S., ed. (1965) *Handbook of Geophysics and Space Environments.* U.S. Air Force Cambridge Research Laboratories, McGraw-Hill, New York.

Venkataraman C., Lyons J. M., and Friedlander S. K. (1994) Size distributions of polycyclic aromatic hydrocarbons and elemental carbon. 1. Sampling, measurement methods, and source characterization. *Environ. Sci. Technol.* **28,** 555–62.

Venkataraman C. and Friedlander S. K. (1994) Size distributions of polycyclic aromatic hydrocarbons and elemental carbon. 2. Ambient measurements and effects of atmospheric processes. *Environ. Sci. Technol.* **28,** 563–72.

Venkatram A. and Viskanta R. (1977) Radiative effects of elevated pollutant layers. *J. Appl. Meteorol.* **16,** 1256–72.

Waggoner A. P., Weiss R. E., Ahlquist N. C., Covert D. S., Will S., and Charlson R. J. (1981) Optical characteristics of atmospheric aerosols. *Atmos. Environ.* **15,** 1891–909.

Waitt R. B. (1985) Case for periodic, colossal jökulhlaups from Pleistocene glacial Lake Missoula. *Geol. Soc. Am. Bull.* **96,** 1271–86.

Wakamatsu S., Ogawa Y., Murano K., Goi K., and Aburamoto Y. (1983) Aircraft survey of the secondary photochemical pollutants covering the Tokyo metropolitan area. *Atmos. Environ.* **17,** 827–36.

Wakimoto R. M. and McElroy J. L. (1986) Lidar observation of elevated pollution layers over Los Angeles. *J. Clim. Appl. Met.* **25,** 1583–99.

Waldman J. M., Munger J. W., Jacob D. J., Flagan R. C., Morgan J. J., and Hoffmann M. R. (1982) Chemical composition of acid fog. *Science* **218,** 677–80.

Wallace L. (2000) Correlations of personal exposure to particles with outdoor air measurements: A review of recent studies. *Aerosol Sci. Technol.* **32,** 15–25.

Wallace L. A. (1991) Personal exposure to 25 volatile organic compounds. EPA's 1987 team study in Los Angeles, California. *Toxicol. Indus. Health* **7,** 203–8.

Wanner H. U. (1993) Sources of pollutants in indoor air. *IARC Sci. Pub.* **109,** 19–30.

Wark K., Warner C. F., and Davis W. T. (1998) *Air Pollution: Its Origin and Control,* Addison-Wesley, Menlo Park, CA.

Weiss S. T. (1986) Passive smoking and lung cancer: What is the risk? *Am. Rev. Resp. Dis.* **133,** 1–3.

Whitby K. T. (1978) The physical characteristics of sulfur aerosols. *Atmos. Environ.* **12,** 135–59.

Williams G. E. (1975) Late precambrian glacial climate and the Earth's obliquity. *Geol. Mag.* **112,** 441–65.

Woodcock A. H. (1953) Salt nuclei in marine air as a function of altitude and wind force. *J. Meteorol.* **10,** 362–71.

World Health Organization (WHO) (2000) http://www.who.int

World Meteorological Organization (WMO) (1995) Scientific assessment of ozone depletion: 1994. Rep. 25, Global Ozone Res. and Monit. Proj World Meterolical Organization, Geneva.

World Meteorological Organization (WMO) (1998) Scientific assessment of ozone depletion: 1998. WMO Global Ozone Research and Monitoring Project – Report No. 44, Geneva.

Worsnop D. R., Fox L. E., Zahniser M. S., and Wofsy S. C. (1993) Vapor pressures of solid hydrates of nitric acid: Implications for polar stratospheric clouds. *Science* **259,** 71–4.

Wuebbles D. J. (1995) Air pollution and climate change. In *Composition, Chemistry, and Climate of the Atmosphere,* Singh H. B., ed. Van Nostrand Reinhold, New York.

Zdunkowski W. G., Welch R. M., and Paegle J. (1976) One dimensional numerical simulation of the effects of air pollution on the planetary boundary layer. *J. Atmos. Sci.* **33,** 2399–414.

Zepp R. G., Callaghan T. V., and Erickson D. J. (1998) Effects of enhanced solar ultraviolet radiation on biogeochemical cycles. *J. Photochem. Photobiol. B: Biology* **46,** 69–82.

Zhang J., Liu S. M., Lu X., and Huang W. W. (1993) Characterizing Asian wind-dust transport to the northwest Pacific Ocean. Direct measurements of the dust flux for two years. *Tellus* **45B,** 335–45.

Zoller W. H., Parrington J. R., and Phelan Kotra J. M. (1983) Iridium enrichment in airborne particles from Kilauea Volcano: January 1983. *Science* **222,** 1118–21.

Zubakov V. A. and Borzenkova I. I. (1988) Pliocene palaeoclimates: Past climates as possible analogues of mid-twenty-first century climate. *Palaeogeog., Palaeoclim., Palaeoecol.* **65,** 35–49.

PHOTOGRAPH SOURCES

COVER

Deer at Golden, Colorado, in front of Denver's "brown cloud," December 14, 1998 (Photo by David Parsons, available from National Renewable Energy Laboratory, U.S. Department of Energy, http://www.nrel.gov)

CHAPTER 1

Figure 1.2. Geber (Edgar Fahs Smith Collection, University of Pennsylvania Library)
Figure 1.3. Libavius (Edgar Fahs Smith Collection, University of Pennsylvania Library)
Figure 1.4. Glauber (Edgar Fahs Smith Collection, University of Pennsylvania Library)
Figure 1.5. Paracelsus (Edgar Fahs Smith Collection, University of Pennsylvania Library)
Figure 1.6. Mayow (Edgar Fahs Smith Collection, University of Pennsylvania Library)
Figure 1.7. Black (Edgar Fahs Smith Collection, University of Pennsylvania Library)
Figure 1.8. Cavendish (Edgar Fahs Smith Collection, University of Pennsylvania Library)
Figure 1.9. D. Rutherford (Edgar Fahs Smith Collection, University of Pennsylvania Library)
Figure 1.10(a). Priestley (Edgar Fahs Smith Collection, University of Pennsylvania Library)
Figure 1.10(b). Priestley oxygen apparatus (Edgar Fahs Smith Collection, University of Pennsylvania Library)
Figure 1.11(a). Scheele (Edgar Fahs Smith Collection, University of Pennsylvania Library)
Figure 1.11(b). Scheele laboratory (Edgar Fahs Smith Collection, University of Pennsylvania Library)
Figure 1.12. The arrest of Lavoisier (Edgar Fahs Smith Collection, University of Pennsylvania Library)
Figure 1.13. Destruction of Priestley's house (Edgar Fahs Smith Collection, University of Pennsylvania Library)
Figure 1.14. Davy (Edgar Fahs Smith Collection, University of Pennsylvania Library)
Figure 1.15. Berzelius (Edgar Fahs Smith Collection, University of Pennsylvania Library)
Figure 1.16. von Liebig (Edgar Fahs Smith Collection, University of Pennsylvania Library)

Figure 1.17. Schönbein (Edgar Fahs Smith Collection, University of Pennsylvania Library)
Figure 1.18. M. Curie (Edgar Fahs Smith Collection, University of Pennsylvania Library)

CHAPTER 2

Figure 2.2. Aurora australis (Photo by David Miller, National Geophysical Data Center, available from NOAA Central Library; http://phlib.hpcc.noaa.gov/lb_images/historic/nws/wea02009.htm)
Figure 2.6. Asteroid Ida (Available from National Space Science Data Center, http://nssdc.gsfc.nasa.gov)
Figure 2.9. Hot sulfur springs (Photo by Alfred Spormann, Stanford University)
Figure 2.10. Hot springs in Yellowstone (Photo by Alfred Spormann, Stanford University)

CHAPTER 3

Figure 3.1. Toricelli apparatus (Edgar Fahs Smith Collection, University of Pennsylvania Library)
Figure 3.6. Robert Boyle (Edgar Fahs Smith Collection, University of Pennsylvania Library)
Figure 3.7. Jacques Charles (Edgar Fahs Smith Collection, University of Pennsylvania Library)
Figure 3.8. Amedeo Avogadro (Edgar Fahs Smith Collection, University of Pennsylvania Library)
Figure 3.9. John Dalton (Edgar Fahs Smith Collection, University of Pennsylvania Library)
Figure 3.10. Forest fire in Yellowstone National Park (Photo by U.S. Forest Service, available from National Renewable Energy Laboratory, U.S. Department of Energy, http://www.nrel.gov)
Figure 3.12. Water pumping at rice paddy in Sundarbans, West Bengal, India, Feb. 25, 1998 (Photo by Jim Welch, available from National Renewable Energy Laboratory, U.S. Department of Energy, http://www.nrel.gov)
Figure 3.13. Thomas Midgley (Edgar Fahs Smith Collection, University of Pennsylvania Library)

CHAPTER 4

Figure 4.1. James Watt (Edgar Fahs Smith Collection, University of Pennsylvania Library)
Figure 4.2a. Panoramic view of Reading, Pennsylvania, c. 1909 (Photo by O. Conneaut, available from Library of Congress Prints and Photographs Division, Washington, DC)
Figure 4.2b. Panoramic view of Youngstown, Ohio, c. 1910 (Photo by O. Conneaut, available from Library of Congress Prints and Photographs Division, Washington, DC)
Figure 4.2c. Panoramic view of Indiana Steel Co.'s big mills, Gary, Indiana, c. 1912 (Photo by Crose Photo Co., available from Library of Congress Prints and Photographs Division, Washington, DC)
Figure 4.3. Noontime view of Donoro, Pennsylvania, during a deadly smog event. Copyright Photo Archive/Pittsburgh Post-Gazette, 2001. All rights reserved. Reprinted with permission
Figure 4.4. Panoramic view of Los Angeles, California, taken from Third and Olive Streets, December 3, 1909 (Photo by Chas. Z. Bailey, available from Library of Congress Prints and Photographs Division, Washington, DC)
Figure 4.5. Backyard incinerator ban, Herald-Examiner Photo Collection, Los Angeles Public Library

Figure 4.6. Smog bothers pedestrians, Hollywood Citizens News Collection, Los Angeles Public Library

Figure 4.7. Arie Haagen-Smit (Courtesy of the Archives, California Institute of Technology)

Figure 4.8. Los Angeles smog 2000 (Photo by M. Z. Jacobson)

CHAPTER 5

Figure 5.3. Saharan dust from satellite (SeaWiFS Project, NASA/Goddard Space Flight Center and ORBIMAGE; http://seawifs.gsfc.nasa.gov)

Figure 5.4. Mount St. Helens eruption (Peter Frenzen, available from Mount Saint Helens National Volcanic Monument Photo Gallery, http://www.fs.fed.us/gpnf/mshnvm/ photo_gallery/images/ hi_res/msh23_lrg.jpg)

Figure 5.5. Biomass burning from satellite (NASA/GSFC/JPL, MISR and Air MISR Teams).

Figure 5.6. SEM particle images (Reid and Hobbs, 1998)

Figure 5.7. Prescribed burn at Horse Creek Mesa, Big Horn National Forest, Wyoming, October 9, 1981 (Photo by U.S. Forest Service, available from National Renewable Energy Laboratory, U.S. Department of Energy, http://www.nrel.gov)

Figure 5.15. TEM particle images (Pósfai et al., 1999)

Figure 5.16. SEM coated soot (Strawa et al., 1999)

CHAPTER 6

Figure 6.6. William Ferrel (National Oceanic and Atmospheric Administration Central Library; http://www.photolib.noaa.gov/historic/c&gs/theb3456.htm)

Figure 6.15. Smog under an inversion in Los Angeles (Photo by M. Z. Jacobson)

Figure 6.17. Dust storm (Mon. Wea. Rev 63, 148, 1935, available from the National Oceanic and Atmospheric Administration Central Library; http://phlib.hpcc.noaa.gov/historic/nws/ wea01422.htm)

Figure 6.18. Satellite image of Atlanta temperatures (NASA-Goddard Space Flight Center Scientific Visualization Studio; http://svs.gsfc.nasa.gov/imagewall)

Figure 6.21. Sunset through elevated pollution layer in Los Angeles, by Gene Daniels, U.S. EPA, May 1972, Still Pictures Branch, U.S. National Archives

Figure 6.22. Elevated pollution layer in Los Angeles (Photo by M. Z. Jacobson)

Figure 6.23. Elevated smoke layer following a greenhouse fire (Photo by M. Z. Jacobson)

Figure 6.26. Lofting of pollution from a power plant (Available from National Renewable Energy Laboratory, U.S. Department of Energy, http://www.nrel.gov)

Figure 6.27. Lofting of pollution from stacks into an inversion (Photo by M. Z. Jacobson)

CHAPTER 7

Figure 7.1. Sir Isaac Newton (Edgar Fahs Smith Collection, University of Pennsylvania Library)

Figure 7.5. Lord Rayleigh (American Institute of Physics Emilio Segrè Visual Archives, Physics Today Collection)

Figure 7.7. Yellow sun before sunset (Photo by M. Z. Jacobson)

Figure 7.8. Red horizon at sunset (Photo by M. Z. Jacobson)

Figure 7.16. Light scattering by a cloud (Photo by M. Z. Jacobson)

Figure 7.17. Rainbow (Commander John Bortniak, NOAA Corps, available from the National Oceanic and Atmospheric Administration Central Library; http://phlib.hpcc.noaa.gov)

Figure 7.22. Maps of visibility in the United States (Schichtel et al., 2001)
Figure 7.23. Haze over Los Angeles, by Gene Daniels, U.S. EPA, May, 1972, Still Pictures
Branch, U.S. National Archives
Figure 7.24. Haze over San Gabriel Mountains, by Gene Daniels, U.S. EPA, May 1972, Still
Pictures Branch, U.S. National Archives
Figure 7.25. Reddish-brown colors in smog (Photo by M. Z. Jacobson)
Figure 7.26. Red sky over transmission lines near Salton Sea, by Charles O'Rear, U.S. EPA,
May, 1972, Still Pictures Branch, U.S. National Archives
Figure 7.27. Purple sky following volcano (Photo by Jeffrey Lew, UCLA)

CHAPTER 8

Figure 8.4. Solar One, solar furnace project (Photo by Sandia National Laboratory, available
from National Renewable Energy Laboratory, U.S. Department of Energy,
http://www.nrel.gov)
Figure 8.5. Wind machines at dusk (Photo by Warren Gretz, available from National
Renewable Energy Laboratory, U.S. Department of Energy, http://www.nrel.gov)
Figure 8.6. Photovoltaics and nuclear power, May 15, 1992 (Photo by Warren Gretz, available
from National Renewable Energy Laboratory, U.S. Department of Energy,
http://www.nrel.gov)
Figure 8.8. Power plant emissions in Moscow, November 28, 1994 (Photo by Roger Taylor,
available from National Renewable Energy Laboratory, U.S. Department of Energy,
http://www.nrel.gov)

CHAPTER 9

Figure 9.1. Henri Becquerel (NBS Archives, courtesy of American Institute of Physics Emilio
Segrè Visual Archives)
Figure 9.2. First evidence of radioactivity, along with Becquerel's notes (American Institute of
Physics Emilio Segrè Visual Archives, William G. Myers Collection)
Figure 9.3. Ernest Rutherford (Edgar Fahs Smith Collection, University of Pennsylvania
Library)
Figure 9.4. Asbestos (Photo by Robert Grieshaber, http://www.uwm.edu/Dept/EHSRM/
ASB/PHOTOS/mineralasb25c.jpg)

CHAPTER 10

Figure 10.1. Leblanc (Edgar Fahs Smith Collection, University of Pennsylvania Library)
Figure 10.2. Robert Angus Smith (Reproduced courtesy of the Library and Information Centre,
Royal Society of Chemistry)
Figure 10.4. Map of rainfall pH in the U.S. National Atmospheric Deposition
Program/National Trends Network (2000) (http://nadp.sws.uiuc.edu/isopleths)
Figure 10.5a. Acidified forest, Oberwiesenthal, Germany, near the border with the Czech
Republic, taken in 1991 (Photo by Stefan Rosengren/Naturbild)
Figure 10.5b. Acidified forest in Czechoslovakia (Photo by Owen Bricker, U.S.G.S).
Figure 10.6. Eroded statue (Herr Schmidt-Thomsen)
Figure. 10.7. Liming of lake (Photo by Tero Niemi/Naturbild)
Figure 10.8. Cathedral of Learning in 1930 and 1934 (Courtesy of University Archives,
University of Pittsburgh)

CHAPTER 11

Figure 11.6. Sydney Chapman (American Institute of Physics Emilio Segrè Visual Archives, Physics Today collection)
Figure 11.8. Mount Pinatubo eruption (Photo by Dave Harlow, USGS; http://geology.wr.usgs.gov/fact-sheet/fs113-97/)
Figure 11.16. Polar stratospheric clouds, NASA

CHAPTER 12

Figure 12.3a. Ultraviolet image of Venus's clouds as seen by the Pioneer Venus Orbiter, February 26, 1979 (Available from National Space Science Data Center, http://nssdc.gsfc.nasa.gov)
Figure 12.3b. Earth, from Apollo 17 spacecraft (Available from National Space Science Data Center, http://nssdc.gsfc.nasa.gov)
Figure 12.3c. Mars, from Hubble telescope. David Crisp/WFPC2 Science Team (Jet Propulsion Laboratory/California Institute of Technology) (Available from National Space Science Data Center, http://nssdc.gsfc.nasa.gov)
Figure 12.4. Greenhouse-gas absorption spectrum (Valley, 1965)
Figure 12.5. John Tyndall (American Institute of Physics Emilio Segrè Visual Archives, E.Scott Barr collection)
Figure 12.6. Svante Arrhenius (Edgar Fahs Smith Collection, University of Pennsylvania Library)
Figure 12.9. Fires in Africa (Data collected by U.S. Air Force Weather Agency; image and data processing by NOAA's National Geophysical Data Center; http://www.ngdc.noaa.gov/dmsp/fires/africa.html)
Figure 12.14. Sea ice, by Rear Admiral Harley D. Nygren, NOAA Corps (ret.), available from NOAA Central Library (a) http://www.photolib.noaa.gov/corps/images/big/corp1098.jpg; (b) http://www.photolib.noaa.gov/corps/images/big/corp1079.jpg
Figure 12.15. Ross Sea ice shelf, by Michael Van Woert, NOAA NESDIS, ORA. available from NOAA Central Library (a) http://www.photolib.noaa.gov/corps/images/big/corp2399.jpg; (b) http://www.photolib.noaa.gov/corps/images/big/corp2551.jpg
Figure 12.21. Island of Tuvalu (Photo by Peter Bennetts)

INDEX

Abiotic synthesis, 41, 42
Absorption
  aerosol particle, 187–190
    cross section, 195
    effect on visibility, 200
    effect on UV radiation, 189, 190, 278
    efficiency, single-particle, 195–197
    extinction coefficient
      population of particles, 195
      single particle, 188
  defined, 181
  gas
    cross section, 182
    effect on visibility, 200
    extinction coefficient, 182, 183
    infrared, 316–318
    ultraviolet and visible, 181–183
Abstraction, hydrogen, 96, 107, 108
Accumulation mode (see modes)
Acetaldehyde, 7, 98, 99, 105, 111, 228
Acetic acid, 258, 260
Acetone, 23, 78, 105, 247
Acetyl radical, 99
Acetylene, 105
Acid, 133, 258
Acid deposition, 254–272
  defined, ix, 254
  dry, 254
  causes
    hydrochloric acid, 254–257, 259, 260
    nitric acid, 259, 263
    sulfuric acid, 9, 254, 255, 259–262
  effects
    buildings and sculptures, 266, 268, 269

forests and crops, 264–266
health, 263
lakes and streams, 263, 264
rainwater pH in the United States, 264
fog, 254, 260
haze, 254
history of, 7, 254–257
locations of damage, 254–257, 263–266
methods of reducing
  ammonia, 270
  ammonium hydroxide, 267
  calcium carbonate, 268–269
  calcium hydroxide, 267–268
  sodium chloride, 269
  sodium hydroxide, 267–268
natural, 65, 258–260
pH (see pH)
rain, 254, 256–258
  first coined, 256
regulation of, 219, 256, 270, 271
wet, 254
Acid fog (see acid deposition)
Acid Precipitation Act, U.S., 270
Acid rain (see acid deposition)
Acidification
  sea spray, 119, 133
  soil-particle, 134
Acidity, 132
Active chlorine, 292, 298, 300
Addition process, 106–108
Adiabatic process
  compression, 158
  expansion, 157
Adsorption, 298

Advection, 54
Aerobic respiration (*see* respiration)
Aerosol (definition), 3
Aerosol particle, 116–142
    absorption (*see* absorption)
    accumulation mode (*see* modes)
    cloud condensation nucleus, 128
    coarse mode (*see* modes)
    coagulation of (*see* coagulation)
    composition, 64, 135, 137–139
    condensation onto (*see* condensation)
    concentration (*see* concentration)
    defined, 3
    deposition of (*see* deposition)
    dissolution into (*see* dissolution)
    effect on climate, 316–319, 339–341, 347–9
        BC-low-cloud-positive feedback loop, 341
        daytime stability effect, 340
        indirect effect, 341
        particle effect through large-scale meteorology, 341
        particle effect through surface albedo, 341
        photochemistry effect, 340
        semidirect effect, 341
        smudge-pot effect, 340
    effect on UV radiation (*see* radiation)
    effect on health, 117, 140–142, 344
    effect on tropospheric ozone, 190
    effect on stratospheric ozone, 283, 344
    emissions of (*see* emissions)
    externally-mixed, 140
    fine (*see* modes)
    internally-mixed, 130, 140
    liquid water content, 133, 136
    long-range transport of, 166, 167
    morphology (*see* morphology, aerosol particle)
    nucleation of (*see* nucleation)
    nucleation mode (*see* modes)
    primary, 118
    regulation (*see* aerosol particle, standards)
    removal processes (*see* air pollution)
    scattering (*see* scattering)
    secondary, 118
    sedimentation of (*see* sedimentation)
    shape (*see* shape, aerosol particle)
    size distribution (*see* size distribution)
    standards, outdoor, 213, 214, 218, 225–238
    visibility effects of (*see* visibility)
Agricola, Georgius, 17
Agriculture, invented, 333
Air
    defined, 2,
    density (*see* density)
    pressure (*see* pressure)
    specific heat (*see* specific heat)
    temperature (*see* temperature)
    thermal conductivity (*see* thermal conductivity)
Air mass, 165
Air pollution (*see also* smog; aerosol particle)
    defined, 4

chemicals in, 64
chemistry (*see* chemical reactions)
dispersion
    plume (*see* plume dispersion)
    vertical, 157–165
effect of clouds on, 167, 168
effect of ground temperatures on (*see* temperature)
effect of regulation on, 220–222
effect of soil liquid water on (*see* soil)
effect of winds on (*see* winds)
effect on health, 63–79, 140–142
effect on UV radiation (*see* radiation)
indoor (*see* indoor air pollution)
long-range transport of (*see* winds)
mortality, 84–86, 141, 226
    Bhopal (*see* Bhopal)
    Donora, 86–88
    Edinburgh, 85
    London, 86, 210, 229, 263
    Meuse Valley, 86
    outdoor (*see* smog)
    regulation, 82–92, 204–233, 210–238
    removal processes
        aerosol particle, 137, 138
        gas, 111, 137
    time of peak pollution during day, 163
    trends (*see* trends, air quality)
Air Pollution Control Act of 1955 (*see* Regulation, United States)
Air Quality Act of, 1967 (*see* Regulation, United States)
Air Quality Control Region (AQCR), 212
Air Quality Criteria (AQC), 212, 213
Aircraft emission standards (*see* regulation of air pollution)
Albedo, 313
    effect on climate, 338, 341
    planetary, 313, 314, 328
Alcohol, 23, 102, 104, 105, 109–111
    ethanol (*see* ethanol)
    fuel program, 227, 228
    methanol (*see* methanol)
Aldehyde, 23, 24, 102, 104, 105
Aleutian low, 153, 155
Algae, 41, 303
    blue-green, 42, 43
Alkali, (*see also* base) 18
Alkali Act, 84, 256, 270
Alkali industry, 254–257
Alkaline acid air (*see* ammonia)
Alkane, 23, 102, 104–106
Alkene, 23, 102, 104–107
Alkyne, 23, 105
Allegheny County, Pennsylvania, 86, 87
Allergen, 64, 242, 247
Allergy, 75, 247
Alpha particle, 245
Alum, potassium, 9, 20
Aluminum, 4, 5, 20, 30, 36, 37, 119, 127, 264, 339
Aluminum hydroxide, 264
Aluminum oxide, 187

American Chemical Society, 76
American Conference of Governmental Industrial
        Hygienists, Inc. (ACGIH), 251
Amino acid, 41
Ammonia
    aqueous, 136, 270
    gas
        as neutralizing agent, 270
        as refrigerant, 77, 286
        chemistry, 23, 41, 42, 46, 257
        discovery, 6, 9, 11, 13
        dissolution, 135, 267
        evolution in atmosphere, 39, 42, 43, 46
        sources and sinks, 19, 135
Ammonification, 46
Ammonium carbonate, 9
Ammonium chloride, 6, 8, 9
Ammonium hydroxide, 258, 267
Ammonium ion (ammonium)
    abundance in accumulation mode, 135–139
    chemistry, 40, 46, 135, 136, 267, 270
    concentrations, 135, 137
    role in air pollution problems, 64
    sources and sinks, 119, 135–139
Ammonium nitrate, 6, 9, 10, 136
Ammonium sulfate, 6, 9, 10, 136, 187
Anaerobic, 42
Anaerobic respiration (*see* respiration)
Aneroid barometer (*see* barometer)
Anion, defined, 132
Anoxygenic photosynthesis (*see* photosynthesis)
Antarctic ice sheets (*see* ice sheets)
Antarctic ozone depletion (*see* ozone)
Antarctic ozone hole (*see* ozone)
Anthracite (*see* coal)
Anthropogenic emissions (*see* emissions)
Anticyclone, 149
Apparent centrifugal force (*see* forces)
Apparent Coriolis force (*see* forces)
Aqueous, 4
Aragonite (*see* calcium carbonate)
Arcanite (*see* potassium sulfate)
Archean period, 327, 328
Arctic ozone dent (*see* ozone)
Area source (*see* emissions)
Argon, 4, 5, 20, 21, 22, 30, 36, 61, 62, 314
Aromatic
    chemistry, 104, 105, 107, 108
    compounds, 23
        nitrated (*see* nitrated aromatics)
    definition, 7, 23, 24
    emissions, 102
Arrhenius, Svante August, 318
Asbestos, 64, 242, 247, 248
Asbestosis, 248
Ash
    from industrial combustion (fly ash), 126, 127, 138, 139
    from biomass burning, 124, 138
Asteroid, 36, 37, 329

Asthma, 72, 75, 140–142, 248–250
Athens, 91
Atmosphere
    composition, 62–79
    history, 36–47
Atmospheric window, 317, 318
Atomic number, 2, 3
Atomic oxygen (*see* oxygen)
Atomic mass, 2, 3
Atom, 2
Attainment area, 213, 216
Aurora, 33
Australia, 32, 84, 125, 238, 345, 347
Autotroph, 40, 41
    lithotrophic, 40, 41
    photo, 40, 41
Autumnal equinox (*see* equinox)
Avogadro, Amadeo, 59
Avogadro's law, 58, 59
Avogadro's number, 59–61, 353

BC (*see* black carbon)
Backscattering (*see* scattering)
Bacteria
    airborne, 242, 248
    cyanobacteria, 40, 41, 43, 44, 303, 328
    decomposition by, 68
    effect on climate, 339
    hydrogen, 41
    iron, 41
    methanogenic, 41, 71
    methanotrophic, 71
    nitrifying, 41, 46
    nitrogen-fixing, 46
    nitrosofying, 46
    photosynthetic, 43
    purple and green, 41
    sulfur, 41
Bacterial decomposition (*see* decomposition, bacterial)
Baking soda (*see* sodium bicarbonate)
Balard, Antoine Jerome, 5, 19
Baldwin, Christopher, 6, 10
Baldwin's Phosphorus (*see* calcium nitrate)
Bangladesh, 342
Barium, 18
Barometer
    aneroid, 52
    mercury, 51
Basal-cell carcinoma (*see* carcinoma)
Basalt, 38
Base (alkaline substances), 133, 258, 266–270
Bay breeze, 170
Beavvais, Vincent de, 6, 9
Becher, Johann Joachim, 12
Becquerel, Antoine Henri, 21, 244, 245
Beijing, 91, 225, 235, 236
Benz, Karl, 110
Benzaldehyde, 107, 108
Benzene, 78, 140, 230

Benzene ring, 23, 24
Benzylperoxy radical, 107, 108
Bering land bridge, 334
Bermuda-Azores high, 155
Berzelius, Jöns Jakob, 5, 18–19
Best available control technology (BACT), 217
Beta particle, 245
Bhopal (*see* India)
Bicarbonate ion, 65, 67, 258
Big Bang, 30
Bimolecular reaction (*see* chemical reaction)
Binary nucleation (*see* nucleation)
Biogenic emissions (*see* emissions)
Biomass energy, 222
Biomass burning, 17, 65–78, 107, 119, 123–125, 321, 322
    composition of smoke from (*see* smoke)
    defined, 123
    indoor, 141
        mortality from, 141, 226
    map of African fires, 322
Biotic evolution, 40–47
Bisulfate ion, 135, 259, 261
Bisulfite, 261
Bisulfite ion, 261, 262
Bituminous coal (*see* coal)
Black carbon (BC) (*see also* soot)
    absorption of light by, 188, 278
        efficiency, 196
    coated, 124, 139, 140
    color (*see* colors)
    defined, 124
    effect on climate, x, 7, 310, 316–319, 339–342, 348, 349
    effect on health (*see* aerosol particle, effect on health)
    effect on visibility, 225
    lifetime, 348
    refractive index, 188
    role in air pollution problems, 64
    scattering efficiency, 196
    shape, 139, 140
    sources and sinks, 119, 124, 126, 138, 139, 319
Blackbody, 33–36, 311
Black, Joseph, 6, 12, 13, 15
Black-lung disease, 140, 141
Blue-green algae (*see* algae)
Blue moon, 185
Blue sky (*see* colors)
Blue Mountains, 185
Blue Ridge Mountains, 185
Bohr, Niels, 2
Boltzmann, Ludwig, 35
Boltzmann's constant, 52, 60, 354
Boundary layer, 54–56
    stable (nocturnal), 55, 56
BoWash corridor, 167, 220
Boyle, Robert, 58, 83
Boyle's law, 58, 59
Brand, Hennig, 5, 10
Brazil, 110, 111, 227, 228
Brimstone, 5

Bromine, elemental, 4, 5, 19
Bromine, atomic, gas, 293, 294, 300
Bromine monoxide, 293, 294, 300
Bromine nitrate, 294
Bromocarbons, 287, 289
Bronchial ailments, 72, 73, 75, 85, 140–142
Brown cloud, Denver, 200
Brownian motion, 128, 129
Buffers (*see* neutralizing agents)
Bussy, Antoine, 5, 20
Butadiene, 78, 89
Butane, 105
Butene, 105
Byron, Lord Percy Bysshe, 336

Cairo, 225, 234
Calcite (*see* calcium carbonate)
Calcined soda (*see* sodium carbonate)
Calcium, 4, 5, 18, 30, 36, 37, 119
Calcium carbonate, 6, 8, 12, 13, 67, 82, 121, 134, 255, 268, 269, 328
Calcium difluoride, 17
Calcium hydroxide, 82, 258, 267
Calcium ion, 67, 119, 134, 267, 268
    concentrations, 135, 137
    produced from weathering, 120, 268
Calcium nitrate, 6, 10
Calcium oxide, 12, 18, 82
Calcium silicate, 67
Calcium sulfate dihydrate (*see* gypsum)
Calcium sulfide, 255
Calcspar (*see* calcium carbonate)
Calcutta, 226, 235
California
    air pollution regulation in, 212–214
    air pollution standards, 213, 214, 219, 220
Canada, 33, 226, 263, 268, 345, 347
Canadian high, 156, 165
Cancer, 78, 248–250 (*see also* carcinoma, melanoma)
Carbohydrate, 43, 44
Carbon, 3–5, 7, 30, 36, 119
    reservoirs, 65
Carbon dioxide
    absorption of radiation by, 181, 182, 317
    buildup on Venus, 315
    characteristics, 65–68
    chemistry, 23, 39, 42–45, 67, 95, 134, 255, 268
    discovery, 6, 11–13, 15
    dissolution in rainwater, 258, 259
    effect of Milankovitch cycles on, 331–334
    effect of UV radiation on, 303
    effect on climate, ix, 316–319, 338, 339, 342–344, 348, 349
    effect on health, 68, 243
    evolution in atmosphere, 39–47, 326–336
    from soil-particle acidification, 134
    lifetime, 26, 348, 349
    mixing ratios, 47, 61–64, 68, 243, 259, 314, 315, 319, 320, 326–336, 342

photosynthesis of, 19, 43–45
regulation of, 345–349
reservoirs (*see* carbon)
solubility, 333, 339
sources and sinks, 39–47, 65–68, 123, 124, 134, 242, 243, 255, 268, 321, 345
structure, 22
Carbon monoxide
characteristics, 68–70, 317
chemistry, 23, 39, 95–97, 104, 107
discovery, 7, 16, 17
effect of catalytic converter on emissions, 216
effect on health, 69, 70
mixing ratios, 63, 69, 243
role in air pollution problems, 64
role in prebiotic atmosphere, 39
smog alert levels for, 220
sources and sinks, 17, 69, 70, 102, 103, 123, 124, 242, 243, 250
standards
emission, 212, 215
indoor, 251, 252
outdoor, 213, 214, 220, 226–238
structure, 22
time of peak mixing ratio during day, 163
Carbon tetrachloride, 287, 289–291, 305, 306, 317, 322
Carbonate ion, 67, 258
Carbonic acid, 67, 133, 258, 259
Carboniferous period, 327, 328
Carbonyl sulfide, 123
Carbonyls, 24
Carcinoma
basal-cell, 302
squamous cell, 302, 303
Catalytic converter, 77, 216, 226–238, 347
Catalytic ozone destruction cycles (*see* ozone)
Cataract, 302
Cation, defined, 132
Caustic potash (*see* potassium hydroxide)
Caustic soda (*see* sodium hydroxide)
Cavendish, Henry, 5, 6, 11, 13, 16, 59
Ceiling concentration, 251
Cellular respiration (*see* respiration)
Celsius, Anders, 52
Cenozoic era, 327, 329
Centrifugal force, apparent (*see* forces)
Centripetal acceleration, 145
CFCs (*see* chlorofluorocarbons)
Chain length, 281, 282, 292, 294
Chaldeans, 4
Chalk, 8, 120, 255
Chapman, Sidney, 279, 280
Chapman cycle, 278–280
Chappuis, M. J., 181
Chappuis bands, 181
Chaptal, Jean-Antoine, 14
Charcoal, 255
Charlemagne, 247
Charles, Jacques, 58, 59

Charles's law, 58, 59
Chemical lifetimes (*see* lifetimes, chemical)
Chemical reactions
air pollution, 99–111
bimolecular, 24
combination, 25
heterogeneous, 298
isomerization, 24, 25
kinetic, 24
nighttime, 95
photolysis, 24, 93–111, 290–300, 340
rate coefficient, 93
reversible, 25, 133
termolecular, 25
thermal decomposition, 24, 25
tropospheric, 93–99
unimolecular, 24
Chemical weathering (*see* weathering)
Chest pains, 72
Chile, 228
Chimney, 166, 167
Chimney effect, mountain, 170, 171
China, 125, 235, 236, 263, 345
Chloride ion (chloride)
chemistry, 133, 259, 269
concentrations, 133, 135, 137
role in air pollution problems, 64
sources and sinks, 119, 138, 139, 259
Chlorine
atomic, 290, 292, 299, 300
elemental, 4, 5, 18
percent of sea water, 119
molecular, 7, 18, 123, 298, 299
reservoirs, 292, 293, 298–300
Chlorine activation, 298
Chlorine gas (*see* chlorine, molecular)
Chlorine monoxide, 292, 299, 300
Chlorine nitrate, 64, 292, 293, 300
Chlorine nitrite, 299
Chlorite, 121
Chlorocarbons, 287, 289
Chlorofluorocarbons
CFC-11, 64, 287, 288, 290–293, 305, 306, 317, 322
CFC-12, 63, 64, 286–288, 290–293, 305, 306, 317, 322
CFC-113, 287, 288, 291, 305
CFC-114, 287
CFC-115, 287
chemistry, 286–301
effect on global warming, 317, 319, 322, 346
effect on stratospheric ozone, 64, 286–301
emissions of, 286–292, 305, 322
invention of, 76, 77, 286–288
lifetimes, 287, 289–291, 294
mixing ratios, 287, 289–291
properties of, 286
regulation of, 218, 303–306, 346
trends in, 288, 305, 306
uses for, 286–289

Chlorophylls, 43, 44
Chloroplasts, 43, 44
Chromosphere (*see* sun)
Cigarettes (*see* smoke and smoking)
Clay, 121 (*see also* specific heat, thermal conductivity)
Clean Air Acts, United Kingdom (*see* Regulation, United Kingdom)
Clean Air Act, United States (*see* Regulation, United States)
Cleve, Per, Theodor, 20, 21
Clevite, 20
Climate change (*see* global warming, greenhouse effect)
Climatic optimum, 335
    Medieval, 335
Cloud condensation nuclei (CCN), 128, 131, 341
Cloud drop formation, 131
Cloud layer, 55, 56
Cloud
    color (*see* colors)
    effect on air pollution, 167, 168
    effect on climate, 338–341
    effect on UV radiation, 167
Coagulation, 117, 128–130, 138, 139
    internal mixing by, 130
Coal
    air pollution from, 82–86, 255, 320–323
    anthracite, 125
    bituminous, 125
    combustion products, 126, 255
    defined, 125
    energy, 222
    first formed, 327
    lignite, 125, 231
    mortality from, 140, 141, 226
    peat, 125
    reserves, 125
    sea, 82, 83, 254
    soft, 85
    sulfur content, 271
Coal Workers' Pneumoconiosis (*see* black-lung disease)
Coarse mode (*see* modes)
Cold air mass (*see* air mass)
Cold cloud (*see* cloud)
Cold front (*see* front)
Collection (*see* gravitational collection)
Collision complex, 25
Colors
    black carbon, 204, 268, 269
    cloud
        black bottom, 197
        white, 202, 203
    haze, 202, 203
    horizon, 186, 187, 204
    nitrated aromatics, 202, 278
    nitrogen dioxide, 74, 202
    ozone, 72, 205
    PAHs, 202
    rainbow (*see* rainbow)
    sky

    blue, 184, 185
    purple, 205
    red, 204
    smog
        black, 204
        red, brown, 202, 203
    soil dust, 204
    sun
        red, 185, 186
        white, 184, 185
        yellow, 185, 186
Combination reaction (*see* chemical reaction)
Compounds
    defined, 3
    history of discovery, 4–21
Concentration
    aerosol particle
        mass, 116
        number, 116–118, 129
        volume, 116–118, 129
    gas, number, 60, 61, 116,
    hydrometeor particle, 116
Condensation, 117, 130–132, 138–139
Conditional instability (*see* stability)
Conduction, 38, 52–54, 327
Conductive heat flux, 53
Conductivity, thermal (*see* thermal conductivity)
Coning (*see* plume dispersion)
Conservation of mass, law of, 15
Contrast ratio, 198, 199
    liminal, 198
Convection, 38, 53, 54, 157, 168, 169
    forced, 53, 54, 157
    free, 53, 54, 157
Convective Brownian diffusion enhancement, 129
Convective mixed layer (*see* mixed layer)
Conventional heterotroph (*see* heterotroph)
Copenhagen Amendments, 304, 305
Copper, 4
Coriolis force, apparent (*see* forces)
Cornea, 302
Corona (*see* sun)
Cough, 72, 86, 87, 140–142
Cresol, 107, 108
Cretaceous period, 327, 329, 342
Cretaceous-Tertiary (K-T) extinction, 329
Criegee biradical, 106, 107
    excited, 25, 106, 107
Criteria air pollutant, 69, 72–75, 213
Crust, Earth's (*see* Earth)
Cryfts, Nicolas, 11
Curie, Marie, 5, 21
Curie, Pierre, 5, 21
Cutaneous melanoma (*see* melanoma)
Cyanobacteria (*see* bacteria)
Cycloalkane, 23
Cycloalkene, 23
Cyclopentene, 23
Cyclone, 149, 165

DNA, 42, 45
Daimler, Gottlieb, 110
Dalton, John, 60
Dalton's law of partial pressure, 60
Davy, Sir Humphry, 5, 18, 20
Decision, 229
Decomposition, bacterial (*see* bacteria)
Deforestation, 321, 322
Dehydration, 119
Denitrification
  bacterial, 40, 43, 46, 74, 280
  stratospheric, 299
Denmark, 346
Denovian period, 327, 328
Density, air, 50–52, 58–62
Denver (*see* brown cloud)
Dephlogisticated air (*see* oxygen, molecular)
Deposition, 111, 137 (*see also* sedimentation),
  dry, 137, 138
  wet (*see* rainout)
Desertification, 321
Des Voeux, Harold Antoine, 85
Deuterium, 3
Diabatic heating processes, 157
Diamonds, 5, 7, 15
Diesel, Rudolf, 126
Diesel fuel, 126, 319
  prevalence of, 126, 230
Diffraction, 192, 193
Dimer mechanism, 300
Dimethyl sulfide (DMS), 73, 260
Diminished nitrous air (*see* nitrous oxide)
Dinitrogen pentoxide, 25, 26, 95
Dioscorides, 5
Directive, 229
Dispersion, light, 192
Dispersive refraction (*see* dispersion, light)
Dissociation, 132
Dissolution, 66, 67, 132–139
  defined, 132
Diterpenene (*see* terpene)
Dizziness, 68, 78, 251
Dobson, Gordon M. B., 274
Dobson unit, 274, 275
Doldrums, 150, 151
Dolomieu, Silvain de, 121
Dolomite, 8, 121
Donnan, Frederick George, 3
Dorn, Friedrich Ernst, 5, 21
Drake, Edwin Laurentine, 109
Drought, 343
Dry deposition (*see* deposition)
Dry ice, 315
Dry adiabatic lapse rate (*see* lapse rate)
Duryea, J. Frank and Charles E., 110

E-folding lifetime (*see* lifetime)
Earth
  core, 38, 39

  inner, 38, 39
  outer, 38, 39
comparison with other planets, 314–316
Crust, 37, 38, 327
density, 13, 38, 39
distance from sun (*see* sun)
evolution of, 36–40
interaction with solar wind, 32, 33
mantle, 38, 39, 67
pressure within, 38, 39
radiation spectrum of, 34–36
radius of, 39, 52, 312, 354
structure, 38, 39
temperature
  equilibrium, 313, 314
  inner, 39
  near-surface air (*see* temperature)
Eccentricity of Earth's orbit, 331–334
Economy, effect of regulation on (*see* regulation, effect
        on economy)
Eddy, 54
Edward I, 83
Egypt, 234
Egyptians, ancient, 4, 5, 7, 8
El Chichón, 283
Electric motor, 85
Electrolysis, 18
Electrolyte, 132
  solid, 136
Electromagnetic wave, 33, 54
Electron, 2, 3
Electrostatic precipitator, 85
Elemental carbon (*see* black carbon)
Elements, 2, 3
  history of discovery, 4–21
Elementary reaction (*see* chemical reaction)
Elevated ozone layers in smog, 173, 174
Elevated pollution layers, 170–174
Emissions
  aerosol particle, 70, 118–127, 250
    regulation of, 212
  anthropogenic, 4, 118
  area source, 103, 118
  biogenic, 108
  biomass burning (*see* biomass burning)
  control techniques, 271
  fossil-fuel combustion (*see* fossil-fuel combustion)
  fugitive dust (*see* fugitive dust)
  greenhouse gas, 102, 320–323, 345–347
  incinerators (*see* incinerators)
  industrial, 119, 127, 138, 139
  meteoric debris (*see* meteoric debris)
  mobile source, 103, 118
    regulations (*see* regulation of air pollution)
  natural, 118
  point source, 103, 118
    regulations (*see* regulations of air pollution)
  power plant (*see* power plant emissions)
  sea spray (*see* sea spray)

Emissions (*cont.*)
  smelters (*see* smelter emissions)
  soil dust (*see* soil dust)
  stationary source, 103
  steel mill (*see* steel mill emissions)
  trends, 221
  volcanic (*see* volcanos)
Emissivity, 35, 311, 313
Emphysema, 72
Energy, 222
  biomass (*see* biomass energy)
  coal (*see* coal)
  geothermal (*see* geothermal energy)
  hydroelectric (*see* hydroelectric energy)
  natural gas (*see* natural gas)
  nuclear (*see* nuclear energy)
  oil (*see* oil)
  renewable (*see* renewable energy)
  solar (*see* solar energy)
  wind (*see* wind energy)
  units, 353
Engine, 109, 110
Engine knock, 76
Entrainment zone, 55, 56
Environmental lapse rate (*see* lapse rate)
Environmental Protection Agency, U.S. (U.S. EPA),
      213, 270
Environmental tobacco smoke (*see* smoke, environmental
      tobacco)
Eocene epoch, 327, 329
Epidermis, 301
Epsom salt (*see* magnesium sulfate)
Epsomite (*see* magnesium sulfate)
Equation of state, 58–62
  for dry air, 61
Equatorial low-pressure belt, 150, 151
Equilibrium temperature (*see* temperature)
Equinox
  autumnal, 311, 312
  vernal, 311, 312
Ethane
  chemistry, 98
  lifetime, 98
  mixing ratios, 63, 98
  structure, 23, 286
Ethanol
  as vehicle fuel, 109–111, 227
  chemistry, 42, 111
  lifetime, 105, 110
Ethanoloxy radical, 106
Ethanyl radical, 106
Ethene, 23, 63, 64, 105–107
Ethene molozonide (*see* molozonide, ethene)
Ethoxy radical, 98, 111
Ethyl (*see* leaded gasoline)
Ethyl radical, 98
Ethylperoxy radical, 98
Eukaryote, 40, 45, 327
European Union, 228–230, 304, 347

Evaporation, 63, 64, 66, 130–132, 168
Evelyn, John, 83
Excited criegee biradical (*see* criegee biradical)
Exothermic, 42
Externally-mixed particle (*see* aerosol particle)
Extinction, species, 329, 343
Extinction coefficient, 182, 199 (*see also* absorption,
      scattering)
Eye irritation, 78, 86–90, 97, 98, 249 (*see also*
      lachrymator)

Fahrenheit, Gabriel Daniel, 52
Faint-Young Sun paradox, 327
Famine, 343
Fanning (*see* plume dispersion)
Far-UV (*see* radiation)
Feedbacks to climate
  aerosol particles (*see* aerosol particle, effect on climate)
  gases, 338, 339
  negative, 338
  positive, 338
Feldspars, 38, 121
Fennoscandian Ice Sheet (*see* ice sheets)
Fermentation, 42, 43, 66
Ferrel, William, 150
Ferrel cell, 150–152
Fertilizer, 46
Fine particles (*see* aerosol particles)
Finland, 268, 346
Fire-air, 11
Fixed air (*see* carbon dioxide)
Fixed gases, 62
Fluor acid (*see* hydrofluoric acid)
Fluorine, 3–5, 17, 123
Fluorite (*see* calcium difluoride)
Fluorspar (*see* calcium difluoride)
Fly ash (*see* ash)
Fog, 86
Forced convection (*see* convection)
Forced ventilation (*see* ventilation)
Forces acting on air
  apparent centrifugal, 147–154
  apparent Coriolis, 146–154
  friction, 147–154
  pressure gradient, 146–154
Ford, Henry, 110
Formaldehyde
  chemistry, 97, 104–106, 109, 110, 181
  effect on health, 97
  lifetime, 105
  mixing ratios, 63, 97, 243
  reactivity, 105
  role in air pollution problems, 64, 97, 242, 243
  sources and sinks, 97, 242, 243
  structure, 23
Formic acid, 260
  excited, 25, 107
Formyl radical, 97, 99, 104
Forward scattering (*see* scattering)

Fossil-fuel combustion, 17, 65–78, 107, 125–127
    aerosol particle products of, 119, 125–127, 138, 139
Fourier, Jean Baptiste, 318
Fowler, Alfred, 181
France, 230, 231, 345
Free convection (*see* convection)
Free radical, 22, 74, 103
Free troposphere (*see* troposphere)
French Revolution, 16, 254
Freon, 288
Friction force (*see* forces)
Front
    cold, 165
    polar, 152
    subtropical, 151
Frontal inversion (*see* inversion)
Fugitive dust, 120–122
Fumarole, 36, 39
Fumigation (*see* plume dispersion)
Fungal spores, 64, 242, 248

Galileo Galilei, 50, 52
Gamma radiation (*see* radiation)
Gary, Indiana, 86, 87
Gas, 3, 11, 15
    concentration (*see* concentration)
    history of discovery, 11–21
    origin of word, 11
Gas constant
    dry air, 61, 354
    universal, 60, 354
Gas pingue (*see* ammonia)
Gas scattering (*see* scattering)
Gas silvestre (*see* carbon dioxide)
Gasoline
    defined, 126
    leaded (*see* leaded gasoline)
    unleaded (*see* unleaded gasoline)
Geber, 6, 8, 9
General Motors, 76, 77, 286
Geneva Convention on Long-Range Transboundary
        Air Pollution, 270
Geometric regime, 196, 197
Geostrophic
    adjustment, 148
    balance, 148
    wind (*see* winds)
Geothermal energy, 222
Germany, 84, 231, 232, 265, 345
Geyser, 36, 39
Glacier, 329, 330, 334, 335, 342
    Alaskan, 327, 330, 342
Glaser, Christopher, 6, 10
Glauber, Johann Rudolf, 6, 9, 10
Glauber's salt (*see* sodium sulfate)
Global ozone reduction (*see* ozone)
Global warming 310–349
    aerosol particle contribution to, 316–319, 339–342,
        348, 349

defined, ix, x, 310, 316
discovered, 318
effects, 342–344
    on ecosystems, 343
    on human health, 344
    on regional climate and agriculture, 343
    on sea levels, 342, 343
    on stratospheric ozone, 344
    on stratospheric temperatures, 324, 325, 344
evidence of (*see* temperature, trends)
greenhouse gas contribution to, 316–319, 338, 339,
        348, 349
pollutants affecting, 64, 316–319
regulation of, 345–349
Glucose, 42, 44, 45
Glycol aldehyde, 106
Gobi Desert, 122, 235
Gold, 4
Goldilox hypothesis, 316
Gossage, William, 256, 271
Gradient wind (*see* winds)
Grand Canyon, pollution in, 167
Granite, 38, 120
Gravitational collection, 129
Gravity, 2, 354
Greeks, ancient, 7, 8, 82
Green vitriol, 9
Greenhouse effect
    natural, 310, 314, 316–318, 328
    runaway, 315
Greenhouse gases, 63, 65, 70, 310, 316–318
    emissions (*see* emissions)
    trends in, 319–323
Greenland ice sheet (*see* ice sheets)
Grew, Nehemiah, 6, 10
Ground temperature, effect on air pollution
        (*see* temperature)
Growth, particle (*see* condensation, dissolution)
Gypsum, 6, 8, 120, 121, 136, 255, 268, 269

Haagen-Smit, Arie, 89–91, 210
Hadley, George, 150, 151
Hadley cell, 150–152
Halite (*see* sodium chloride)
Halons, 218, 287, 289, 291, 293, 305
    H-1211, 287, 289, 293
    H-1301, 287, 289, 293
    H-2402, 287, 289
Hartley bands, 181
Hazardous air pollutants (HAPs), 78, 79, 215, 218, 249
Haze, 202, 203
HCFCs (*see* hydrochlorofluorocarbons)
Headaches, 68, 69, 78, 251
Health advisory, 220
Health, Education, and Welfare, U.S. Department of,
        211–213
Health effects (*see* acid deposition, aerosol particle, air
        pollution, global warming, radiation, smoke,
        smoking, and individual chemicals)

Heart ailments, 79, 141

Heat island effect, urban, 169

Hebrews, ancient, 7

Helium, 3–5, 20–22, 30, 31, 36–38, 62, 314, 315

Hematite, 121, 187

Hemiterpene (*see* isoprene)

Heterogeneous nucleation (*see* nucleation)

Heterogeneous reactions (*see* chemical reaction)

Hemoglobin, 69

Heterotroph, 40

   conventional, 40, 41

   lithotrophic, 40, 41

   photo, 40, 41

High pressure systems (*see* pressure systems)

HFCs (*see* hydrofluorocarbons)

Holocene epoch, 327, 335, 336

Holocene maximum, mid-, 335

Homogeneous nucleation (*see* nucleation)

Homomolecular nucleation (*see* nucleation)

Homosphere, 54

Horace, 82

Horizon, color of (*see* colors)

Horse latitudes, 150, 152

Howard, Luke, 169

Huggins bands, 181

Huygens's principle, 192

Hydration, 133

Hydrobromic acid, 294

Hydrocarbon

   alkane (*see* alkane)

   alkene (*see* alkene)

   alkyne (*see* alkyne)

   aromatic (*see also* aromatic)

      polycyclic (PAH), 124, 127, 140, 242, 339

         color (*see* colors)

         UV absorption by, 187–190, 272

   cycloalkane (*see* cycloalkane)

   cycloalkene (*see* cycloalkene)

   defined, 23

   emission standards for, 211, 212, 215

   nonmethane, 24

   oxygenated, 24

   terpene (*see* terpene)

Hydrochloric acid

   adsorbed, 298, 299

   aqueous

      chemistry, 18, 133, 134, 259

      discovery of, 6, 9, 13

   gas

      acid deposition from (*see* acid deposition)

      discovery of, 7, 16, 18

      chemistry, 255, 269

      dissolution of 133, 134, 259

      lifetime, 287, 290, 293

      mixing ratios, 287

      regulation of, 84, 256

      removal by rainout, 291

      role in acid deposition, 64, 254

      role in stratospheric ozone loss, 64, 291–293, 299–301

      sources, 17, 119, 123, 133, 134, 255, 260, 269, 291, 292

Hydrochlorofluorocarbons (HCFCs), 287, 289, 290, 305, 306

   HCFC-22, 287, 289–291, 305, 306, 317, 322

   HCFC-141b, 287, 305, 306

   HCFC-142b, 287, 306

Hydroelectric energy, 222, 228

Hydrofluoric acid, 7, 17, 18

Hydrofluorocarbons (HFCs), 287, 289, 305, 306, 347

   HFC-134a, 287, 289, 305, 306

Hydrogen

   atomic, 42, 95, 97, 104, 292, 314, 315

   elemental, 3–5, 19, 21, 30, 31, 36, 37, 38

   ion, 67, 132–136, 257–262, 267–270

   molecular (gas), 6, 11, 13, 23, 39–42, 47, 59, 97, 107, 123, 282, 292, 314

Hydrogen convection zone (HCZ) (*see* sun)

Hydrogen oxides, 282

Hydrogen peroxide

   aqueous, 262

   gas, 292, 294

Hydrogen sulfide, 39, 40, 42, 43, 45, 73, 214, 255

Hydrologic cycle, 39, 63

Hydrometeor, 3

Hydrometeor particle, 3, 116

Hydroperoxy radical

   chemistry, 95–107, 244, 281, 282, 292, 294

   concentrations, 105, 244

Hydroxide ion, 133, 257

Hydroxyl radical

   chemistry, 23, 25, 39, 94–111, 261, 263, 281, 282, 299, 315

   concentrations, 94, 105

   lifetime, 26, 244

   sources, 103, 104

   structure, 22

Hygroscope, 11

Hygroscopic, 139

Hypochlorous acid, 7, 19, 298, 299

Ice core record, 320, 326, 331, 335, 337

Ice age

   last, 334

   Little, 335, 336

Ice sheets

   Antarctic, 327, 329, 342

      East, 329, 342

      West, 329, 342

   defined, 329

   Greenland, 327, 330, 342

   Fennoscandian, 334, 335

   Laurentide, 334, 335

Icelandic low, 155

Ideal gas law, 58–60

Igneous rocks (*see* rocks)

Illite, 121

Imaginary refractive index (*see* refractive index)

Immune system ailments, 302, 303

Incinerators

backyard, 89, 90
industrial, 127
emissions from, 127
India, 84, 125, 235, 345
Bhopal, 235
Indirect effect (*see* aerosol particle, effect on climate)
Indonesian forest fires, 167
Indoor air pollution, ix, 64, 242–252
Industrial Revolution, 84, 254
Infiltration, 242, 243
Inflammable air (*see* hydrogen, molecular)
Infrared radiation (*see* radiation)
Inner core (*see* Earth)
Inorganic compound, 23
Intermediate interior (*see* sun)
Internal variability in Earth's climate, 338
Internally-mixed particle (*see* aerosol particle)
Intertropical Convergence Zone (ITCZ), 150, 152
Inversion, 55, 56, 160–165
base height, 160, 161
defined, 56, 160
effect on air pollution, 56, 86, 91, 161
frontal, 165
large-scale subsidence, 162–164
in Los Angeles, 163, 164
in Mexico City, 227
in Santiago, 228
in Tehran, 234
marine, 164
popping, 163
radiation (nocturnal), 162
small-scale subsidence, 165
strength, 160, 161
thickness, 160, 161
Involatile, 134
Ion, 2, 22
anion (*see* anion)
cation (*see* cation)
Iran, 234
Iron, 4, 5, 30, 36, 37, 39, 119, 127, 264, 339
Isobar, 148
Isomerization reaction (*see* chemical reaction)
Isoprene
chemistry, 104, 108, 109
lifetime, 105, 109
sources, 102, 108
structure, 23
Isotope, 2, 244–246, 326
Israel, 233, 234

Janssen, Pierre, 5, 20
Japan, 84, 236, 237, 263, 345, 347
Jet streams
polar front, 152, 275, 297
subtropical, 151
Johannesburg, 91, 237
Junge layer, 123
Jupiter, 314, 315
Jupiter Ammon, Temple of, 9

Jurassic period, 327, 328

Kaolinite, 121
Keeling, Charles David, 320
Kerosene, 126
Ketones, 23, 24, 104, 105
Koschmieder equation, 199
Krypton, 4, 5, 20, 21, 22, 62
Kyoto Protocol, 347–349

Lachrymator, 97, 98 (*see also* eye irritation)
Lake breeze, 170
Lakes of Natron (*see* Natron, Lakes of)
Land breeze, 170, 171
Landcover, 169
Langlet, Nils Abraham, 20
Lapse rate,
adiabatic
dry, 157–160
wet, 157–160
environmental, 157–160
Large-scale subsidence inversion (*see* inversion)
Last ice age (*see* ice age)
Latent heat, 13, 151, 157
Laughing gas (*see* nitrous oxide)
Laurentide ice sheet (*see* ice sheets)
Lavoisier, Antoine Laurent, 13, 15, 16, 18
Lead
characteristics, 19, 75–79, 245, 246
concentrations, 78
discovery, 4, 5
effect on catalytic converter, 216
effect on health, 76, 78, 79
emissions, 70
poisoning, 76, 78
role in air pollution problems, 64, 234
sources and sinks, 70, 75–77, 127, 245, 246
standards
emission, 212, 226–238
outdoor, 213, 214, 226–238
Leaded gasoline, 75–79, 226–238, 286 (*see also* lead)
Leblanc, Nicolas, 254–256
Lenoire, Étienne, 110
Libavius, Andreas, 6, 9, 255
Liebig, Justus von, 19, 20
Life, earliest, 327
Lifetime, chemical, 26
Lightning, 24, 74
Lignite coal (*see* coal)
Lime
quick- (*see* calcium oxide)
slaked (*see* calcium hydroxide)
Lime kilns, 82
Limestone, 8, 82, 268, 269
Limonene, 108, 247
Liquids, 4, 15
history of discovery, 4–10
Liquid water content
aerosol particles (*see* aerosol particles)
soil (*see* soil)

Lithotroph, 40
Lithotrophic autotroph (*see* autotroph)
Lithotrophic heterotroph (*see* heterotroph)
Little Ice Age (*see* ice age)
Lockyer, Joseph Norman, 5, 20
Lofting (*see* plume dispersion)
Lognormal distribution (*see* size distribution)
London, 82, 86, 210, 225, 229
London Amendments, 304, 305
London-type smog (*see* smog)
Long-range transport of air pollution (*see* winds)
Looping (*see* plume dispersion)
Los Angeles air pollution, 88–92, 101–105, 109, 135–137,
        163, 170–173, 202–204, 221, 225
    regulation of, 88–92, 210
    trends, 221
Los Angeles Air Pollution Control District
        (LAAPCD), 210
Loschmidt, Joseph, 24, 59
Low pressure systems (*see* pressure systems)
Lowest achievable emissions rate (LAER), 217
Lung ailments, 72, 73, 75, 85, 97, 141, 246, 248, 250
Lye (*see* sodium hydroxide)

Madrid, 231
Magma, 120, 123
Magma oceans, 38, 327
Magnesia, city of, 5, 20
Magnesia (*see* magnesium oxide)
Magnesia alba (*see* magnesium carbonate)
Magnesite (*see* magnesium carbonate)
Magnesium, elemental, 4, 5, 20, 30, 36, 37, 119
Magnesium carbonate, 6, 12, 134
Magnesium ion, 119
    concentrations, 135, 137
Magnesium oxide, 20
Magnesium sulfate, 6, 10, 121
Mainstream smoke (*see* smoke, environmental tobacco)
Manganese dioxide, 18
Manganic oxide, 15
Mantle, Earth's (*see* Earth)
Marble, 8, 13, 120, 268, 269
Marcus, Siegfried, 110
Marine acid air (*see* hydrochloric acid, gas)
Marine inversion (*see* inversion)
Mars, 314–316
Mascagnite (*see* ammonium sulfate)
Mass, 2
    atomic (*see* atomic mass)
    of an air molecule, 52, 354
Mass concentration (*see* concentration)
Maximum Achievable Control Technology (MACT), 218
Mayow, John, 11, 12
Mean free path, 31
Mechanical turbulence (*see* turbulence)
Medieval climatic optimum (*see* climatic optimum)
Melanin, 301
Melanocyte, 302
Melanoma

    cutaneous, 302
    ocular, 302
Mendeleev, Dmitri, 2
Mephitic air (*see* nitrogen, molecular)
Mercuric oxide, 14, 15
Mercury (element), 4, 14, 19, 127
Mercury (planet), 314, 315
Mercury barometer (*see* barometer)
Meridianal, 150
Mesopause, 55, 57
Mesosphere, 54, 55, 57
Mesothelioma, 248
Mesozoic era, 327, 328
Metals from industrial emissions, 126, 127, 138–141
Metamorphic rocks (*see* rocks)
Meteoric debris, 127
Meteorite, 36, 37, 327
Meteorological effects on air pollution,
Meteorological range (*see* visibility)
Methacrolein, 109
Methane
    bacterial production, 42
    characteristics, 23, 70–72
    chemistry, 23–25, 39, 41, 42, 96, 282, 283, 292
    constituent of natural gas, 126
    effect of Milankovitch cycles on, 333, 334
    effect on climate, ix, 64, 317–319, 339, 348, 349
    effect on health, 72
    emissions, 102, 321
    formation of chlorofluorocarbons from, 286
    lifetime, 71, 96
    mixing ratios, 63, 71, 96, 319, 320, 331
    regulation of, 347
    sources and sinks, 70, 71, 124, 282
    structure, 22
Methanogenic bacteria (*see* bacteria)
Methanol, 23, 110, 111
Methanotrophic bacteria (*see* bacteria)
Methoxy radical, 96, 97, 99, 110
Methyl bromide, 287, 289, 293
Methyl chloride, 63, 287, 289, 291, 317, 319
Methyl chloroform, 287, 289, 291, 305, 306
Methyl radical, 25, 96, 99, 282, 292, 293
Methylene chloride, 78
Methylethylketone, 78, 247
Methylperoxy radical, 96
Methylvinylketone, 109
Mexico, 226, 227
Mexico City, 91, 225–227
Midgley, Thomas, 76, 77, 286–288
Midlatitudes, 152, 195
Mie, Gustav, 196
Mie regime, 196, 197
Milankovitch, Milutin, 331
Milankovitch cycles, 331–334, 337
Miller, Stanley, 41, 42
Mineral
    defined, 120
    in soil, 121

Miocene epoch, 327, 330

Mites, dust, 247

Mixed layer, 55, 56

Mixing depth, 160, 161, 163, 164, 168, 169, 173

Mixing ratio

saturation, 64

volume, 61, 62, 93

Mobile source (*see* emissions)

Modes, 116, 117

accumulation, 116, 117, 138, 197

importance for visibility, 197

coarse, 116, 117, 121, 138

fine, 117

nucleation, 116, 117, 138

Moissan, Henri, 17

Mojave Desert, 122, 155, 166

Molarity, 132, 257

Molecular hydrogen (*see* hydrogen)

Molecular mass (*see* mass, of an air molecule)

Molecular nitrogen (*see* nitrogen)

Molecular oxygen (*see* oxygen)

Molecular weight

atomic (*see* mass, atomic)

dry air, 61, 354

Molecule, 3

Molozonide, ethene, 106

Monoterpene (*see* terpene)

Monsoon, 156

Montgolfier brothers, 54

Montreal Protocol, 304, 305

Moon, 314, 315

blue (*see* blue moon)

Morey, Samuel, 109

Morphology, aerosol particle, 139–140

Morveau, Louis de, 20

Moscow, 233

Motor Vehicle Air Pollution Control Act of 1965
     (*see* Regulation United States)

Motor Vehicle Pollution Control Board, 211

Mount Pinatubo, 283, 284

Mount St. Helens, 123

Mountain breeze, 171

Mountain chimney effect (*see* chimney effect)

Muriatic gas (*see* hydrochloric acid, gas)

Nail polish remover, 78

National Ambient Air Quality Standards (NAAQS), 69–74,
     213–220, 251

National Atmospheric Deposition Program (NADP), 270

National Emission Standards for Hazardous Air Pollutants
     (NESHAP), 215, 217

National Institute for Occupational Safety and Health
     (NIOSH), 251

Natrite (*see* sodium carbonate)

Natron, Lakes of, 7

Natural gas, 70, 126, 222

Nausea, 68, 78, 87, 251

Near UV radiation (*see* radiation)

Near-infrared radiation (*see* radiation)

Negative-feedback mechanism (*see* feedbacks to climate)

Neon, 3–5, 20, 21, 22, 30, 36, 62, 314

Neptune, 314, 315

Nerve damage, 78, 79

Netherlands, 346

Neutral stability (*see* stability)

Neutralizing agents, 266–270

Neutron, 2, 3

New Delhi, 225, 235

New Source Performance Standards (NSPS), 215, 217

Newcomen, Thomas, 83

Newton, Sir Isaac, 180, 192

Nickel, 36, 37, 39, 127

Niton (*see* radon)

Nitrammite (*see* ammonium nitrate)

Nitrate ion (nitrate)

abundance in coarse mode, 134–139

chemistry, 43, 46, 134, 259

concentrations, 135, 136, 190

dissociation reaction producing, 134, 259

effect on climate, 339

effect on visibility, 225

role in air pollution problems, 64

sources and sinks, 74, 119, 138, 139, 259

Nitrate radical

absorption of UV radiation by, 181, 182

chemistry, 25, 26, 95, 103, 105, 108

concentrations, 105

Nitrated aromatics

color (*see* colors)

imaginary refractive indices of, 189

UV absorption by, 187–190, 339

Nitre (*see* potassium nitrate)

Nitric acid

adsorbed, 298, 299

aqueous

chemistry, 15, 95, 133, 134, 259

discovery of, 6, 9

gas

acid deposition from (*see* acid deposition)

chemistry, 94, 181, 259, 263, 281

discovery of, 7, 16

dissolution of, 94, 119, 134, 259

lifetime, 94

role in air pollution problems, 64, 254

sources, 74, 134, 260, 263, 281

Nitric acid dihydrate, 297

Nitric acid trihydrate (NAT), 297

Nitric oxide

characteristics, 74

chemistry, 23, 24, 46, 93–109

destruction of ozone with, 93, 173, 174, 280, 281

discovery of, 7, 16

effect on health, 74

lifetime, 26

mixing ratios, 63, 74, 93, 100, 101

role in air pollution problems, 64

sources and sinks, 17, 74, 255, 280, 281

time of peak mixing ratio during day, 163

Nitric oxide (*cont.*)
    structure, 22
Nitrification, 40, 46
Nitrite ion, 40, 43, 46
Nitrocalcite (*see* calcium nitrate)
Nitrogen
    atomic, 42
    elemental, 3–5, 30, 36
    molecular
        absorption of UV radiation by, 58, 181, 182, 277, 279
        chemistry, 23, 25, 39, 42, 43, 45–47, 103
        discovery of, 7, 13, 14
        evolution in atmosphere, 40–47
        lifetime, 26
        mixing ratios, 47, 50, 61, 62, 314
        sources and sinks, 45–47, 123
        structure, 22
Nitrogen cycle, 45–47
Nitrogen dioxide
    absorption extinction coefficient, 182, 183
    absorption of UV radiation by, 181, 182, 278
    characteristics, 74, 75
    chemistry, 23–26, 39, 93–109, 263, 280, 281, 292, 294
    color (*see* colors)
    discovery of, 6, 16
    effect on health, 74, 75
    lifetime, 26
    mixing ratios, 63, 75, 93, 183, 199, 243
    role in air pollution problems, 64
    role in prebiotic atmosphere, 39
    smog alert levels for, 220
    sources and sinks, 17, 74, 75, 242, 243, 250, 263
    standards
        indoor, 251, 252
        outdoor, 213, 214, 220, 226–238
    structure, 22
    visibility reduction because of, 199
Nitrogen fixation, 40, 46, 303
Nitrogen oxides
    catalytic ozone destruction cycle, 280, 281
    chemistry, 89, 94, 100, 101, 111
    defined, 94
    emissions, 70, 102, 103, 124, 260, 281
        effect of catalytic converter on, 216
        standards, 211, 212, 215
    mixing ratios, 102, 103
Nitrous acid, 103, 104
Nitrous air (*see* nitric oxide)
Nitrous oxide
    absorption of radiation by, 317
    chemistry, 46, 280
    discovery, 7, 16, 17
    effect on climate, ix, 317–319
    mixing ratios, 280, 319, 320
    regulation of, 347
    role in air pollution problems, 64
    sources and sinks, 17, 280, 281, 321
Nitrum (*see* potassium nitrate)
Noble elements, 20, 22

Nocturnal boundary layer (*see* boundary layer)
Nonattainment area, 213, 216
Nonmethane hydrocarbon (NMHC) (*see* hydrocarbons)
Nonmethane organic carbon (NMOC), 24
Northern Lights (*see* Aurora)
Norway, 268, 346
Nuclear energy, 222, 224, 230
Nucleation, 128, 138, 139,
    binary, 128
    heterogeneous, 128, 130
    homogeneous, 128
    homomolecular, 128
    ternary, 128
Nucleation mode (*see* modes)
Nucleus, atomic, 2
Number concentration (*see* concentration)

Obliquity, 312, 333
Occupational Safety and Health Administration
        (OSHA), 251
Oceans, formation of, 39, 327
Ocular melanoma (*see* melanoma)
Oil (petroleum) 109, 126, 222
    characteristics of, 126
    discovery of, 109
    formation of, 126
Oil of sulfur (*see* sulfurous acid)
Oil of vitriol (*see* sulfuric acid)
Oligocene epoch, 327, 330
OM (*see* organic matter)
Organic chemistry, founding of, 19
Organic compound, 19, 23, 40–45, 64
Organic matter (OM)
    defined, 124
    effect on climate, with black carbon (*see* black carbon)
    in soil, 121
    particulate
        effect on visibility, 225
        refractive index, 188
        role in air pollution problems, 64
        sources, 119, 124, 126, 138, 139
Otto, Nikolaus, 110
Outer core (*see* Earth)
Outgassing, 36, 39, 327
Oxidants, 213
Oxidation, 23, 42, 44
Oxygen
    atomic
        excited, 94, 104, 278–280
        ground state, 24, 93–105, 278–282, 292, 293
        isotopically-enriched, 326
    elemental, 3–5, 15, 19, 30, 36, 37, 39
    molecular
        absorption of radiation by, 58, 181, 182, 277, 278, 317
        aerobic respiration, 45
        chemistry, 13, 23, 25, 43, 44, 93–111, 278–280, 292–294
    discovery of, 7, 12, 14–16

evolution in atmosphere, 43–47
lifetime, 26
mixing ratios, 47, 50, 61, 62, 314
photosynthesis production, 43–45
sources and sinks, 43–46
structure, 22
Oxygenated hydrocarbons (*see* hydrocarbons)
Oxymuriatic acid (*see* chlorine gas)
Ozone
  aqueous, 262
  gas
    absorption extinction coefficient, 182, 183
    absorption of UV radiation by, 181, 182, 277, 278
    chemistry
      stratosphere, 280–282, 290–300
        catalytic destruction cycles, 280–282, 292, 293, 299, 300
      troposphere, 93–111
    color (*see* colors)
    column abundance, 274–276, 283–285
    dent, Arctic, 275, 276, 285, 286, 295–301
    discovery of, 7, 20
    effect of aerosol particles on, 190, 283, 344
    effect of climate on, 344
    effect of UV radiation on, 190, 303
    effect on climate, 317, 318
    effect on UV radiation, 181, 274, 277, 278, 286
    formation of elevated layers in smog, 173, 174
    health, plant, and material effects, 72, 73, 89, 266, 274
    hole, Antarctic, ix, 275, 284, 285, 295–301
      area of, 295, 296
      defined, 276
      map of, 275, 296
    indoor, 242, 243
    isopleth, 100, 101
    lifetime, 26
    long-range transport of, 166, 167, 219
    mixing ratios, 63, 72, 89, 93, 100, 101, 243, 276
      peak stratospheric, 276, 277, 281
    photostationary state (*see* photostationary state)
    regeneration, 294, 300, 301
    role in different air pollution problems, 64
    smog alert levels for, 220
    sources and sinks, 72, 242, 243
    standards
      indoor, 251, 252
      outdoor, 213, 214, 218, 220, 226–238
    stratospheric, 55, 57, 181, 182, 274–306
      effect of bromine on, 293, 294
      effect of chlorine on, 292, 293, 298–300
      effect of global warming on, 344
      effect of hydrogen on, 281, 282
      effect of nitrogen on, 280, 281
      evolution of, 45, 274
      regulations relating to, 303–306
      trends, 283–286, 305, 306
    structure, molecular, 22, 24
    time of peak mixing ratio during day, 163
Ozone Transport Assessment Group (OTAG), 219

PM_{10}, 70, 140, 213, 214, 217
  emissions, 70
  standards, outdoor, 213, 214
$PM_{2.5}$, 141, 213, 214, 218
  standards, outdoor, 213, 214, 218
Pacific high, 153, 155
  effect on air pollution, 162–164
Paired electrons (*see* electron)
Paleocene epoch, 327, 329
Paleozoic era, 327, 328
Pangea, 327, 328
Papin, Denis, 83
Paracelsus, 5, 6, 11, 59
Paralysis, 78
Paris, 230
Parthenon,
Partial pressure, 60, 61
  dry air, 61
  water vapor, 61
Particles
  aerosol (*see* aerosol particles)
  defined, 3
  hydrometeor (*see* hydrometeor particles)
Pascal, Blaise, 50, 51
Path length, 183
Peak ozone densities (*see* ozone)
Peak stratospheric temperatures (*see* temperature)
Pearl ash (*see* potassium carbonate)
Peat coal (*see* coal)
Penumbra, 338
Pentane, 105
Perfluorocarbons, 287, 289, 347
Perfluoroethane, 287, 289, 322
Periodic table of the elements, 2, 3
Permafrost, 339
Permian period, 327, 328
Permissible exposure limit (PEL), 251
Peroxy radical, 100
Peroxyacetyl nitrate (PAN)
  chemistry, 98, 99, 101, 110, 111
  health and plant effects, 98, 99, 266
  lifetime, 99
  mixing ratios, 98
  role in air pollution problems, 64, 228
  time of peak mixing ratio during day, 163
Peroxyacetl radical, 99
Peroxynitric acid, 281
Personal cloud, 142
Petroleum (*see* oil)
pH, 132–136, 257–260
  acid fog, 258, 260
  acid rain, 258, 260
  defined, 132, 257
  distilled water, 258, 259
  map, United States, 264
  natural rainwater, 258, 259
  scale, 258
  sea water, 258, 269
Philosopher's stone, 20

Phlogiston, 12, 13, 15
Phosphorus, 4, 5, 10, 36
  Baldwin's (*see* calcium nitrate)
Phosphorus of Brand (*see* phosphorus)
Photoaging, 302
Photoautotrophs (*see* autotroph)
Photochemical smog (*see* smog)
Photodissociation reaction (*see* chemical reaction)
Photoheterotrophs (*see* heterotroph)
Photokeratitis, 302
Photolysis reaction (*see* chemical reaction)
Photon, 24, 33, 54
Photosphere (*see* sun)
Photostationary-state relationship, 93–96, 102
Photosynthesis
  anoxygenic, 43
  bacterial, 43, 66
  discovery of, 16
  oxygen-producing, 40, 43, 47, 327
  sulfur-producing, 43
Phototroph, 40, 43
Physical weathering (*see* weathering)
Phytoplankton, 260, 303
Pigment, 44
Pinene
  alpha, 108
  beta, 108
Pittsburgh, Pennsylvania, 86, 87, 268, 269
Plagioclase feldspar (*see* feldspars)
Planetesimal, 36
Plant debris, 127
Plants
  effect on climate, 40, 44, 45, 339
  first, 327, 328
Plaster of paris, 8
Pleistocene epoch, 327, 330, 331
Pliny the Elder, 4, 5, 8
Pliocene epoch, 327, 330
Plumbism, 78
Plume dispersion
  coning, 174, 175
  fanning, 174, 175
  fumigation, 174, 175
  lofting, 174, 175
  looping, 174, 175
Pluto, 314, 315
Point source (*see* emissions)
Poland, 7, 21, 345, 347
Polar cell, 150–152
Polar easterlies (*see* winds)
Polar front (*see* front)
Polar front jet streams (*see* jet streams)
Polar high, 150, 154
Polar stratospheric cloud (PSC), 297–301, 344
  Type I, 297–299, 344
  Type II, 297–299, 344
Polar vortex, 275, 297, 300
Pollen, 64, 127, 247
Pollutant

primary, 100, 101, 118, 163, 171
  secondary, 100, 101, 118, 163, 171
Polonium, 4, 5, 21, 245, 246
Polycyclic aromatic hydrocarbons (PAHs)
              (*see* hydrocarbons)
Popping the inversion (*see* inversion)
Positive-feedback mechanism (*see* feedbacks to climate)
Potash (*see* potassium carbonate)
  caustic (*see* potassium hydroxide)
Potassium, 4, 5, 18, 119
Potassium alum (*see* alum, potassium)
Potassium carbonate, 13
Potassium feldspar (*see* feldspars)
Potassium hydroxide, 13, 18
Potassium ion, 119
Potassium nitrate, 6, 8, 9, 14, 15, 255
Potassium sulfate, 6, 10
Power plant emissions, 127
Precambrian era, 327
Precession, 333
Precipitation, solid, 136
Pressure, air, 50, 51, 58–62
  partial (*see* partial pressure)
  sea-level, 50
  units, 353
Pressure gradient force (*see* forces)
Pressure systems
  low, 149, 155, 156
  high, 149, 154, 155, 156, 164
Prevailing visibility (*see* visibility)
Prevent significant deterioration (PSD), 216, 217
Priestley, Joseph, 5, 7, 12–18
Primary colors, 180
Primary particles (*see* aerosol particles)
Primary pollutant (*see* pollutant)
Primary rainbow (*see* rainbow)
Primary standards (*see* standards)
Prokaryote, 40, 42, 45
Propane, 98
Proterozoic period, 327, 328
Proton, 2, 3, 132
Pseudoadiabatic lapse rate (*see* wet adiabatic lapse rate)
Public Health Service, U.S., 76, 77, 211
Pyrolusite (*see* manganese dioxide)

Quartz (*see* silicon dioxide)
Quaternary period, 327
Quicklime (*see* calcium oxide)

Radiation, 33–35, 54
  defined, 33
  gamma, 33, 245
  infrared (IR), 34, 35
    solar (near)-infrared, 35, 180
    thermal (far)-infrared, 35, 63, 162, 310, 313–318
  radio, 34
  spectrum
    Earth, 34, 35
    Sun, 34, 35

visible light, 180
solar, 34, 35, 310
ultraviolet (UV), ix, 34, 35, 45, 180
absorption by aerosol particles, 187–190
absorption by gases, 181–183
effect of clouds on (*see* cloud)
effect of ozone on (*see* ozone)
effect of air pollution on, 189, 190
effect on health and ecosystems, 301–303
effect on photolysis rates, 303
effect on tropospheric ozone (*see* ozone)
far-, 34, 35, 45, 277, 278, 290
near-, 34, 35, 277, 278
UV-A, 34, 35, 277, 278
UV-B, 34, 35, 45, 277, 278, 301–303
UV-C, 34, 35, 45, 277, 278, 301
variation with altitude, 189, 277, 280
visible, 34, 35, 44, 74, 180, 277
absorption by aerosol particles, 187–189
X-, 34
Radiation (nocturnal) inversion (*see* inversion)
Radioactive decay, 244–246
Radioactive gases, 21, 244–246
Radiosonde temperature record (*see* temperature)
Radium, 21, 245, 246
Radium emanation (*see* radon)
Radius of the sun (*see* sun)
Radon
concentrations, 246
discovery, 4, 5, 21
effect on health, 246
progeny, 246
role in air pollution problems, 64, 242, 244–246
sources and sinks, 242, 244–246
Rainbow, 193–195
primary, 193
secondary, 193
Rainout, 111, 138, 167
Ramazzini, Bernardo, 6, 16
Ramsay, Sir William, 5, 20, 21, 184
Rate coefficient (*see* chemical reaction)
Rayleigh, Lord Baron, 5, 21, 184
Rayleigh regime, 196, 197
Rayleigh scattering (*see* scattering)
Reactive organic gas (ROG)
chemistry, 89, 100–111
defined, 24
emissions, 70, 102, 103
lifetimes, 104, 105, 111
mixing ratios, 100–102
time of peak mixing ratio during day, 163
Reading, Pennsylvania, 86, 87
Real index of refraction (*see* refractive index)
Reasonably achievable control technology
(RACT), 217
Receptor region, 101, 102, 171
Red horizon (*see* colors)
Red nitrous vapor (*see* nitrogen dioxide)
Red sky (*see* colors)

Reduction, 23
Reflection, 190, 191
Reflectivity (*see* albedo)
Refraction, 186, 191, 192
Refractive index
imaginary, 188, 189
real, 188, 191
Refrigerant, 77, 288
Regulation, 82–91, 210–239, 251, 252, 270, 271, 303–306,
345–349
acid deposition, 270, 271
air pollution
indoor, 251, 252
outdoor, 82–91, 210–239
Australia (*see* Australia)
Brazil (*see* Brazil)
Canada (*see* Canada)
Chile (*see* Chile)
China (*see* China)
Denmark (*see* Denmark)
effect on economy, x, 222, 347
effectiveness, 220–222
Egypt (*see* Egypt)
European Union, 229
Finland (*see* Finland)
France (*see* France)
Germany (*see* Germany)
global warming, 345–349
Greece, ancient, 82
India (*see* India)
Iran (*see* Iran)
Israel (*see* Israel)
Japan (*see* Japan)
Mexico (*see* Mexico)
Netherlands (*see* Netherlands)
Norway (*see* Norway)
ozone reduction, stratospheric, 303–306
Rome, ancient, 82
Russia (*see* Russia)
South Africa (*see* South Africa)
Spain (*see* Spain)
Sweden (*see* Sweden)
United Kingdom, 84, 229, 230
Alkali Act (*see* Alkali Act)
Clean Air Act of 1956, 229
Clean Air Act of 1968, 229
Environment Act of 1995, 229, 230
Environmental Protection Act of 1990, 229
London, 82–85
United States, 84, 85, 210–225, 347
Acid Precipitation Act (*see* Acid Precipitation Act)
Air Pollution Control Act of 1955, 210, 211
Air Quality Act of 1967, 212, 213
aircraft, 215
Boston, 85
California (*see* California)
Chicago, 85
Cincinnati, 85
Clean Air Act of 1963, 211

Regulation (*cont.*)
Amendments of 1970, 69–74, 78, 213–216, 270, 347
Amendments of 1977, 216, 217, 237
Amendments of 1990, 217, 218, 244, 248, 249, 270, 271, 304
Revision of 1997, 218, 219
Los Angeles (*see* Los Angeles air pollution)
mobile source, 211, 212, 215
Motor Vehicle Air Pollution Control Act of 1965, 211, 212
Pittsburgh, 85–87
point source, 211
St. Louis, 85
Relative humidity, 131, 132
Renewable energy, 222–224, 231, 232, 347, 348
Residual layer, 55, 56
Respiration
aerobic, 45, 66
anaerobic, 42, 45, 66
cellular, 45
Respiratory ailments, 68, 73, 75, 89, 141, 248–250
Reversible reaction (*see* chemical reaction)
Rice paddy, 70, 71
Rock-forming elements, 36
Rocks
defined, 120,
igneous, 120
metamorphic, 120
oldest, 327
sedimentary, 120
Romans, ancient, 8, 82
Rubber, 72, 89
Rubey, William W., 39
Ruhr region, 231
Runaway greenhouse effect (*see* greenhouse effect)
Russia, 84, 125, 232, 233, 347
Rusting, 15, 46
Rutherford, Daniel, 5, 7, 13, 14
Rutherford, Ernest, 245

S(IV) family, 261
conversion to S(VI) family, 261, 262
S(VI) family, 261
Sahara desert, 122
Sal ammoniac (*see* ammonium chloride)
Sala, Angelus, 6, 9
Salt, common (*see* sodium chloride)
Salt cake (*see* sodium carbonate)
Saltpeter (*see* potassium nitrate)
Sand
specific heat (*see* specific heat)
thermal conductivity (*see* thermal conductivity)
Santa Ana winds (*see* winds)
Santiago, 91, 228
Sao Paolo, 225, 226, 227
Satellite temperature record (*see* temperature)
Saturation mixing ratio (*see* mixing ratio)
Saturation vapor pressure (SVP), 130–132

high molecular-weight organic gases, 132
sulfuric acid, 132, 261
water over ice, 131
water over liquid, 130, 131, 339, 344
Saturn, 314, 315
Savery, Thomas, 83
Scattering
aerosol particle, 190–197
back-, 193, 339
cross section, 195
defined, 190, 193
effect on visibility, 200
efficiency, single-particle, 195–197
extinction coefficient, 195
forward, 193, 194
efficiency, single-particle, 196, 197
geometric (*see* geometric regime)
Mie (*see* Mie regime)
Rayleigh (*see* Rayleigh regime)
side-, 193
Tyndall (*see* Tyndall absorber, scatterer)
gas, 183, 184
atmospheric colors resulting from, 184–187
cross section, 184
effect on visibility, 199, 200
extinction coefficient, 184
meteorological range due to, 199
Rayleigh, 184, 199
Scattering cross section (*see* scattering)
Scheele, Karl Wilhelm, 5, 7, 14–18
Schönbein, Christian Friederich, 7, 20
Scintillate, 192
Scrubber, 256, 271
Sea breeze, 101, 170, 171
effect on pollution, 101, 170, 171
Sea coal (*see* coal)
Sea ice, 327, 329, 330, 336, 342
Sea level, changes in, 327–335, 342, 343
Sea spray, 64, 117–119, 138, 139
acidification (*see* acidification)
Sea water composition, 119
Second-hand smoke (*see* smoke, environmental tobacco)
Secondary pollutant (*see* pollutant)
Secondary rainbow (*see* rainbow)
Secondary standards (*see* standards)
Secret sal ammoniac (*see* ammonium sulfate)
Sediment analysis, ocean floor, 326
Sedimentary rocks (*see* rocks)
Sedimentation
gas, 137
particle, 117, 137
time required for, 121, 122
Semipermanent pressure system, 154–156
Sesquiterpenes (*see* terpene)
Shape, aerosol particle, 139, 140
Shear, turbulent (*see* turbulent shear)
Shelley, Mary Wollstonecraft, 336
Shells, 8, 326, 327
Short radio wavelengths (*see* radiation)

Short-time exposure limit (STEL), 251, 252
Short ton, defined, 69
Siberian high, 156
Sick-building syndrome, 251
Sidescattering (*see* scattering)
Sidestream smoke (*see* smoke, environmental tobacco)
Silicon, 4, 5, 18, 19, 30, 36, 37, 119, 123
Silicon dioxide, 19, 38, 67, 121, 126, 187
Silver, 4
Sintering, 326
Size distribution, 116–118, 135–139
    lognormal, 116–118
Skeletons, 8, 120
Skin cancer (*see* carcinoma, melanoma)
Skin color, cause of, 301
Skin irritation, 78, 247–249, 251
Sky color (*see* colors)
Slaked lime (*see* calcium hydroxide)
Small-scale subsidence inversion (*see* inversion)
Smectite, 121
Smelter emissions, 127
Smith, Robert Angus, 256, 257
Smog (*see also* air pollution)
    chemicals in, 64
    defined, ix, 85
    effect on health (*see* air pollution)
    London-type, 82, 85, 86, 88
    photochemical, 82, 87–92, 99–111
    regulation (*see* regulation, air pollution)
Smog alerts, 219–221
Smoke
    biomass burning (*see also* biomass burning)
        composition of, 119, 123–125, 138, 139
    environmental tobacco, 64, 248–250
        concentrations, 249
        effect on health, 248–250
        mainstream, 248–250
        sidestream, 248–250
Smokestacks (*see* chimneys; plume dispersion; Sudbury)
Smoking, effect on health, 246, 248–250
Smudge-pot effect (*see* aerosol particle, effect on climate)
Snell's law, 191
Snowblindness (*see* photokeratitis)
Soap, 7, 84, 85, 255
Soda ash (*see* sodium carbonate)
Sodium
    elemental, 4, 5, 18, 30, 36, 37
        percent of sea water, 119
    ion, 134, 135, 137, 269
Sodium bicarbonate, 258
Sodium carbonate, 6, 7, 85, 86, 254, 255
Sodium chloride, 6, 8, 9, 19, 119, 187, 255, 269
Sodium hydroxide, 6, 18, 258, 267
Sodium ion, 119, 135, 137, 138
Sodium sulfate, 6, 9, 10, 255
Soil
    composition, 121
    defined, 120

liquid water, 168, 169
    effect on air pollution, 168, 169
Soil dust
    absorption of UV radiation by, 187, 278
    color (*see* colors)
    composition, 119, 121, 138
    defined, 121
    effect on visibility, 225
    effect of winds on, 165, 166
    emissions, 117, 119–122, 165, 166
    light absorption by, 187, 188
    role in air pollution problems, 64
    source regions, 122
    storms in China, 235
Soil-particle acidification (*see* acidification)
Solar albedo (*see* albedo)
Solar constant, 311, 312, 354
Solar energy (from photovoltaics), 222–224
Solar nebula, 30, 36
Solar radiation (*see* radiation)
Solar spectrum (*see* spectrum, solar)
Solar wind (*see* sun)
Solid electrolyte (*see* electrolyte)
Solids, 15
    history of discovery, 4–10
Solstice
    summer, 311, 312
    winter, 311, 312
Solubility, 132, 333, 339
Solute, 4, 132
Solution, 4, 132
Solvay, Ernst, 257
Solvent
    defined, 4, 132
    industrial, 78
Soot (*see also* black carbon, organic matter)
    blackened buildings because of, 268, 269
    coated, 124, 139, 140
    composition of, 124–126
    defined, 124
    effect on climate (*see* black carbon)
    effect on plants, 266
    health effects (*see* aerosol particle, effect on health)
    size of aged particles, 139, 140
    size of emitted particles, 125, 139, 140
    morphology, 139, 140
    shape, 139, 140
    sources, 12, 66, 124, 125, 126, 139, 255
Source region, 101, 102, 171
South Africa, 84, 237, 238
South Coast Air Quality Management District (SCAQMD), 89, 220, 221
Southern Lights (*see* Aurora)
Spain, 231
Specific heat, 13, 155, 168, 169
    air, 155
    clay, 155
    liquid water, 155
    sand, 155

Spectrum (*see* radiation)
Speed of light, 191, 353
Spirit of nitre (*see* nitric acid, aqueous)
Spirit of salt (*see* hydrochloric acid, aqueous)
Spores, 127
Spray cans, 288, 304
Spume drops, 119
Squamous cell carcinoma (*see* carcinoma)
Stability, atmospheric
    absolutely stable, 159, 160
    absolutely unstable, 159, 160
    conditionally unstable, 159, 160
    criteria, 159
    defined, 158–160
    effect on climate, 340, 341
    neutral, 159
    stable, 158–160
    unstable, 158–160
Stable boundary layer (*see* boundary layer)
Stahl, Georg Ernst, 12
Standards
    primary, 213–215
    secondary, 213–215
State Implementation Plan (SIP), 213, 215
Stationary source (*see* emissions)
Steam engine, 83–85
Steam wells, 36, 39
Steel mills, 86, 87, 127
Stefan, Josef, 35
Stefan-Boltzmann constant, 35, 313, 354
Stefan-Boltzmann law, 35, 311, 313
Stratopause, 55, 57
Stratosphere, 54, 55, 57
    color following volcano, 205
    effect of global warming on, 324, 325, 344
Stratospheric ozone (*see* ozone)
Stromatolites, 328
Strong acid (*see* acid)
Strong electrolyte (*see* electrolyte)
Strutt, John William (*see* Rayleigh, Lord Baron)
Strutt, Robert John, 181
Styrene, 78, 247
Subcloud layer, 55, 56
Sublimation, 132
Subpolar-low-pressure belts, 150, 152, 153
Subtropical front (*see* front)
Subtropical high-pressure belts, 150, 152
Subtropical jet stream (*see* jet streams)
Sudbury, 167, 271
Sulfate ion (sulfate)
    abundance in accumulation mode, 135–139, 197
    chemistry, 120, 134, 135, 259, 261, 262, 268
    concentrations, 135, 137
    effect on climate, 339
    effect on visibility, 197, 225
    role in air pollution problems, 64
    role in limestone erosion, 268, 269
    sources, 118, 120, 126, 134–139, 261, 262
    standard for, outdoor, in California, 214

structure, 22
Sulfite ion, 261, 262
Sulfur, elemental, 4, 5, 9, 30, 36, 37, 43, 119
Sulfur dioxide
    aqueous, 262
    gas
        characteristics, 73
        chemistry, 23, 261
        discovery of, 7, 16
        dissolution in rainwater, 262
        effect on health, 73
        effect on visibility, 205, 224, 225
        emission control techniques, 271
        long-range transport of, 166, 167, 261
        mixing ratios, 63, 73, 86, 244
        role in air pollution problems, 64
        role in prebiotic atmosphere, 39
        sources and sinks, 5, 17, 70, 73, 102, 103, 123, 242,
            244, 254, 260, 261
        standards
            emission, 212
            indoor, 251, 252
            outdoor, 213, 214, 226–238
        structure, 22
Sulfur hexafluoride, 287, 289, 322, 347
Sulfur oxides, 102, 103
Sulfur triangle, 231
Sulfur trioxide, 102, 261
Sulfuric acid
    aqueous
        absorption of solar radiation by, 187, 188
        acidification by (*see* acidification)
        chemistry, 11–13, 41, 133, 135, 255, 259, 261
        discovery of, 6, 9
        health effects, 73
        refractive index, 188
        role in acid deposition (*see* acid deposition)
        role in air pollution problems, 64, 254
        sources and sinks, 5, 73
    gas
        acid deposition from (*see* acid deposition)
        condensation of, 117, 132, 134, 135, 259
        nucleation of, 128
        sources and sinks, 73, 123, 255, 260, 261
Sulfurous acid, 6, 8, 12, 261, 262
Summer solstice (*see* solstice, summer)
Sun
    chromosphere, 31, 32
    color (*see* colors)
    core, 31
    corona, 31, 32, 37
    distance from earth, 32, 311, 354
    hydrogen convection zone (HCZ), 31, 32
    intermediate interior, 31
    mass, 31
    origin, 30
    photosphere, 31–36, 310, 311
    radiation spectrum, 34–36
    radius 31, 311, 354

solar wind, 31, 32, 36, 37
  structure, 31, 32
  temperature, 30–32, 310, 354
Sunburn, 303
Sunspot, 338
Supercooled liquid water, 297
Surface (heterogeneous) reaction (*see* chemical reaction)
Surface layer, 55, 56
Surface winds (*see* winds)
Surgeon General, U.S., 76, 77, 211
Swartout, H. O., 89
Sweden, 263, 267, 346
  acidified lakes, 263

T-Tauri stage, 37
Tambora, 336
Taxes, environmental, 231, 232, 345, 346
Tehran, 226, 234
Tel Aviv, 233
Temperature
  air, 50, 52, 56–62
  effect of Milankovitch cycles on, 331–334
  equilibrium, of planets, 310–316
  ground, effect on pollution, 168
  inner Earth (*see* Earth)
  peak stratospheric, 57
  structure of the atmosphere, 54–58
  sun (*see* sun)
  trends
    historical, 326–337
    recent
      from radiosonde, 323–326
      from satellite, 323–326
      from surface stations, 323–326
      tropospheric, 323–326
      stratospheric, 323–326
Temperature inversion (*see* inversion)
Termolecular reaction (*see* chemical reaction)
Ternary nucleation (*see* nucleation)
Terpene, 23, 108, 109
  diterpene, 23, 108
  hemiterpene (*see* isoprene)
  monoterpene, 23, 108
  sesquiterpene, 23, 108
Tertiary period, 327
Tetraethyl lead (*see* leaded gasoline)
Thenardite (*see* sodium sulfate)
Theophrastus, 6, 8
Thermal, 54
Thermal conductivity, 53
  air, 53
  clay, 53
  liquid water, 53
  sand, 53
Thermal decomposition reaction (*see* chemical reaction)
Thermal pressure system, 155, 156, 170
Thermal speed of an air molecule, 52
Thermal turbulence (*see* turbulence)
Thermal-IR radiation (*see* radiation)

Thermometer, liquid-in-glass, 52
Thermoscope, 52
Thermosphere, 54, 55, 58
Thorium, 18, 245, 246
Time-weighted average threshold limit values
      (TWA-TLVs), 251, 252
Tin, 4
Tire particles, 64, 127, 138
Titanium, 124, 127
Tobacco smoke (*see* smoke, environmental tobacco)
Tokyo, 91, 173, 237
Toluene
  chemistry, 107, 108, 132
  health effects, 78
  lifetime, 105
  mixing ratios, 63, 107
  reactivity, 105
  role in air pollution problems, 64
  sources, 78, 107, 247
  structure, 23, 24
Torricelli, Evangelista, 50–52
Total organic gas (TOG), 24
Total suspended particulates (TSP), 213
Trade winds (*see* winds)
Tragedy of the commons, x
Transboundary air pollution, 167, 219, 235, 261, 270
Travers, M. W., 5, 21
Trends, air quality, 220–238
Triassic period, 327, 328
Tritium, 3
Tropic of Cancer, 312
Tropic of Capricorn, 312
Tropopause, 55–57
Troposphere, 54–57, 63
  boundary layer (*see* boundary layer)
  free, 54–57
Tucker, Raymond R., 89
Turbulence, 54
  mechanical, 54
  thermal, 54
Turbulent inertial motion, 129
Turbulent shear, 129
Tuvulu, 342, 343
Twilight, 186, 187
Twinkle, 192
Tyndall, John, 196, 318
Tyndall absorber, scatterer, 196

Ultraviolet (UV) radiation (*see* radiation)
Umbra, 338
Unimolecular reaction (*see* chemical reaction)
United Kingdom, air pollution laws (*see* Regulation,
      United Kingdom)
United Nations Framework Convention on Climate
      Change, 346
United States
  Air pollution laws (*see* Regulation, United States)
  Dept. of Health, Education and Welfare (*see* Health,
      Education, and Welfare, U.S. Dept. of)

Environmental Protection Agency (*see* Environmental Protection Agency, U.S.)

Public Health Service (*see* Public Health Service, U.S.)

Surgeon General (*see* Surgeon General, U.S.)

Unleaded gasoline, 77, 216

Unstable atmosphere (*see* stability, atmospheric)

Uranium, 244–246

Uranus, 314, 315

Urban air pollution (*see* smog)

Urey, Harold, 41, 42

Vacuum, 33, 50, 191

Valley breeze, 171

Van Helmont, John Baptist, 6, 11, 13

Van Niel, Cornelius, 45

Vapor deposition,

Variable gas, 62, 63

Vehicle emission regulations, 211, 212, 215

Ventilation
  forced, 243
  natural, 243

Venus, 314–316

Vermiculite, 121

Vernal equinox (*see* equinox)

Vienna Convention for the Protection of the Ozone Layer, 304

Vinci, Leonardo da, 11, 185

Vinegar (*see* acetic acid)

Vinyl chloride, 78, 214

Viruses, 127, 242, 248

Visibility, 197–202
  defined, 197
  effect of aerosol particles on, 196, 197, 199, 200
    accumulation mode, 196, 197
  effect of gases on, 199, 200
  meteorological range, 198–200, 225
  prevailing, 197, 200–202, 224, 225
  standards, 224, 225
  trends, 200, 201, 224, 225
  visual range, 197

Visible spectrum (*see* radiation)

Visual range (*see* visibility)

Vitamin D, 301

Vitriolic acid (*see* sulfuric acid)

Vitriolic acid air (*see* sulfur dioxide)

Volatile organic compounds (VOCs), 24, 242, 246, 247, 251

Volcanos
  effect on climate, 329, 336
  effect on stratospheric color, 205
  effect on stratospheric ozone, 283, 284, 291, 292
  El Chichón (*see* El Chichón)
  emissions, 17, 36, 39, 66, 73, 117, 119, 123, 260, 291
  Mount Pinatubo (*see* Mount Pinatubo)
  Mount St. Helens (*see* Mount St. Helens)
  origin of name, 123
  Tambora (*see* Tambora)

Volume concentration (*see* concentration)

Volume mixing ratio (*see* mixing ratio)

Warm cloud (*see* cloud)

Washing soda (*see* sodium carbonate)

Water
  aerosol liquid water content (*see* aerosol particles)
  hydration (*see* hydration)
  ice (*see also* hydrometeor particles and Polar stratospheric clouds)
    heterogeneous chemistry on, 298, 299
    sublimation, 132
  liquid (*see also* hydrometeor particles)
    absorption efficiency, 197
    chemistry, 41–44, 67, 95, 120, 134, 257, 258, 262, 267
    percent of sea water, 119
    refractive index, 188
    scattering efficiency, 197
    specific heat (*see* specific heat)
    thermal conductivity (*see* thermal conductivity)
  soil liquid water content (*see* soil)
  vapor (gas)
    absence on Venus, 315
    absorption of radiation by, 181, 182, 317
    characteristics, 63, 64
    chemistry, 19, 23, 25, 39, 41, 94–111
    condensation of, 117, 128, 130–132, 261, 282, 283
    constituent of early atmosphere, 39
    depositional growth of, 132
    effect on climate, 315–319, 338, 339
    effect on health, 64
    measurement of, 11
    mixing ratios, 63, 64, 130–132, 182, 314
    sources and sinks, 39, 63, 64, 123, 282, 283, 316
    structure, 22

Waterwheel, 83

Watt, James, 83, 84

Wavelength, 24, 33

Weak acid (*see* acid)

Weathering
  chemical, 66, 67, 120
  physical, 120

Weichselian (*see* ice age, last)

Weight, 2

Wells, H. G., 21

Westerly winds (*see* winds)

Wet adiabatic lapse rate (*see* lapse rate)

Wheezing, 73

White night, 187

Wien, Wilhelm, 33

Wien's law, 33

Wind energy, 222, 223, 232, 347, 348

Winds
  bay (*see* bay breeze)
  causes of, 146–154
  effect on pollution, 165–167, 170–175
    long-range transport, 166, 167, 219, 261, 270
  geostrophic, 148
  gradient, 149
  lake (*see* lake breeze)
  mountain (*see* mountain breeze)
  Polar easterlies, 150, 154

Santa Ana, 165, 166
sea (*see* sea breeze)
southwesterly, 150
surface
    along curved isobars, 149, 150
    along straight isobars, 148, 149
trade, 150, 152
valley (*see* valley breeze)
westerly, 150, 151, 154, 195
Winter solstice (*see* solstice)
Wisconsin (*see* ice age, last)
Wöhler, Friedrich, 5, 20
Würm (*see* ice age, last)

X radiation wavelengths (*see* radiation)
Xenon, 4, 5, 20, 21, 22, 62

Xylene
    chemistry, 108, 132
    health effects, 78
    mixing ratios, 63, 108
    reactivity, 104, 105
    role in air pollution problems, 64
    sources, 78, 108, 247

Year without a summer, 336
Younger-Dryas period, 334, 335, 337
Youngstown, Ohio, 86, 87

Zinc emissions, 87, 127, 138
Zonal, 150
Zonally averaged, 56, 57, 150, 274